12th Topical Conference on Gas Utilization 2012

Topical Conference at the 2012 AIChE Spring Meeting and 8th Global Congress on Process Safety

Houston, Texas, USA
1-5 April 2012

ISBN: 978-1-62276-566-9

Printed from e-media with permission by:

Curran Associates, Inc.
57 Morehouse Lane
Red Hook, NY 12571

Some format issues inherent in the e-media version may also appear in this print version.

Printed by Curran Associates, Inc. (2012)

For permission requests, please contact AIChE
at the address below.

AIChE
3 Park Avenue
New York, NY 10016-5991

Phone: (203) 702-7660
Fax: (203) 775-5177

www.aiche.org

Additional copies of this publication are available from:

Curran Associates, Inc.
57 Morehouse Lane
Red Hook, NY 12571 USA
Phone: 845-758-0400
Fax: 845-758-2634
Email: curran@proceedings.com
Web: www.proceedings.com

TABLE OF CONTENTS

() Future Fuels and Chemicals: Grand Challenges and Opportunities

Joseph B. Powell, Shell Projects & Technology, Houston, TX

To insure a Sustainable Future, CO_2emissions must be reduced despite strong future growth in demand. "Easy oil" from conventional production will be less abundant, though coal and natural gas, as well as unconventional hydrocarbon reserves will remain in strong supply for decades to come. Carbon footprint, cost, and convenience will be critical drivers for future energy choices. Basis recent advances in horizontal drilling and hydraulic fracturing, enhanced production of tight gas, shale gas, and coal bed methane will enable natural gas to be both an affordable and abundant option for meeting short and long-term sustainable energy goals. Sustainable production of biofuels, with consideration of well-to-wheels carbon intensity and other resource impacts, will play a major role in reducing carbon intensity in the transportation sector. Carbon capture and sequestration along with renewable energy sources and enhanced energy efficiency will play increasing roles in reducing future greenhouse gas emissions across the diversity of resources that will be deployed to meet local and global stakeholder needs for energy and security. Advantaged routes to chemicals will be impacted by new feedstocks and performance requirements for fuels, providing opportunities for innovation in gas to chemicals, biobased chemicals, and selective conversions associated with hydrocarbon portfolio balancing. Significant time and investment will be required to bring forth new alternative energy systems to materiality, driving the development of new paradigms for efficient de-risking and deployment of technology.

This presentation will address opportunities for the chemical engineering profession to differentiate and excel in sustainably meeting future needs in fuels and chemicals.

(24a) CO2 Capture Via Potassium Carbonate/MEA Solution

Yee Soong, National Energy Technology Laboratory, U.S. Department of Energy, Pittsburgh, PA, Robert Dilmore, U.S. Department of Energy, Pittsburgh, PA, Sheila Hedges, U.S. DOE/NETL, Pittsburgh, PA and D. E. Allen, Salem State College, Salem, MA

A great deal of concern has been expressed with regard to global climate change and its link to growing atmospheric concentrations of carbon dioxide. To decrease the impact of anthropogenic CO_2on global climate, several strategies are under development that will potentially remove $CO_2$from the atmosphere or decrease CO_2emission. One such strategy involves the capture of $CO_2$from large point sources (such as pulverized coal (PC) plants) and the long-term storage of CO_2underground. The**post-combustion**capture of $CO_2$is a challenging application because of (1) the low pressure and dilute concentration (10 to 15 %) of the carbon dioxide dictate a high volume of gas to be treated and (2) trace impurities in the flue gas tend to reduce the effectiveness of the CO_2absorbing processes.

Much work has been studied in this area and has included development of amine-based liquid sorbents, solid amine-based sorbents, metal carbonate sorbents, ionic solution, chemical looping, membrane system and other novel technologies. The aqueous amine system is one of the commercial technologies for CO_2capture for PC power plants.

The commonly used amine in scrubbing solutions contains monoethanolamine (2-aminoethanol) ($NH_2CH_2CH_2OH$) (MEA). The disadvantages of using MEA solutions include energy-intensive, amine degradation and corrosive. The typical reaction is illustrated as follows:

2 MEA + H^++ HCO_3^-¬> MEA& Middot;H^++ MEA& Middot;HCO_3^-

Work also has been conducted in the area of using carbonate sorbents to remove carbon dioxide from flue gas. A carbon dioxide capture process using regenerable sorbents such as an alkali metal oxide or an alkali metal carbonate has been reported. Potassium carbonate (K_2CO_3) can be supported on a high surface area of activated alumina (Al_2O_3). The product formed from absorption of carbon dioxide is taken aside and heated to regenerate the sorbent. No significant chemical degradation is observed, but there is physical attrition. Loading occurs between from about 60^0C to 100^0C according to the following reaction equation.

K_2CO_3+ H_2O + CO_2¬> 2 $KHCO_3$

Thermal desorption of potassium bicarbonate ($KHCO_3$) and the alumina support takes place at temperatures around 145^0C to 150^0C.

In this paper, a CO_2capture process (US patent 7,842,126) is presented where $CO_2$from combustion flue gas dissolves into an aqueous solution of amine and soluble potassium carbonate. Potassium carbonate rapidly reacts with amine-associated carbamate and water to form less-soluble potassium bicarbonate that precipitates from solution. This reaction serves both to chemically regenerate the amine solution by removing the carbamate and to separate CO_2in the form of solid bicarbonate precipitate. Precipitated potassium bicarbonate is further physically separated from bulk solution and heated to temperatures greater than approximately 150 ºC to thermally regenerate back to solid potassium carbonate and release concentrated CO_2gas. Because of the absence of bulk aqueous solution, regeneration enthalpy requirements for separated bicarbonate solid is predicted to be low as compared to those reported for regeneration in traditional MEA-based absorption processes. Following thermal regeneration of the separated solid, potassium carbonate is re-dissolved in potassium carbonate-lean aqueous amine solution, and the reactive solution is re-exposed to flue gas to close the process loop. This process concept has the potential to achieve high CO_2capture efficiency with relatively low regeneration energy requirements.

In addition to details description of capture methodology, preliminary laboratory results will be presented.

(24b) Where Are the Opportunities for CO_2 Conversion?

Lubo Zhou, UOP LLC, a Honeywell Company, Des Plaines, IL

Emission of green house gas CO_2 has caught the attention of people in most recent years. Conversion of CO_2 to other products has been proposed by many scientists to reduce the CO_2 emission to atmosphere. However, there is a big challenge for the CO_2 conversion to any product: large amount of energy or hydrogen will be used during CO_2 conversion. In this paper, thermodynamics and energy efficiency will be analyzed for CO_2 conversion reactions. Life cycle CO_2 emission will also be analyzed. The objective of the paper is to identify the opportunities of CO_2 conversion that can actually reduce CO_2 emission in a life cycle of the whole process, which includes energy and hydrogen production.

CO_2 Reforming for Variable Chemical Process

Osamu Hirohata, Tomoyuki Mikuriya and Mitsunori Shimura
Chiyoda Corporation, Yokohama, Japan

1. Introduction

Production of synthesis gas from natural gas and conversion of synthesis gas to fuels and chemicals would become increasingly attractive from the viewpoint of energy resource diversification and increasing price of crude oil. The technology of synthesis gas production by reforming of natural gas with CO_2 is one of the effective solutions for the issues of energy resource and environmental preservation.

Chiyoda CO_2 reforming catalyst can generate synthesis gases by CO_2 and H_2O reforming under lower H_2O/Carbon and CO_2/Carbon ratios in feed gas compared with the conventional steam reforming catalyst. Therefore, this catalyst produces synthesis gases with valuable H_2/CO ratios more efficiently. This is because this catalyst has higher resistance against carbon deposition under the severe condition. The commercial size CO_2 reforming catalyst has been demonstrated its activity and stability in the GTL demonstration plant tests in Japan.

In this paper, a feasibility study on applying Chiyoda CO_2 reforming to methanol plant and CO production plant has been conducted. CO_2 reforming can increase production of methanol and produce CO as by product. These increased methanol and CO can be utilized for the production of acetic acid and MMA.

2. Characteristics and performance of Chiyoda CO_2 reforming catalyst

(1) Favorable conditions of Chiyoda CO_2 reforming

Fig.1 shows a carbon limit diagram which is usually utilized for the operation of the steam reformer. The bold line in the center of this figure represents the carbon limit line at 600 ℃. On the left side of this line, the process gas has high thermodynamic potential for carbon formation at 600 ℃. Operation near this line cannot be carried out in conventional reforming process. Usually, conventional catalyst (Ni based catalyst) is operated under the condition of steam carbon ratio (S/C) in excess of around 2.0, in the absence of pre-reformer.

On the other hand, Chiyoda's CO_2 reforming catalyst (Noble metal catalyst) has higher resistance for carbon formation. So this catalyst can be used in low S/C conditions in addition to co-feeding of CO_2 on the left side of carbon limit line. High energy efficiency operation can be performed for the synthesis gas production by Chiyoda's CO_2 reforming process.

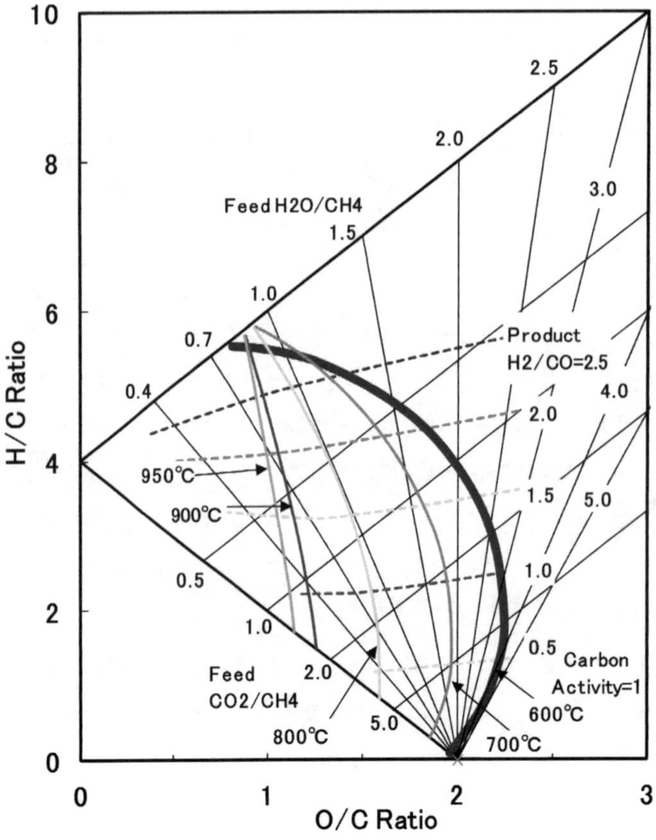

Fig.1 Carbon Limit Diagram

(2) Reaction mechanism

It is considered that the simplified reaction sequence for CO_2 reforming of methane may involve following two irreversible steps, namely, activation of methane followed by the surface reaction with adsorbed oxygen atoms[1)]:

$$CH_4 + * = CH_3* = CH_2* = CH* = C* \quad \cdots\cdots (1)$$

$$CHx* + O* = CO + X/2\ H_2 + 2* \quad \cdots\cdots (2)$$

The adsorbed oxygen atom is formed by the dissociation of CO_2 on the catalyst support as shown in Eq(3).

$$CO_2 + * = O* + CO \quad \cdots\cdots (3)$$

These reaction mechanisms are schematically represented in Fig.2

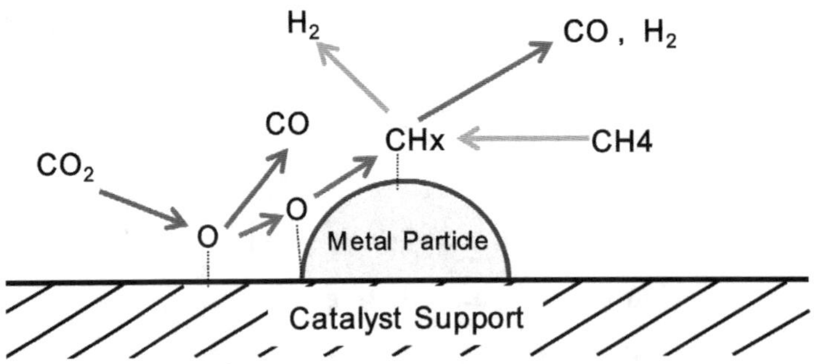

Fig.2 Reaction mechanism on CO_2 reforming catalyst

From this reaction mechanism, to prevent the carbon deposition on the catalyst surface that causes catalyst deactivation, it is important that the x-value of CHx can be maintained high and the rapid oxidation of CHx should be performed by O＊ species.

It was reported that the x-value of CHx species is related to the acidity of the catalyst support as shown in Table 1.[2)] The basic support that is used for this catalyst can keep the x-value high.

Table.1 CHx species observed on Ni and Co during carbon dioxide reforming of methane

Catalyst	x-value
Ni/MgO	2.7
Ni/ZnO	2.5
Ni/Al_2O_3	2.4
Ni/TiO_2	1.9
Ni/SiO_2	1.0
Co/Al_2O_3	0.75

The rate of oxidation of CHx species is considered to be crucially dependent on the particle size of active metal. [3)] If metal particles are small and interact with catalyst support, oxygen atom that is generated by the dissociation of CO_2 moves quickly to the interface of metal and support and is activated. The oxygen atom (O＊) is rapidly supplied to hydrocarbon species (CHx) on the metal surface as shown in the Fig.2.

Based on the consideration from these investigation results, our CO_2 reforming catalyst was designed, in which highly dispersed noble metals are supported on the basic catalyst support .

(3) Performance

The demonstration plant operation of Chiyoda CO_2 reforming catalyst was conducted. Average synthesis gas production rate is around 20,000Nm3/hr with hydrogen to carbon monoxide ratio of 2.0. Feed natural gas contains around 90% of methane and the rest are almost higher hydrocarbon than methane. In the normal operation condition, the temperature of catalyst bed outlet was set around 870-880 ℃ as inlet temperature of waste heat boiler. Operation pressure was 1.94-2.09MPaG. And ratios of carbon dioxide to carbon and steam to carbon of the feed stream were controlled to 0.43-0.50 and 1.1-1.32 respectively. As shown in Fig.1, these conditions cause carbon formation with conventional catalyst. As shown in Fig.3, the stable operation for over 10,000 hours has been confirmed under the target reaction condition.

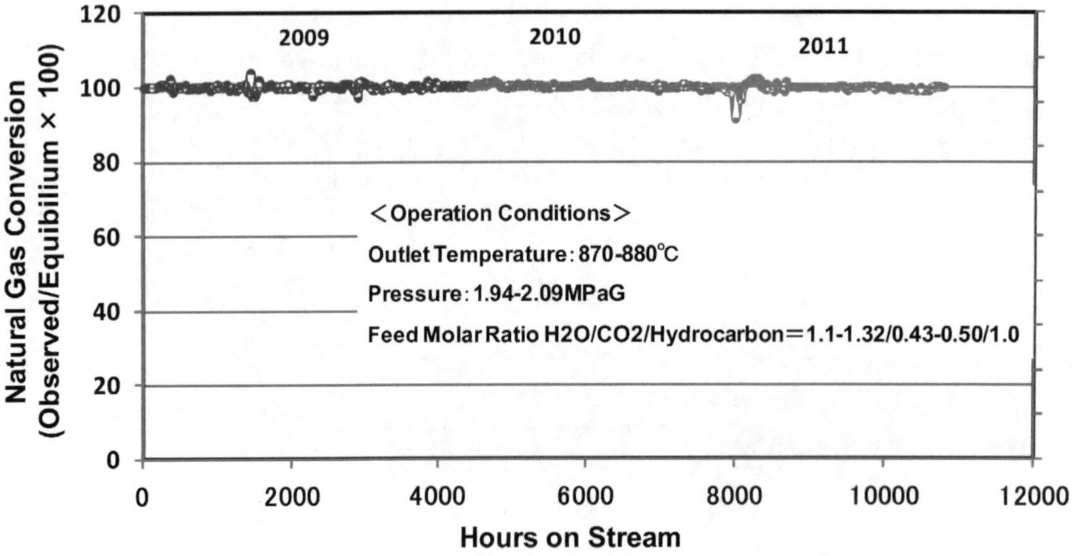

Fig.3 Demonstration Plant Result of Chiyoda CO_2 Reforming Catalyst

3. Study Results

(1) Case-1; CO_2 reforming for methanol production (Revamping case)

(a) Study basis

In Fig.4, the conventional methanol plant flow scheme is shown based on a plant with approx. 3000 t/d methanol production.

Compared with conventional steam reforming, CO_2 reforming case (Fig.5) has two additional systems; ie , import CO_2 line and PSA system. In revamping case, CO_2 is utilized as additional feed for reforming. PSA system can reduce methane concentration and increase H_2 concentration in the recycle stream of the methanol reactor to promote methanol synthesis reaction. As a result of CO_2 import and hydrogen recycle, energy for the reformer to produce excess hydrogen can be reduced and synthesis gas composition can be adjusted to the ideal composition for methanol synthesis.

Process study is conducted on same methanol capacity plant with these two flows.

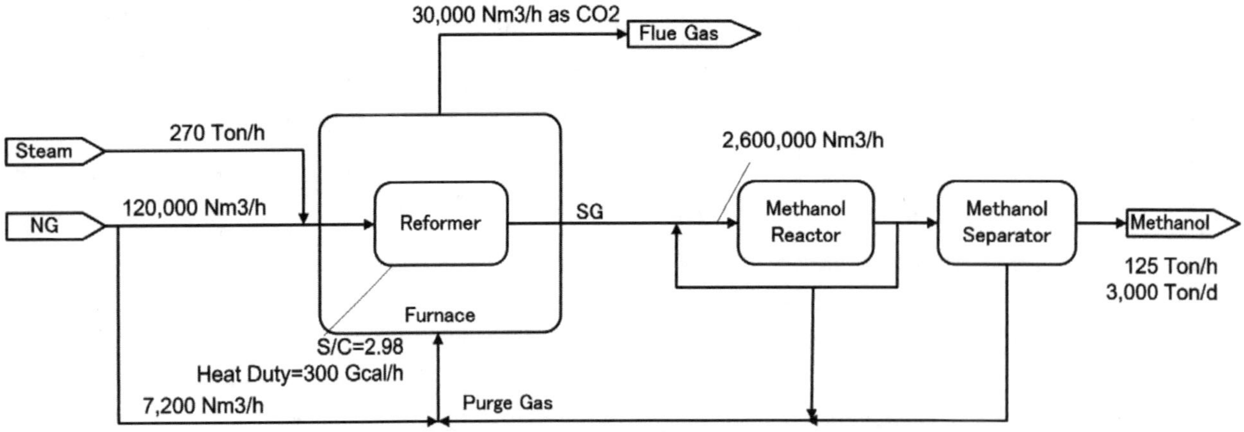

Fig.4 Conventional Case (Steam Reforming)

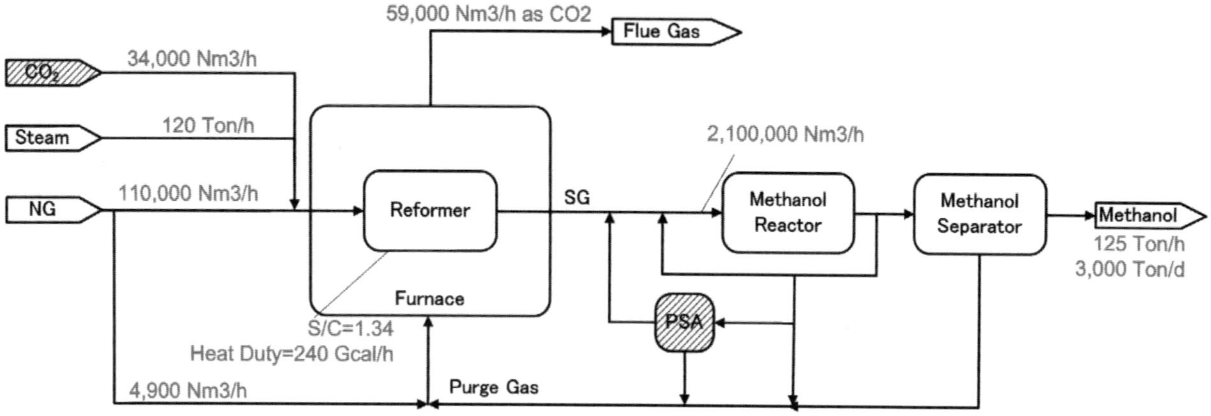

Fig.5 Revamping Case (CO_2 Reforming with PSA)

(b) CO_2 utilization and NG consumption

Methanol plant exhausts the CO_2 from reformer furnace. However, in the revamped plant with CO_2 import, CO2 is effectively utilized NG reforming to produce CO. Therefore, the net exhaust CO_2 can be reduced by approx.18% under the same methanol production capacity (Fig.6). NG consumption in revamped case is 4.4% lesser than conventional case by utilizing CO_2 as feedstock (Fig.7).

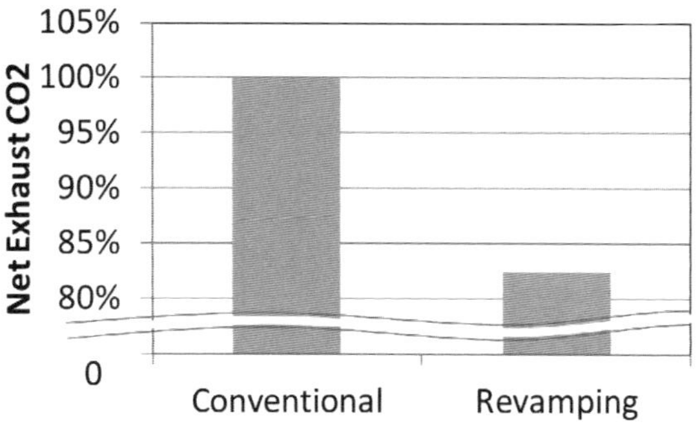

Fig.6 Net Exhaust CO_2

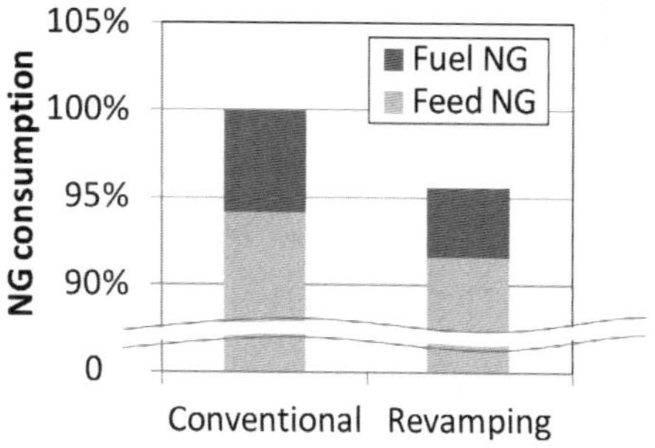

Fig.7 NG consumption

(c)Possibility of increased methanol production

As shown in Fig.8, in the revamped case reformer duty is 20% less than conventional case. Therefore, 25% increase in product can be expected in the revamped case at current reformer capacity (Fig.9).

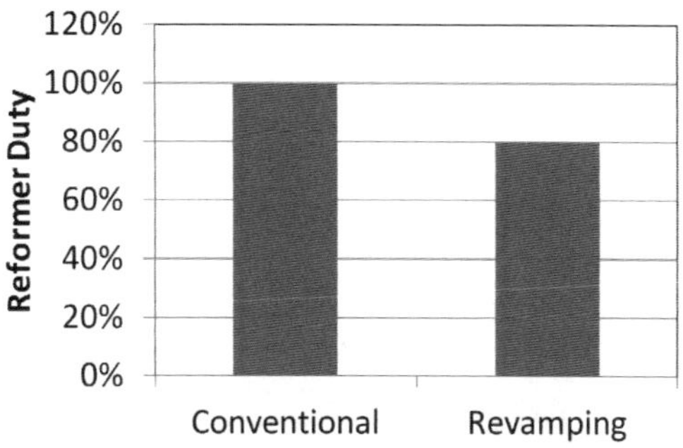

Fig.8 Reformer Duty

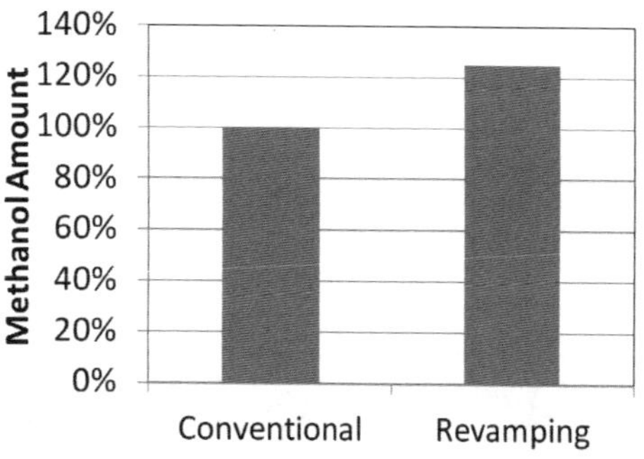

Fig.9 Product Methanol Amount (Same Reformer Duty)

(2) Case-2; CO_2 reforming for methanol and chemical production with one reformer (Revamping case)

(a) Study basis

From the study of case-1 (CO_2 reforming for methanol production plant), methanol product can be increased with CO_2 reforming. In case-2, this increase in production is used for another chemical synthesis.

The revamped flow is shown in Fig.10. After CO_2 reforming, some of the synthesis gas is fed to CO separation unit to produce CO for chemical synthesis. Separated CO_2 and H_2 are utilized in the reformer and methanol synthesis. Even though some synthesis gas is separated to produce CO, there still would be increase in methanol production. This increased methanol and separated CO are used as feedstock for chemical synthesis such as acetic acid or MMA.

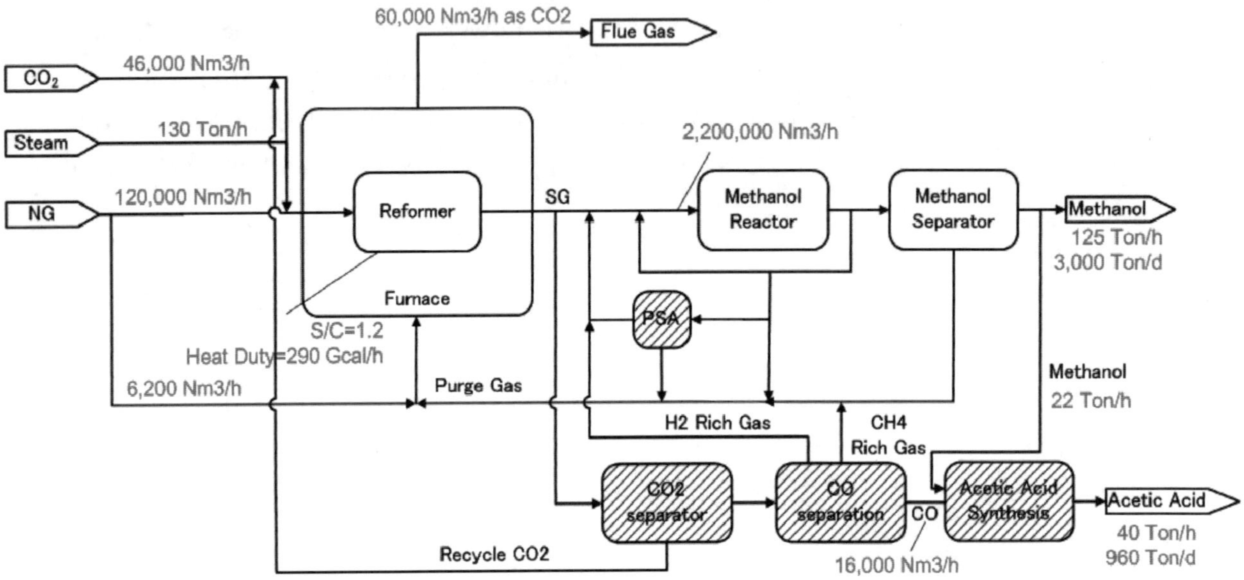

Fig.10 Revamping Case(Methanol and CO production with one reformer)

(b) Possibility of increased methanol production

From Fig.10, revamped case can produce 960 Ton/d acetic acid, with maintaining 3,000 Ton/d methanol production. Therefore, catalyst replacement and new construction of CO separation unit and chemical synthesis unit can produce new chemical products while maintaining current methanol production.

(3) Case-3; CO_2 reforming for CO production (Grass roots case)

(a) Study basis

Partial oxidation of NG is assumed as conventional case for CO production (Fig.11). With partial oxidation (POX), H_2/CO of synthesis gas is adjusted to 2.0. NG and O_2 which is separated from air are fed into POX reactor. After CO_2 removal, synthesis gas is fed into CO separation unit such as cryogenic unit

The block flow of CO_2 reforming case is shown in Fig.12. With CO_2 reforming, H_2/CO of synthesis gas is adjusted to 0.5, which means it can reduce energy required to produce excess hydrogen. After CO_2 is removed from synthesis gas, synthesis gas is fed into CO separation, such as cryogenic unit. Separated CO_2 is recycled to reformer, and off gas from CO separation unit is used as reformer fuel.

In this balance, the plant capacity is assumed as 16,000 Nm3/h CO which is the same amount as CO in Case-2.

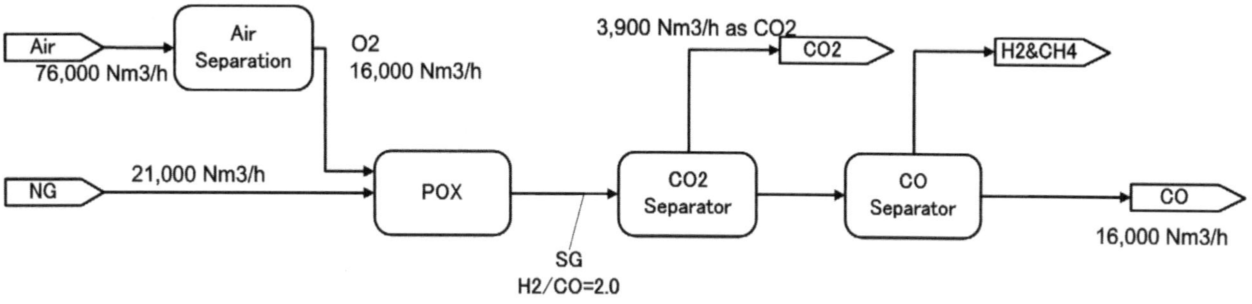

Fig.11 Conventional case (Partial oxidation of NG)

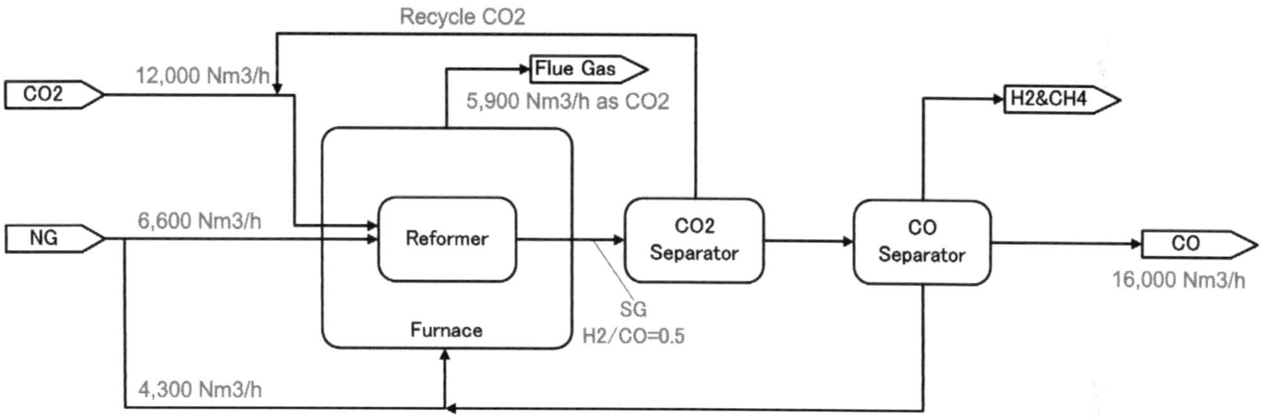

Fig.12 CO_2 reforming case

(b) CO_2 utilization and NG consumption

While CO2 is exhausted from CO_2 removal unit in conventional case, CO_2 is exhausted from reformer furnace in CO_2 reforming case. However, in CO_2 reforming case, because much CO_2 is utilized as feedstock for CO production, amount of import CO_2 is higher than exhaust CO_2 (This means zero emission because net exhaust CO_2 is a negative value (Fig.13)).

NG consumption in CO_2 reforming case is 48% lesser than conventional case by utilizing CO_2 as feedstock (Fig.14). The effect is much more than case-1, (CO_2 reforming for methanol production case). In general, lesser the H_2/CO of the synthesis gas, lesser the NG consumption in CO_2 reforming.

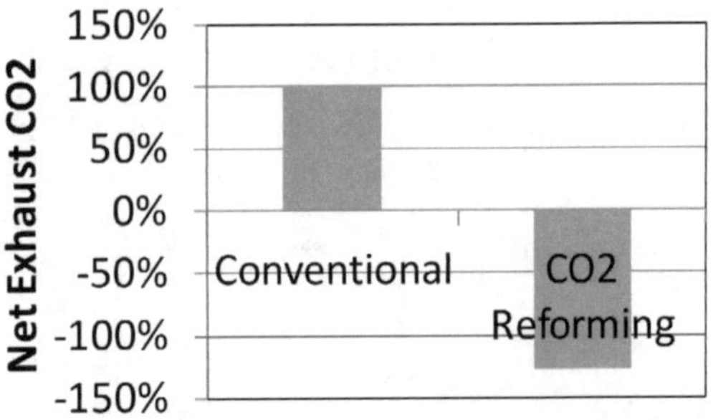

Fig.-13 Net Exhaust CO_2

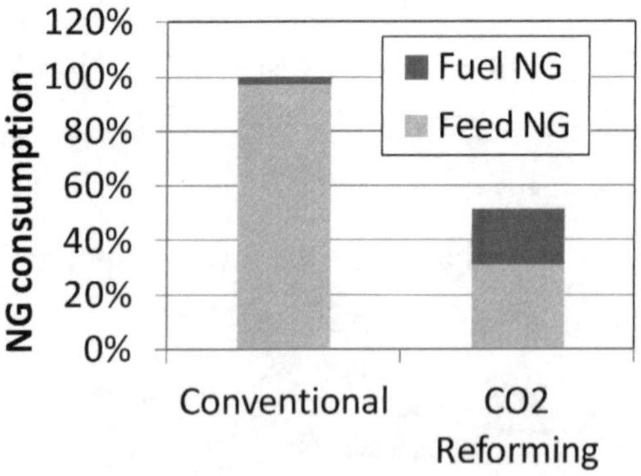

Fig.-14 NG consumption

4. Conclusion

Methanol product amount can be increased by 25% by revamping of the existing methanol plant with CO_2 reforming catalyst. Furthermore, with this increase and CO separation from synthesis gas, new chemical product can be produced while maintaining current methanol production.

In CO production case, CO_2 is mostly utilized as feedstock and 48% of NG can be saved compared to CO production by Partial Oxidation of NG.

5. References

1) J.R.Rostrup-Nielsen ,J-H. Ban Hansen, J.Catal. 144,38-49(1993)
2) T.Osaki, H.Masuda, T.Horiuchi, T.Mori, Catal.Lett., 34,59(1995)
3) K.Tomishige, K.Fujimoto, Catal. Surveys from Japan 2,3-15(1988)

(24d) Spectroscopic Studies of Gas-Liquid Reactions Using Microfluidics for the Study of CO_2 Conversion to High-Value Products

Jesse Greener[1], Rachelle Choueiri[1], Ethan Tumarkin[1], Milad Abolhasani[2], Axel Guenther[3]and Eugenia Kumacheva[1], (1)Department of Chemistry, University of Toronto, Toronto, ON, Canada, (2)Department of Mechanical and Industrial Engineering, University of Toronto, Toronto, ON, Canada, (3)Dept. of Mechanical and Ind. Engineering, Inst. of Biomaterials and Biomedical Engineering, University of Toronto, Toronto, ON, Canada

We present a new concept for spectroscopic studies of the kinetics of fast gas-liquid reactions involving CO_2. The strategy relies on the microfluidic generation of highly monodisperse gas bubbles in the liquid reaction medium. Characterization of the reaction system was achieved by monitoring time-dependent changes in (i) vibrational spectra of the liquid phase and (ii) changes to bubble dimensions. These measurements were achieved using*in situ*attenuated total reflection Fourier transform infrared (ATR-FTIR) spectroscopy and from the analysis of optical micrographs collected by a high speed camera, respectively. The method can be exploited to study important CO_2(g) reactions including: reactions with frustrated Lewis pairs, secondary amines, Grignard reactions and enzyme catalysis. This talk will demonstrate that the method enables rapid determination of reaction rate constants and the observation of transient species. The proposed strategy offers new possibilities in studies of rapid multiphase reactions in microfluidic reactors, and can also be combined with high-throughput optimization of reaction conditions for synthesis of high-value products from waste CO_2gas.

(24e) A Technoeconomic Analysis of Biodegradable Polymers Production From $CO_2$and Waste Plastics

Michael Hartmann, Chemistry and Chemical Engineering Department, Fuels and Energy Development Section, Southwest Research Institute, San Antonio, TX, Maoqi Feng, Div. of Chemistry and Chemical Engineering, Southwest Research Institute, San Antonio, TX and Eloy Flores III, Chemical Engineering Department, Southwest Research Institute, San Antonio, TX

This paper evaluates the economic viability of manufacturing biodegradable polymers (mostly polycarbonates) from $CO_2$and waste plastics. Waste plastics recycling are beneficial to environment and economy as well. Thermal processes have been used for feedstock recycling from waste plastics, such as polyethylene, polypropylene and polystyrene. Olefins are very useful feedstock in industry; however, it's expensive to produce olefins by using conventional hydrocarbons cracking technology. In our simulation, olefins are generated from pyrolysis of waste plastics with thermal energy supplied from a solar concentrator, thus no net $CO_2$is produced in the pyrolysis. Biodegradable polymers are then produced from the polymerization of $CO_2$and olefins directly in a catalytic reactor. This method provides an alternative method for CO_2sequestration and utilization. A detailed economic analysis using VMGSim software is presented.

25a

Mixed Fluid Cascade, Experience and Outlook

Heinz C. Bauer, Linde Engineering, Pullach, Germany

Background

Starting in 1996 Linde AG and Statoil ASA developed a new process for natural gas liquefaction. The Mixed Fluid Cascade (MFC®) process was successfully commercialized in the Snøhvit plant with 4.3 mtpa LNG capacity, which is located in Hammerfest/Norway. After some teething problems, which will be discussed in some depth, the plant is running since early 2009 beyond its design capacity and has passed the performance test in all aspects successfully.

The main features of the Hammerfest LNG concept include

- MFC® process with three cascaded mixed refrigerant cycles
- Direct sea water cooling for all compressor coolers and the main refrigerant condenser
- All electric concept with power generation in highly efficient aero-derivative gas turbines
- Full sequestration of CO_2 contained in the plant feed gas
- High purity nitrogen (99.99 vol%) rejection from end-flash gas
- Waste heat recovery from gas turbines

In total the actual energy consumption of the liquefaction plant is as low as 243 kWh/t_{LNG}, which is unsurpassed by any other base load LNG plant worldwide. Together with the low carbon foot print this marks an impressive achievement with respect to overall plant efficiency.

Plant concept

Since the early 1980ies large natural gas reservoirs had been identified in the Tromsø patch, which is located in the Barents Sea north of Norway (see Fig. 1). The recoverable resources include 190 BCM (billion cubic meters) of natural gas and 34 million cubic meters of condensate. A discovery of oil would have been more welcome at that time as a production concept for natural gas in arctic regions was not readily available. It took almost 20 years[1] until the offshore

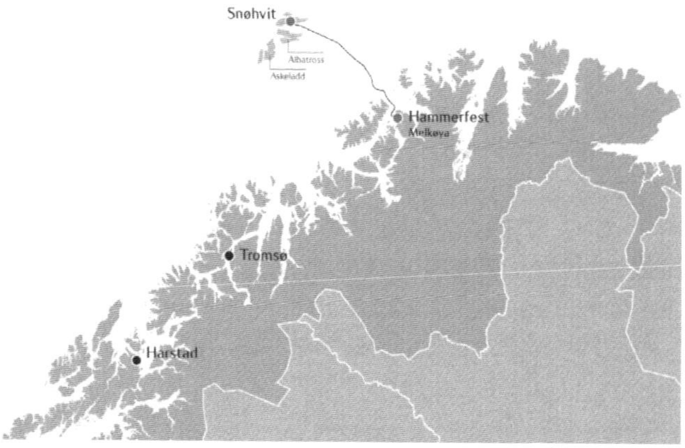

Fig. 1 Location of the Snøhvit field north of Norway
Courtesy Statoil

production technology matured sufficiently to face all challenges successfully.

As the Barents Sea is an ecologically very sensitive area, Snøhvit is the first major development on the Norwegian continental shelf without a fixed or floating unit. Instead, a subsea production system on the seabed feeds a land-based plant on the north-west coast of Melkøya, at the entrance to the shipping channel into Hammerfest via a 26" ID, 143 km multiphase pipeline. In addition two chemical lines, an umbilical and a separate pipeline for transporting carbon dioxide were laid in 2005. Both the subsea production system located on the field and pipeline transport are monitored and controlled from a control room at Melkøya via signals transmitted along fiber-optic cables and high-voltage electrical and hydraulic power lines.

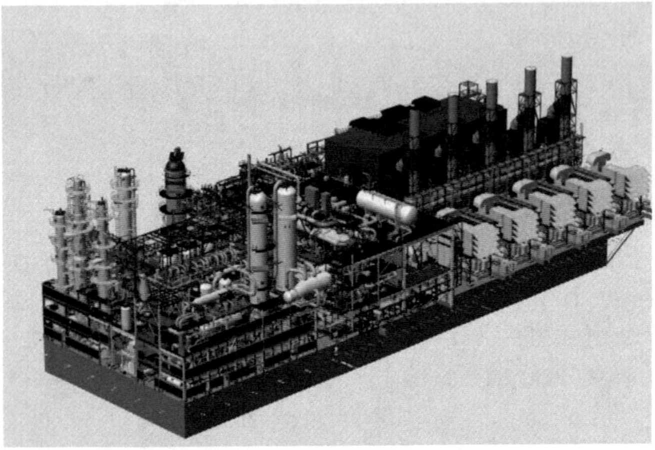

Fig. 2 Process barge

A stick-built concept for the process plant was not applicable as the available plot on Melkøya Island was very limited and as extensive construction work at site would have been very inefficient. Therefore, a modular design with the maximum possible prefabrication was selected. A process barge (see Fig. 2) with a 154x54 m footprint and a shipping weight of 35,000 tons accommodates pre-treatment, all large compressors and power generation with 5 LM6000 aero-derivative gas turbines.

When Statoil started to select the liquefaction technology, all relevant licensors had been invited to a detailed design competition. Eventually, the Mixed Fluid Cascade (MFC®) process[2] (see Fig. 3), a co-development by Statoil and Linde within a technology alliance, was selected for several reasons. (1) The use of three independent mixed refrigerant cycles provided the best conditions for excellent heat integration with a fair power distribution to the cycle compressors. In arctic conditions the temperature range of traditional propane pre-cooling would have been very much restricted by a sea water temperature below 10 degree C all over the year. (2) At that time the desired nameplate capacity of 4.3 mtpa LNG could not be reached by proven concepts, even if modern Mega Trains reach almost twice the capacity today.

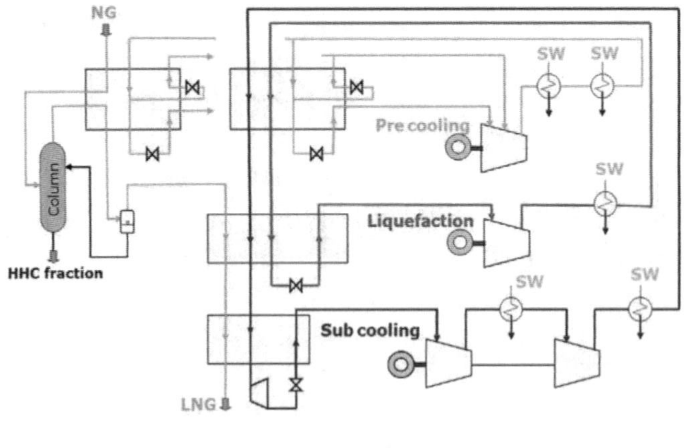

Fig. 3 Mixed Fluid Cascade (MFC®) process Courtesy Statoil

The Norwegian government was very clear with its request to authorize only the best available technique at a location like Melkøya. Hence, further actions had to be taken to minimize the environmental footprint. This includes

- direct sea water cooling for all compressor inter- and aftercoolers and the main refrigerant condenser with titanium tubes and tube sheets,
- all electric concept with power generation in highly efficient LM6000 aero-derivative gas turbines, connection to the national grid as back-up, waste heat recovery from gas turbines
- reinjection and sequestration of 700,000 tons per year of CO_2 from feed gas,
- high purity nitrogen (99.99 vol%) rejection from end-flash gas to the atmosphere,
- minimized flaring.

The result of this effort convinced the Norwegian Ministry of Petroleum and Energy and the Plan for Development and Operation (PDO) was sanctioned in early 2002. It took until September 2007 to complete the construction of the entire plant and to be ready for start-up.

Maturing phase

Shortly after the initial start-up of the plant, a major challenge surfaced. All sea water cooled heat exchangers had been designed in a special way as helix type exchangers[3] in order to save plot and weight on the process barge. However, the actual shell side pressure drop of gas coolers had been significantly beyond the guaranteed figures provided by the licensor[4] and a predicted helical flow did not develop in the main condenser of the pre-cooling refrigerant. Rather, differential condensation[5] of the mixed refrigerant components had to be observed. These shortcomings could be compensated only by an increased discharge pressure of the refrigerant cycle compressors and a corresponding increase of the energy consumption. Even more concerning was the fact that several of the HELIXCHANGER®s failed after a few months due to tube leakages. The failure mode was wear of the tubes against baffle plates caused by flow induced vibration. The resulting sea water ingress into the refrigeration cycles caused several times an extended shut-down of the plant. After provisional plugging of the leaking tubes all gas coolers have been replaced meanwhile by heat exchangers with a different design. Only the replacement main condenser (see Fig. 4) is still a HELIXCHANGER® in a 'robustified' version.

Fig. 4 Replacement of the main condenser Courtesy Statoil

The pre-cooling cycle uses vacuum brazed aluminum plate-fin heat exchangers for heat transfer. Temperature probes on the individual cores of the parallel exchangers showed significant temperature gradients perpendicular to the flow direction, which should not occur. A detailed root-cause analysis revealed uneven blockages of passages caused by solid particles and ambiguous fluid distribution due to the complex hydraulics[6] of mixtures in a two-phase regime. In 2009, all plate-fin heat exchangers have been exchanged by clean ones with an optimized hydraulic design.

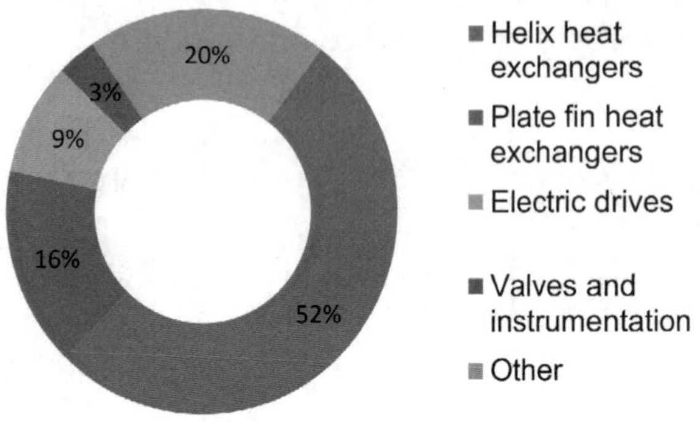

Fig. 5 Distribution of downtime causes since start-up
Courtesy Statoil

The all electric concept had been selected to improve the overall availability of the plant. In essence, this goal has been reached, but some lessons had to be learnt. For example, some details of the exciter machine for the motor of the end-flash gas compressor had to be optimized to tolerate the stress caused by high centrifugal forces and heavy thermal load. Further, the quantitative interaction between mechanical and electrical resonances (see Campbell diagrams[7]) in a compressor train with a variable speed drive (VSD) had to be evaluated. In the context of a short circuit damage part of the large (65 MW) drives had to be replaced by optimized components.

Further causes for downtime (see Fig. 5[8]) include the two coil wound heat exchangers for liquefaction and sub-cooling, which experienced a minor number of leaking tubes. In essence, all challenges of an arctic location with many pioneering concepts have been accepted and have been surmounted as per today.

In addition to equipment related improvements also the operation of the plant became more and more sophisticated. For example, the CO_2 emission from warm restart flaring could be reduced by about a factor 30, as cool-down is now based on the use of vaporized LNG instead of the use of feed gas.

Performance test

About four years after the initial start-up the plant was ready for the official performance test. The long interval between these dates was used to overcome all hurdles, which might have jeopardized the success of the project.

The main results from the plant performance test are

- The guaranteed plant capacity is exceeded by 4%. This translates into an absolute production quantity of 13,200 t/d of LNG (rising in tank).

- All guaranteed product qualities are fulfilled (including the CO_2 routed to offshore and the N_2 Tail Gas of the NRU released to atmosphere).

- The specific refrigeration power consumption is 5.4% lower than guaranteed value. The achieved level of 243 kWh/t_{LNG} is the lowest level for all LNG plants worldwide. Based on the thermal efficiency of the LM6000 gas turbines of 39% (measured at site) only about 5.5 wt% of the feed gas are used to generate the required power for the whole LNG plant.

- The measured total power consumption of the plant is 9.9% less than the guaranteed total power consumption.

- The calculated fuel gas consumption is less than the guaranteed volume.

- All CO_2 separated from the feed stock is routed to offshore. This CCS (carbon capture and sequestration) effort reduces the total CO_2 emission of the plant by about 45% to only 110 t/h CO_2.

- All N_2 separated from the feed stock is released to atmosphere as the remaining methane content is well below the specified 100 mol-ppm.

- The guaranteed levels of noise are deemed to be met.

- The guaranteed NO_x levels in the gas turbine exhaust gas are met.

- Furthermore, there wasn't any flaring and the operation of the plant was quite smooth and stable.

Outlook

Despite of the successful implementation of the MFC process in the arctic region other projects in a more moderate climate require different solutions. With this goal in mind a 2nd generation[9] of the MFC® concept with optional propane pre-cooling (MFC®3) has been developed, which will be described underneath (see Fig. 6). Traditionally, all heavier hydrocarbons (C_{5+}) and to some extent natural gas liquids (C_2 to C_4) are removed from the feed gas in a scrub column in order to avoid freezing of heavies, especially aromatic material, to meet residue specifications (C_4 content) of the LNG, and to provide make-up components (C_2 and C_3) for the refrigeration cycles.

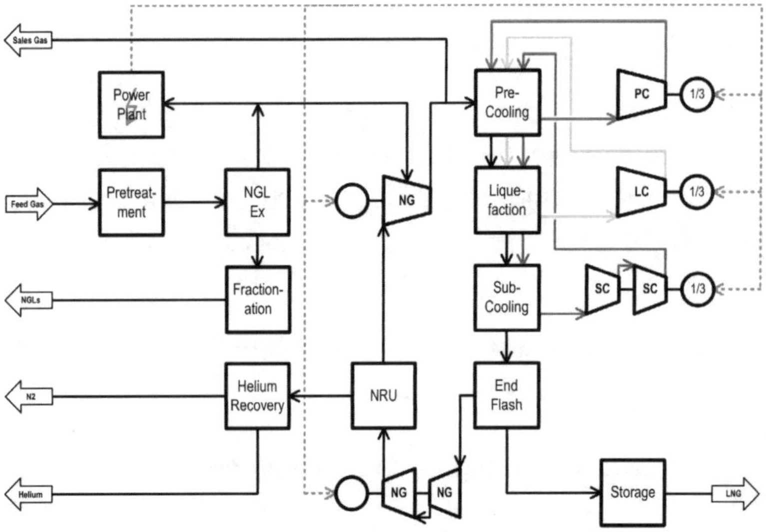

Fig. 6 Overall concept for a 2nd generation MFC® plant

If, however, deep LPG recovery or even C_{2+} recovery is economically recommendable, conventional scrub column configurations hit their limits. Close to the critical point high NGL recovery rates involve large quantities of co-recovered liquid methane in the scrub column bottoms due to the low relative volatility (k-value) between methane and ethane and/or propane. A more efficient and cost effective concept is a dedicated NGL recovery step between pre-treatment and liquefaction. It involves a cryogenic column (demethanizer or deethanizer), whose operating pressure is limited by the critical pressure of the bottoms to a range between 30 and 35 bar. Lean natural gas downstream NGL recovery is not only a perfect feedstock for an LNG plant, but can be sent also to a gas pipeline as sales gas. This requires recompression of the lean sales gas to typically about 100 bar. A common header for lean natural gas feeding both the pipeline grid and the LNG plant provides an attractive means to balance demand on either side.

Linde's base load LNG process is characterized by three independent refrigeration cycles. The intermediate (liquefaction) cycle and the coldest (sub-cooling) cycle use mixed refrigerants in any case, while the type of refrigerant in the warmest (pre-cooling) cycle depends mostly on the ambient temperature. The Mixed Fluid Cascade (MFC®) process is suffixed with a figure 3 in case propane is used for pre-cooling. Fig. 7 shows a simplified process sketch for the MFC®3 process with propane pre-cooling. As discussed above NGL extraction is not implemented via a scrub column,

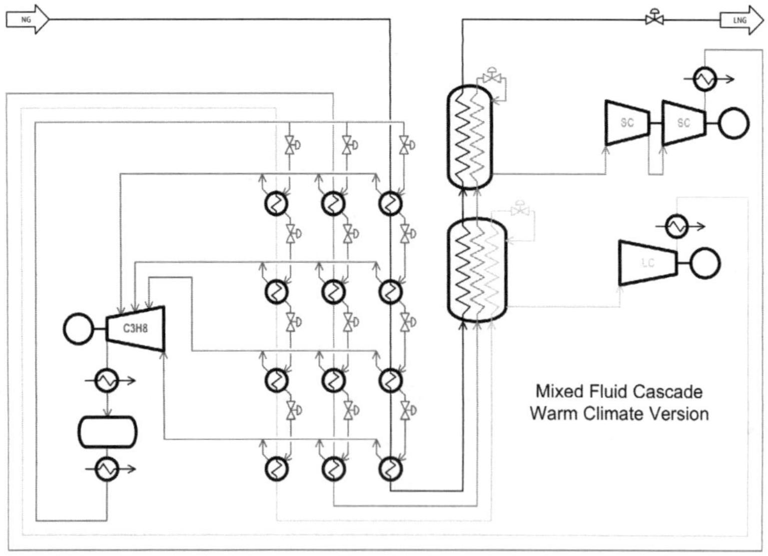

Fig. 7 Mixed Fluid Cascade (MFC®3) process for warm climate

but rather with an independent process step, which is not shown in Fig. 7. As a result of the upstream NGL extraction the lean and compressed natural gas does not cross a two-phase region. Now, there is no specific need to meet a certain process temperature by any of the three refrigeration cycles. Therefore, the load between the three refrigeration cycles can be balanced perfectly. That means the shaft power of all three compressor trains is identical. With such a configuration, which is unique amongst the present Base Load LNG technologies, the largest LNG capacities can be achieved with a given set of main compressor drives.

This load balanced concept is compatible with the use of propane pre-cooling only as long as the ambient temperature is not too low. In a warm Middle East climate a load balanced MFC®3 design results in a lowest operating temperature of the pre-cooling cycle with about -30 to -35 deg C. A lower limit for an economic propane pre-cooling temperature is -35 deg C. Below this limit propane compressor suction line sizes would grow significantly and the risk of air ingress into the closed cycle increases, when the propane system is operated too close to the atmospheric pressure. As soon as the relevant ambient temperature is consistently lower than above mentioned, either the propane pre-cooling compressor no longer can contribute with one third of the overall shaft power to the overall duty or the choice of the pre-cooling refrigerant needs to be reconsidered.

In a conventional plant design the end flash gas from the tail end of the LNG production frequently changes in quality and quantity, when operation requires process adjustments to external influences like plant load, ambient temperature, loading/holding mode, and so on. The result is modifications in the Wobbe index, which cannot be tolerated by highly efficient and clean burning (LoNox) gas turbines. Therefore, the end flash gas is not the preferred fuel source for modern gas turbines in any configuration. An alternative concept uses lean natural gas as gas turbine fuel and reprocesses the end flash gas in a nitrogen rejection unit (NRU). The NRU splits the end flash gas into clean nitrogen with less than 1 vol% methane, which can be sent to the atmosphere without further processing. The remaining hydrocarbon fraction will be re-liquefied either jointly with the feed stream to the LNG plant or separately within the NRU itself. The described method provides good quality fuel to the gas turbines at any required flow rate and maintains low nitrogen levels in the LNG even at high nitrogen concentrations in the natural gas. This enhanced operational flexibility does not impact the plant efficiency, but rather increases its availability.

As soon as the natural gas contains at least 500 vol ppm helium a commercial production can be considered. The best source for helium extraction in an LNG plant is the nitrogen fraction of the NRU. Technology for helium purification and liquefaction is widely available and has been implemented several times successfully.

Conclusions

Linde has successfully developed and implemented a new contender in the arena of world scale Base Load LNG processes. The new concept results in outstandingly low energy consumption, which is just 243 kWh/t_{LNG} in case of the Hammerfest LNG plant. This excellent efficiency can be achieved only with the implementation of three independent mixed refrigerant cycles.

The integration of carbon capture and sequestration (CCS) features reduces the overall CO_2 emission to about 110 t/h at 550 t/h LNG production, which is an exceptionally low value. Clever concepts for a warm restart with vaporized LNG further contribute to a very low carbon footprint. Nitrogen from the feed gas, which cannot be exported as LNG component, is cleaned up to less than 100 ppm hydrocarbons and can be vented without environmental concerns.

More recent concepts with propane pre-cooling (MFC®3) and perfectly balanced shaft power of all three refrigeration cycles have been applied to several projects and are in different phases of execution.

References

---

[1] Alf R. Jacobsen, Snøhvit, The history of oil and gas in the Barents Sea, Statoil publication, 2010

[2] e.g. US patent 6,253,574

[3] US Patent 6,827,138

[4] http://www.cbi.com/technologies-services/helixchanger-heat-exchanger/

[5] Jostein Pettersen, Experience with water cooled mixed refrigerant condensation, Proceedings of the 16th International Conference & Exhibition on Liquefied Natural Gas, Poster PO1-3, 2010, Oran, Algeria

[6] G. Skaugen, G. A. Gjøvåg, P. Nekså, P. E. Wahl, Use of sophisticated heat exchanger simulation models for investigation of possible design and operational pitfalls in LNG processes, Journal of Natural Gas Science and Engineering, Volume 2, Issue 5, November 2010, pages 235-243

[7] http://en.wikipedia.org/wiki/Campbell_diagram

[8] Jostein Pettersen, Øivind Nilsen, Snøhvit, A Successful Frontier LNG Project, paper presented at the 6th Annual LNG Tech Global Summit, 2011, Rotterdam, Netherlands

[9] Heinz Bauer, Modular Design of a Base Load LNG Plant, Proceedings of the International Gas Union Research Conference, Poster P4-11, 2011, Seoul, South Korea

AIChE 2012 Spring Conference
Control Panel - 12th Topical Conference on Gas Utilization (T6A), T6A00 LNG simulations and controls.

Fuel Gas System Design – Key to Reliable and Efficient LNG Plant Operation

Pei-Han (Jason) Yong, P.E.
Process Engineer
Bechtel Oil, Gas & Chemicals, Inc.

Christine Teh
Process Engineering Specialist
Bechtel Oil, Gas & Chemicals, Inc.

Bindupriya Bandari
Process Engineering Specialist
Bechtel Oil, Gas & Chemicals, Inc.

Robert A. Zinsmeister, P.E.
Control Systems Senior Engineer
Bechtel Oil, Gas & Chemicals, Inc.

Abstract

A well designed fuel gas system in an LNG liquefaction plant ensures a continuous, reliable, and properly specified fuel gas supply to all users while optimizing the thermal efficiency of the plant. Fuel gas system design development starts with identifying the fuel gas users, their needs, the applicable design standards or constraints, and Owner preferences. Fuel gas system design is also influenced by various plant operating modes such as plant start-up, normal operation, turn down, and upset conditions.

This paper will identify typical high pressure and low pressure fuel gas users in an LNG plant. Generally, the users with the most stringent fuel gas specifications are the gas turbines with their complex burner design. Therefore, the high pressure fuel gas is usually treated to meet the original equipment manufacturer (OEM) requirements. The fuel gas specification for several commercially available gas turbines used in LNG industry and its impact on fuel gas system design will be analyzed in this paper. The sources of fuel gas generally available within an LNG plant and their impact on overall thermal efficiency will be also be discussed. The process parameters used to tune the performance of the unit including supply pressure, temperature, composition, specific gravity, wobbe index and fuel gas heating value will be evaluated and described in detail.

A few case studies to demonstrate the robustness of the fuel gas system, advanced controls and analytical measurements required to facilitate various plant operating modes will also be addressed.

1.0 Introduction

An LNG liquefaction plant is designed for stable baseload production under different operating conditions. The continuous availability of fuel gas is very important for operating an LNG plant since it is used as fuel for the gas turbine mechanical drivers to drive the refrigeration compressors, gas turbine generators to generate electrical power, and fired heater to heat up the heating medium. The fuel gas system also supplies other miscellaneous users, such as flare pilots and purges.

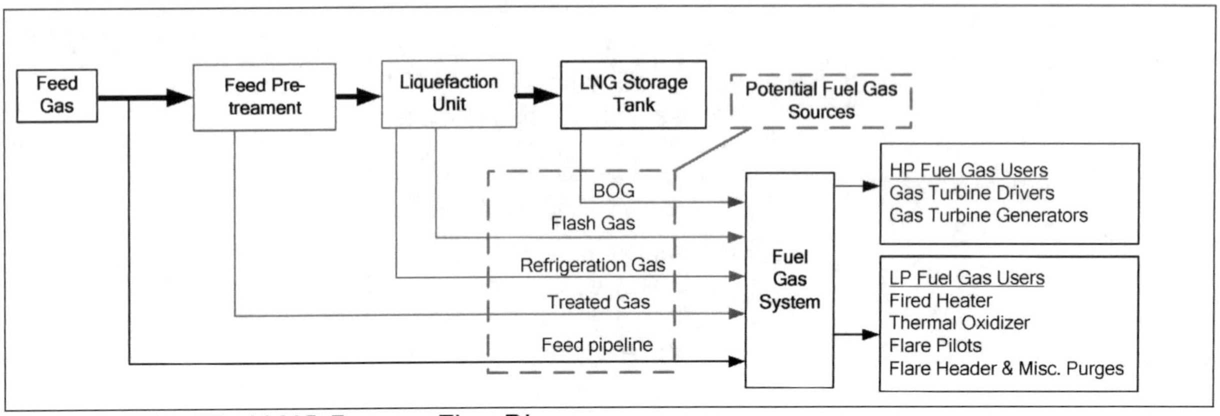

Figure 1. Simplified LNG Process Flow Diagram

Fuel gas for a stand-alone LNG liquefaction plant is supplied internally. There are several fuel gas sources within an LNG facility as shown in Figure 1. The typical fuel gas sources are from the incoming feed pipeline, feed pre-treatment, liquefaction unit, and BOG from LNG storage tank. The amount of heavy hydrocarbons sent to the fuel gas system are typically limited by the gas turbine combustor's specification, such that they need to be blended with the normal fuel gas supply to bring the heating value within the allowable range and allowable rate of change.

Fuel gas composition varies based on the various feed sources. The typical feed gas stream composition is approximately 90 mol% methane and the rest are inerts and heavy components.

2.0 Fuel Gas System Design Development

In general, the fuel gas system in an LNG plant is designed for at least two pressure levels: high pressure (HP) and low pressure (LP). The main users for HP fuel gas are the gas turbine mechanical drivers and gas turbine generators. Whereas, the main users for the LP fuel gas vary depending on the design of heating medium system and CO_2 / H_2S emission requirements. If the plant does not have a waste heat recovery (WHR) system to recover heat energy from turbine exhaust, the main user for the LP fuel gas is the fired heater providing heat to the heating medium. Another potential user for LP fuel gas is the flare system, which is governed by the emission requirements for CO_2 and H_2S. If the plant design includes a CO_2 sequestration unit, there will be no LP fuel gas demand from CO_2 and H_2S thermal oxidation.

The typical breakdown of HP and LP fuel gas users during normal operation are shown in Table 1. This table shows the typical breakdown for HP and LP fuel gas users in an LNG plant without a waste heat recovery unit. The Refrigeration Gas Turbines and Gas Turbine

Generators are the main users whereas only a small fraction of fuel gas is sent to the LP fuel gas system. During normal operation, the only LP users are the flare pilots and fired heater. LP fuel demand to the flare system can increase slightly during flaring events.

Table 1. Fuel Gas Breakdown in Absence / Presence of WHR

	Fuel Gas Breakdown	
	Without WHR	With WHR
LP Fuel	5%	2%
HP Fuel	95%	98%

Table 1 also shows the typical fuel gas breakdown in an LNG plant with a waste heat recovery unit. The LP fuel gas demand decreases slightly in terms of overall fuel gas demand due to the absence of a fired heater during normal operation. This translates to fuel gas savings and more LNG production. Technical and economical studies are performed for the selection of the waste heat recovery. The use of a fired heater is beneficial during startup when the gas turbines are unavailable to supply waste heat. Once the refrigeration gas turbine reaches the stable operating mode, the fired heater can be turned down gradually while ramping up the waste heat recovery system. The heating medium duty ultimately depends on the inlet gas components such as carbon dioxide and heavy hydrocarbons and also plant configuration. A higher carbon dioxide concentration will require a higher duty in the Amine Regeneration Unit to strip off the dissolved carbon dioxide from the solvent. The presence of heavier hydrocarbons will require a higher duty in the NGL Fractionation Units. These conditions may justify the use of the waste heat recovery system due to additional fuel cost savings and reduction in CO_2 emission.

Since the primary HP fuel gas users are the gas turbines for mechanical and power generation applications, there are specific compositional limits specified by the gas turbine manufacturers. Some of those limitations are supply pressure, temperature, MWI, and contaminant levels.

- The typical HP fuel gas supply pressure is between 25 barg to 45 barg.
- The HP fuel gas stream is superheated to at least 28°C (50°F) above the dew point temperature at the gas turbine fuel delivery pressure.
- A strict MWI range can limit the range of fuel gas composition and temperature to the combustors.
- Any contaminants (i.e. particulate, sulfur, liquid droplets) should be removed before the fuel gas is sent to the combustor.

These specifications, which generally vary from gas turbine types and manufacturers, will be discussed further in the next section.

Conversely, LP fuel gas compositional requirements are not as stringent as that for the HP fuel gas. Being one of the main LP fuel gas users, the flare unit is supplied with stable, dry and, continuous fuel gas to ensure smooth operation. The main source of LP fuel gas is from HP let down, when the pressure is let down to approximately 3.5 barg (50 psig).

Gas turbines with advanced combustor design have been developed to ensure that emissions meet the strict criteria for pollutants such as NO_x and CO. Specific emission requirements depend on the prevailing application and the local environmental permitting requirements and is checked on a project by project basis.

In order to lower NOx formation, DLN and DLE type combustors must operate at low flame temperature, shorter residence time in the flame zone, and lean fuel conditions as shown in Figure 2. The fuel is premixed with air to avoid forming hot spots in the flame during combustion [6].

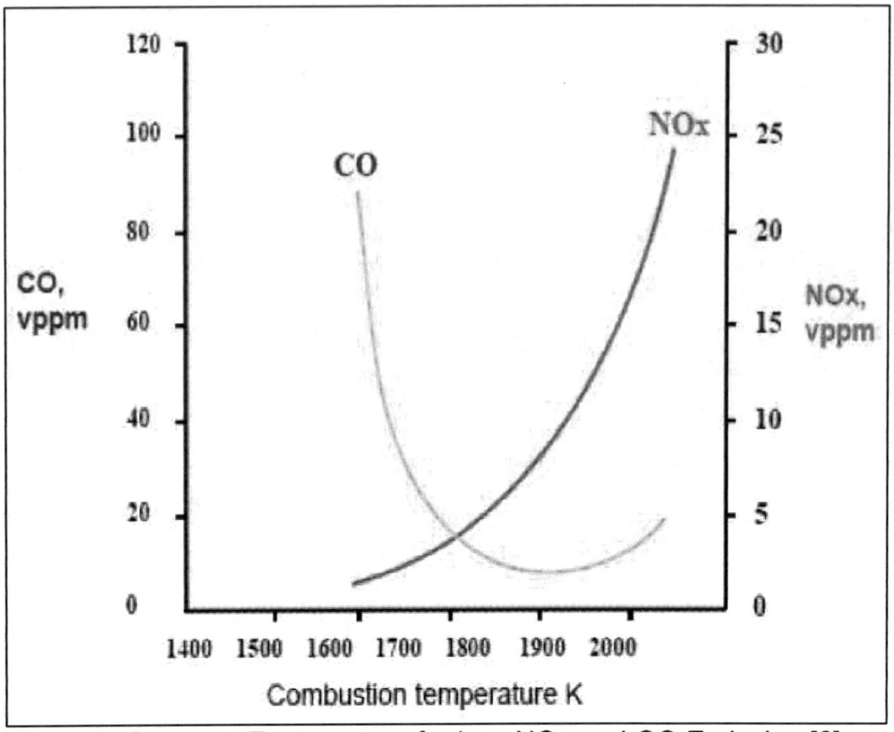

Figure 2. Optimum Temperature for Low NOx and CO Emission [6].

Since DLN and DLE combustor configurations are more sensitive to fuel gas composition variations, the following fuel gas specifications need to be met for uninterrupted operation:

2.1 Temperature

Fuel gas temperature at the fuel gas skid should be maintained at least 28°C (50°F) plus the higher of moisture or hydrocarbon dew point temperature at the gas turbine fuel delivery pressure to compensate for temperature drop when the pressure is reduced between the fuel gas filtration skid and the combustion chamber. Both hydrocarbon and moisture superheat requirements should be met to prevent any liquid droplet formation that can affect gas turbine performance and/or cause damage to combustor hardware. Superheat requirements should be analyzed on a case-by-case basis. For example: when water saturated regeneration fuel gas from the dehydration bed is used as the high pressure fuel gas source, moisture superheat will govern the superheat requirement for the fuel gas stream. Hydrocarbon superheat can become the governing case when there are heavy hydrocarbons present in the fuel gas stream, such as C6+. It is important to obtain a detailed fuel analysis and not just calculate the dew point based on "C4+" type analyzes. The dew point can be very sensitive to the heavier hydrocarbons if they are present.

2.2 Modified Wobbe Index (MWI)

MWI is a function of fuel LHV, temperature, and fuel gas molecular weight. MWI is defined as below:

$$MWI = \frac{LHV}{\sqrt{SG_{gas} x T_{gas}}}$$

(Equation 1)

where:

MWI	= Modified Wobbe Index (Btu/scf-°$R^{1/2}$)
LHV	= Lower Heating Value of fuel gas (Btu/scf)
SGgas	= specific gravity of fuel gas with respect to air
Tgas	= Fuel gas temperature (°R)

The supplied fuel gas MWI range must meet the gas turbines OEMs specifications. If two different fuel gases have the same MWI at the same pressure, the energy output is similar. The lower range of Modified Wobbe Index limits the amount of inerts in the fuel gas. If too many inerts are present in the fuel gas, this can affect the combustion and the energy output. The upper range of MWI limits the amount of heavy hydrocarbons (C2+) in the fuel gas. Fuel gas with heavier hydrocarbons produces more NOx during combustion. In addition, the presence of heavy hydrocarbons can lower the dew point of the fuel gas stream, which is undesirable considering liquid droplets are not allowed in the gas turbines.

The required MWI range for DLN combustors is between 40 and 54 Btu/scf-°$R^{1/2}$ and while the MWI range for DLE and SAC combustors ranges from 40 to 60 Btu/scf-°$R^{1/2}$ [4]. In addition to meeting the allowable limits, the rate of change in MWI should also be discussed with the gas turbines manufacturer. As shown in GE presentation slides, presented at the CORE Symposium, Houston, May 2011, the allowable MWI rate of change for GE Heavy Duty gas turbine is 1% per second [3].

2.3 Pressure

HP fuel gas supply pressure requirements are governed by the gas turbine burner, site minimum ambient temperature, and fuel gas composition. The typical range of HP fuel gas supply pressure is between 25 barg to 45 barg. The fuel gas supply pressure for Aeroderivatives gas turbines is generally higher than Heavy Duty gas turbines.

2.4 Contaminants

Fuel gas to the gas turbine should not contain any significant concentration of liquid droplets or solids. The presence of contaminants that include but not limited to particulates, liquids, and sulfur can damage the combustor and gas turbine hardware. Normally, there is a fuel gas filter located downstream of the fuel gas KO drum, to remove the solid particulates up to 5 microns in size. Stainless steel piping between the fuel gas filter and inlet flange to gas turbines can avoid corrosion products from entering the combustors [6]. Liquid contaminants can cause performance related issues such as firing temperature excursions, re-ignition and flashback, and damages to the hardware [4]. Liquid contaminants can be avoided by meeting the superheat requirements of the fuel gas. The presence of sulfur in the fuel gas stream can cause severe corrosion due to sulfuric acid condensation when used in combination with an HRSG unit [4]. Therefore, the allowable sulfur concentration should be discussed with the gas turbine vendor. The presence of trace amounts of diolefins can cause fouling and should be removed from the fuel gas [4]. H_2S in the presence of oxygen can form elemental sulfur. Elemental sulfur deposition downstream of the control valve can potentially plug the valve and lead to poor plant controllability and possible plant shutdown.

Feed to the liquefaction unit can contain a high mol% of nitrogen. The fuel gas system typically serves as a purge for nitrogen that cannot be condensed in the liquefaction unit. The maximum amount of nitrogen to the fuel gas system is restricted by the gas turbine combustor design. Nitrogen Rejection Units may be required to remove the additional nitrogen that cannot be sent to the fuel gas system. Tables 2 and 3 summarize the GE fuel gas specifications for DLN and DLE combustors.

Table 2. Summary of Fuel Gas Specification of GE DLN Combustor [4]

		Minimum	Maximum
Pressure		(Note 1)	
Temperature		(Note 2)	
LHV	Btu/scf	775	1160
MWI	Btu/(scf . $R^{1/2}$)	40	54
Variation of MWI		- 5%	+ 5%
Constituent Limits, %mole			
Methane		85	100
Ethane		0	15
Propane		0	15
Butane + C_4+		0	5
Hydrogen		0	Trace
Carbon Monoxide		0	Trace
Oxygen		0	Trace
Total Inerts (N_2 + CO_2 + Ar)		0	15

<u>Notes:</u>

1. *Fuel gas supply pressure varies based on the gas turbine model and combustor configuration.*
2. *Fuel gas temperature should meet both hydrocarbon and moisture superheat requirement. Whereas, maximum allowable fuel gas temperature varies with fuel gas pressure and is also limited by metallurgy.*

Table 3. Summary of Fuel Gas Specification of GE DLE Combustor [5]

		Minimum	Maximum
Pressure		(Note 1)	
Temperature		(Note 2)	
LHV	Btu/lb	18000	
MWI	Btu/(scf . $R^{1/2}$)	40	60
Constituent Limits, % mole			
Methane		50	100
Ethane (C2+ paraffins)		0	(Note 3)
Hydrogen		0	5
Diolefins*		0	None allowed
Other Combustibles**		0	Consult GE

* Diolefins - butadiene, propadiene
** Other Combustible - CO, olefin hydrocarbons, etc.

Notes:

1. *Fuel gas supply pressure varies based on the gas turbine model and combustor configuration.*
2. *Fuel gas temperature should meet both hydrocarbon and moisture superheat requirement. Whereas, maximum allowable fuel gas temperature varies with fuel gas pressure and is also limited by metallurgy.*
3. *Maximum allowable C_2+ paraffins for DLE combustor varies based on the type of gas turbine.*

In summary, there are several parameters that are controlled to maintain a stable supply of fuel gas to the high pressure fuel gas users. These parameters are:

1. Fuel gas pressure at the pre-mixing chamber to ensure smooth combustion process.
2. Fuel gas temperature entering the combustor to avoid any liquid droplets in the combustion chamber.
3. Fuel gas system should be designed to ensure that the fuel gas MWI and its rate of change are within the allowable range.

There are several ways to control each parameter in ensuring a continuous and reliable fuel gas supply to the gas turbines. One method is to install a robust control scheme that combines temperature and composition analysis to mitigate the rate of change of MWI. Another method is to install a fuel gas mixing drum, which will provide smooth transition of MWI due to sudden changes in the fuel gas source (i.e. fuel gas composition). The next section discusses a few design features and control schemes that can be used to handle various operating scenarios.

3.0 Fuel Gas System Design Features for Various Operating Scenarios

3.1 Temperature control to meet fuel gas requirements

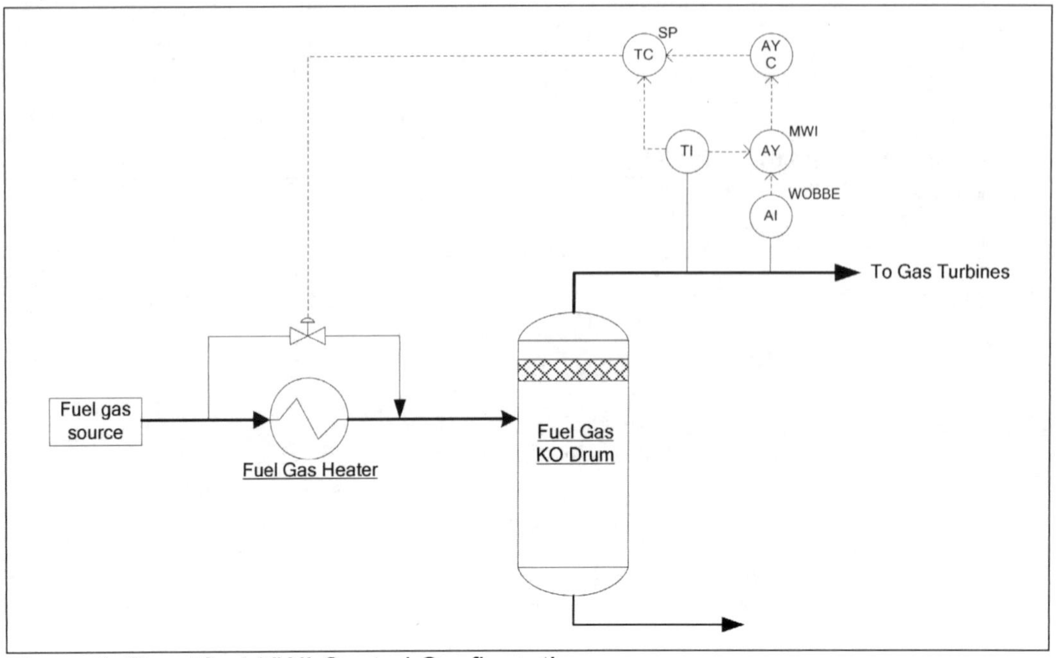

Figure 3. Simplified MWI Control Configuration

Figure 3 shows a simplified MWI control configuration. Fuel gas flows through a fuel gas heater, fuel gas KO drum, and fuel gas filter before it reaches gas turbine skids. Fuel gas temperature is adjusted to minimize the rate of change in MWI and also to ensure that the fuel gas is well above its dew point. A Wobbe analyzer provides the set point to the temperature controller. Any liquid droplets in the fuel gas should be knocked out at the upstream KO drum and there should be no compositional or MWI change between Fuel Gas KO Drum and inlet to the gas turbine skids.

3.1.1. Loss of fuel gas heater

The normal fuel gas source to the gas turbine from the liquefaction unit is dry and at least 28°C superheated. Therefore, the fuel gas heater duty is used primarily to ensure stable temperature to the users and the heating demand can be expected to be minimal. If the heating medium supply valve fails closed, the fuel gas temperature will be reduced. In the absence of fuel gas compositional changes, the MWI usually stays within the allowable limits and significant impact to the gas turbine performance is not anticipated.

If the alternate fuel gas source is instead used and gas temperature drops below its dew point, the fuel gas sources will be shut down to protect downstream gas turbine skids. One of the mitigation measures is to install a standby parallel electric heater to ensure heating duty is always available during plant operation.

3.1.2. Temperature spike from heating medium supply problem

If heat input to the fuel gas heater is accidentally increased, the temperature will rise and the bypass valve on the fuel gas heater will open to maintain a fixed temperature. If the temperature continues to rise, it will raise an alarm on the DCS panel and eventually shut down the fuel gas source to protect the gas turbines.

3.2 Equipment configurations to meet fuel gas requirements

When the fuel gas supply is switched to an alternate source, the gas turbine will most likely experience a MWI change due to compositional variation. The following mitigations is used to meet the rate of change MWI requirements and also to ensure that the MWI is still within the allowable limits:

i) If the compositional variation is significant and falls outside the MWI allowable range, the temperature can be adjusted to meet the target MWI using control configuration shown in Figure 3. A fuel gas mixing drum may be used as an alternate mitigation to dampen the step change of MWI.

ii) If the compositional variation is insignificant provided the superheat requirement is met, installing sufficient piping volume between Fuel Gas KO Drum and Gas Turbine skids can enhance fuel gas mixing and dampen the rate of change of MWI. The control configuration is simpler, in which the Wobbe analyzer is not connected to the temperature controller for this mitigation and is instead sent to the gas turbine controller.

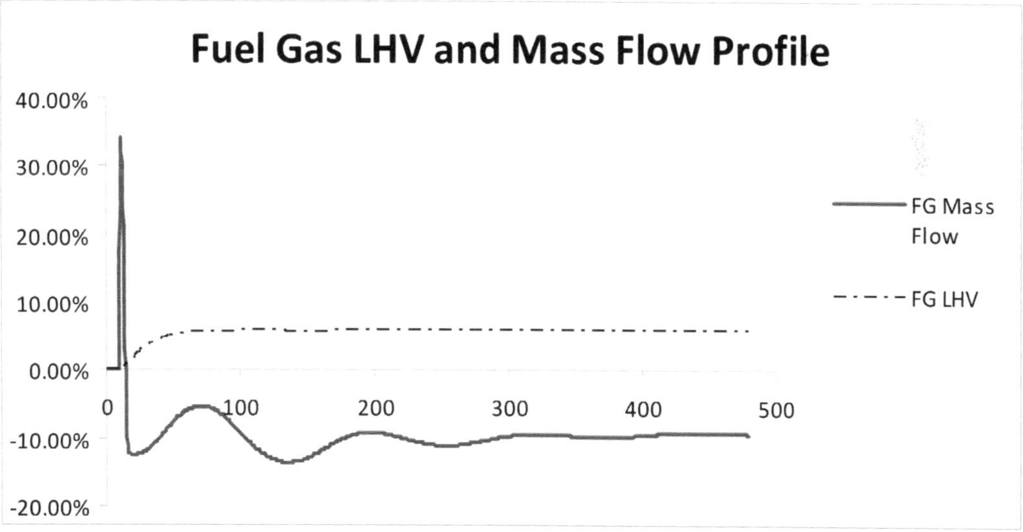

Figure 4. Fuel Gas LHV and Mass Flow Rate Profile

Figure 4 illustrates a typically step change in MWI and composition when switching from a lower LHV fuel source to a higher LHV fuel source. The MWI rate of change is shown to be approximately 0.2% per second, which is less than GE Heavy Duty gas turbine maximum specification of 1% per second [3]. The fuel gas flow is reduced when sending a richer fuel gas source to the gas turbine skids provided all other gas turbine operating parameters (e.g. ambient temperature, gas turbine speed, power output) are constant. An initial spike from the sudden opening of the control valve on the alternate fuel gas source can be noticed in Figure 4. Flow control is however restored in less than 350 seconds.

3.3 Pressure fluctuation in fuel gas system

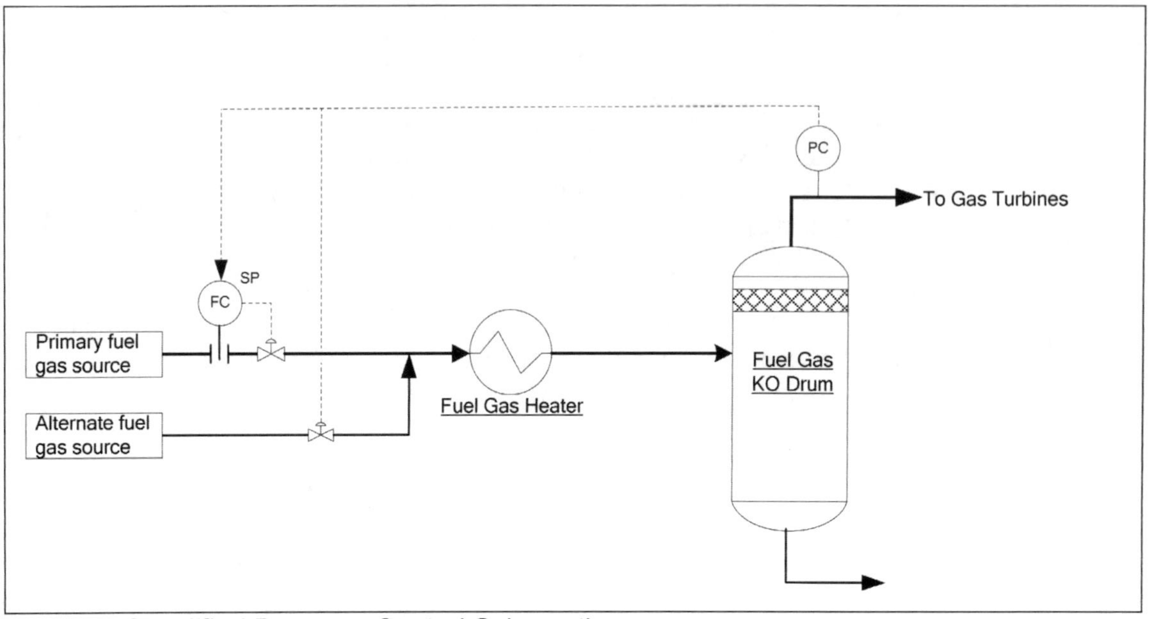

Figure 5. Simplified Pressure Control Schematic

Fuel gas supply for both high and low pressure systems need to be maintained at relatively constant pressure. Any significant pressure fluctuation can affect the fuel gas supply and overall plant performance. If the fuel gas supply losses pressure, the possible consequences are tripping of the gas turbines, the flare system, and the fired heater. In order to ensure stable fuel gas pressure, a combination of pressure and flow controller is used as shown in Figure 5. This figure illustrates a simplified pressure control schematic where a normal bleed from the primary fuel gas source along with a fast acting control valve is used to ensure smooth transfer when fuel sources are switched.

3.4 Fuel gas system design consideration during Start-up / Planned Shutdown

The typical fuel gas source for the GTGs during initial startup is the feed pipeline since most of the units are not ready. The pipeline feed gas may not only be in saturated condition, but the pressure let down from pipeline pressure to the fuel gas system pressure may lead to formation of liquid particles. Therefore, it is very important that the electrical startup fuel gas heater is in operation. During plant start-up, fuel gas will initially supply the flare units and the gas turbine generator to generate sufficient electric power to start the remainder of the facilities.

The fuel gas supply source is switched from pipeline feed gas to the fuel gas from the liquefaction unit once normal operation is established. The composition, MWI, degree of superheat requirement, temperature and pressure conditions are different for these two sources. These factors must be considered in the fuel gas system design to ensure that manual and automated fuel source switching will not trip the gas turbine operability and reliability.

For a planned shut down, the fuel gas demands will gradually reduce. However, a stable and continuous fuel gas purge to the flare unit should be maintained to ensure plant safety at all times.

4. Summary

This paper identifies the fuel gas users in an LNG Liquefaction plant and explains the importance of a well designed fuel gas system to meet all requirements such as, degree of superheat, pressure, MWI, LHV, contaminant particle size, and the composition limits. These factors are used to design a robust fuel gas system and to maintain stable and continuous fuel gas supply during various plant operation modes. A few real operating scenarios are shown to demonstrate how the fuel gas control system can react to meet OEMs requirements during startup or other upset conditions.

5. Acknowledgements

The authors would like to show their gratitude to Vinod Rajkumar, Himanshu Patel, and David Messersmith for their time to give us advice and motivation in preparing this paper. The authors would also like to thank Cyrus Meher-Homji for his technical guidance on gas turbine and combustor fuel requirement and James Kernan for his expertise on LNG plant commissioning and start up. Lastly, the authors would like to thank Dipanjan Bhattacharya for his help and support in developing the simulation for case studies.

Abbreviations

BOG – Boil Off Gas
CO – Carbon monoxide
DCS – Distributed Control System
DLE – Dry Low Emissions
DLN – Dry Low NO_x
GE – General Electric
GTG – Gas Turbine Generator
HP – High Pressure
HRSG – Heat Recovery Steam Generator
KO – Knock Out
LHV – Lower Heating Value
LNG – Liquefied Natural Gas
LP – Low Pressure
LPG - Liquefied Propane Gas
MWI – Modified Wobbe Index
NGL – Natural Gas Liquids
NO_x – Nitrogen oxides
OEM – Original equipment manufacturer
PPMV – Parts per million by volume
SAC – Single Annular Combustor
WHR – Waste Heat Recovery

References

[1] Badeer, G. H. (2005, September). *GE's LM2500+G4 Aeroderivative Gas Turbine for Marine and Industrial Applications*. Retrieved January 4, 2012, from http://site.ge-energy.com/prod_serv/products/tech_docs/en/all_gers.htm

[2] Davis, L. B., & Black, S. H. (2000, October). *Dry Low NOx Combustion Systems for GE Heavy-Duty Gas Turbines*. Retrieved January 4, 2012, from http://site.ge-energy.com/prod_serv/products/tech_docs/en/all_gers.htm

[3] Evulet, A. T. (2011, May). *Gas Turbine Combustion Technologies* [Presentation Slides]. Retrieved December 10, 2011, from http://www.coresymposium.com/2011_COREPresentations/Proven%20Solutions%20GE-Evulet.pdf

[4] GE Energy. (2009, January). *Specification for Fuel Gases for Combustion in Heavy-Duty Gas Turbines* (GEI 41040k).

[5] GE Energy. (2009, September). *Process Specification Fuel Gases for Combustion in AeroDerivative Gas Turbines* (MID-TD-0000-1).

[6] Meher-Homji, C. B., Zachary, J., & Bromley, A. (2010). *Gas Turbine Fuels - System Design, Combustion and Operability*. Proceedings of the 39th Turbomachinery Symposium. Turbomachinery Laboratory, Texas A&M University.

[7] Physical Properties. (n.d.). In *GPSA Engineering Data Book* (12th ed., Vol. 2, pp. 23-5). (Original work published 2004)

[8] Wien, S., Beres, J., & Richani, B. (2005, August). *Air Emissions Terms, Definitions and General Information*. Retrieved January 4, 2012, from http://site.ge-energy.com/prod_serv/products/tech_docs/en/all_gers.htm

Authors Biography

Bindupriya Bandari

Bindu Bandari is a Process Engineering Specialist for Bechtel's LNG. She also has design and field experience in LNG Re-gasification facility; and Safety Relief System for the gas processing facilities and refineries. She holds a Masters of Science, Chemical Engineering degree from Texas A&M University.

Pei-Han (Jason) Yong, P.E.

Pei-Han (Jason) Yong is a Process Engineer for Bechtel with experience on LNG liquefaction, gas processing and treating, and IGCC coal gasification projects. Mr. Yong holds a BS degree in Chemical Engineering from University of Texas at Austin and Master of Chemical Engineering from University of Houston. He is a licensed engineer in the state of Texas.

Christine Teh

Christine Teh is a Process Engineering Specialist for Bechtel working on LNG Liquefaction projects. She holds a Bachelor of Science in Chemical Engineering degree from the University of Texas at Austin.

Robert A. Zinsmeister, P.E.

Robb Zinsmeister is a Senior Control Systems Engineer for Bechtel's LNG Technology Group located in Houston, Texas. He has design experience in LNG, gas processing and refining. He also has field experience in construction, commissioning, startup and plant operation/maintenance. He is responsible for providing support to LNG projects and Control Systems discipline design standards.

Robb has ten years of experience in the Oil and Gas industry. He holds an B.A.Sc. degree in Electrical Engineering from Clarkson University and is a registered Professional Engineer in the state of Texas. He has previous work experience with ExxonMobil at one of their gulf coast refineries.

WASTE HEAT RECOVERY CONSIDERATIONS IN LNG PLANTS

DOUGLAS WILES; STANLEY KWAN; VINOD RAJKUMAR

ABSTRACT

Thermal integration in an LNG plant improves its overall plant thermal efficiency. Thermally integrated plants tend to have lower operating costs and reduced environmental emissions including, CO, CO_2, SO_x and NO_X.

Waste Heat Recovery Unit (WHRU) viability will depend on project specific factors including the availability of waste heat, type, quality, quantity and cost of fuel, type of heating medium and environmental regulations. The scope of the project, for example whether it is a green-field or retrofit construction, will determine the relative importance of many of these factors.

WHRU installation at the gas turbine driver exhaust requires higher initial capital. The economic viability will depend on the magnitude of the accrual of operating cost savings, and their ability to counteract the initial capital outlay. Savings can be in the shape of reduced fuel gas costs and reduced carbon dioxide taxes. Ultimately the impact of these savings will depend on the owner's measurement of the value of fuel gas; whether fuel usage is accounted for as lost feed or lost product. The negative impacts include the reduction in nitrogen rejection that occurs with reduced fuel gas usage and the power restrictions imposed on gas turbine drivers due to the increased back-pressure caused by the presence of the WHRU.

This paper investigates the factors impacting the viability of WHRU in an LNG plant and identifies scenarios where WHRU equipment can add value to a plant. A real world case study will be used to quantify the impact of these variables on the financial considerations of the WHRU.

1. INTRODUCTION

The cost of operating an LNG plant can be shown to be closely related to the plant's overall thermal efficiency, and its impact on the environment. LNG plants have sometimes been criticized for emitting greenhouse gases. The recovery of waste heat can minimize these concerns by integrating plant energy; however the technologies available to recover waste heat can be associated with higher initial capital.

2. WASTE HEAT RECOVERY UNIT DESIGN INTENT

The installation of a waste heat recovery unit (WHRU) enables the overall plant thermal efficiency of an LNG plant to be increased and consequently the use of fuel gas to be reduced. Lower fuel gas consumption enables a reduction in environmental emissions and a reduction of operating costs related to fuel use. The monetary impact of these savings will depend on the owner's fuel gas valuation methodology; whether fuel usage is accounted for as lost feed or lost product.

The commonly accepted source of waste heat in an LNG plant is from the exhaust of a gas turbine driver. Gas turbine drivers are used extensively in LNG plants to drive both refrigeration compressors and power generators. Gas turbine driver exhaust provides a high quantity of waste heat, as typical exhaust temperatures are in excess of 450°C (840°F). The waste heat can be recovered using heat transfer coils, placed in the exhaust stacks. A heat transfer medium such as hot oil or boiler feed water/steam can be circulated within the coil in a closed loop system to recover and distribute the heat to process users

The decision to implement waste heat recovery will be influenced by the Plant Owner's corporate goals, or may be a response to local environmental regulatory emission requirements such as carbon dioxide taxes. The Plant Owner must also consider the slight reduction in overall power available from a gas turbine driver when a waste heat recovery unit (WHRU) is installed on its exhaust stack. The installation of a WHRU also increases the total installed cost (TIC) of the plant and the operating complexity of the process.

3. WASTE HEAT RECOVERABLE

The maximum amount of waste heat available for recovery is dependent on gas turbine driver selection. In general, more waste heat is available from larger gas turbines due to increased exhaust flow rates.

Table 1.0 and Figure 1.0 show the exhaust heat recoverable for a series of gas turbine drivers typically used in the LNG industry. Both hot oil and steam applications have been summarized. Note that the steam analysis does not include co-generation and is purely heat recovered for the process users. Exhaust stack gas temperatures must be kept above a certain temperature to prevent moisture or acid gas condensation and subsequent corrosion. At the same time gas temperatures must be selected such that sufficient heat can be delivered to process users. Hot Oil temperatures in the 200°C to 250°C range are often required.

The data in Table 1 is based on a stack gas temperature of 250°C with the use of waste heat recovery, inlet air conditions of 1.013 bara @ 26°C and fuel gas conditions of 34.47 bara @ 25°C.

Table 1.0: Heat Recoverable from Turbine Exhausts for Hot Oil and Steam Recovery Mediums

	GAS TURBINE DRIVER TYPE			
	GE Frame 6B	GE Frame 7EA	GE LM2500+G4	GE LM6000
Fuel Gas Flow Rate (kg/s)	8,867	17,622	5,436	6,502
	Turbine Operating with Hot Oil Waste Heat Recovery			
Exhaust Temperature (°C)	557	545	527	475
Exhaust Pressure (bara)	1.03	1.03	1.03	1.03
Exhaust Flow (kg/hr)	503,640	1,026,000	294,840	394,200
Heat Recoverable (kW)	50,278	98,056	26,389	28,333
	Turbine Operating With Steam Waste Heat Recovery			
Exhaust Temperature (°C)	558	546	529	476
Exhaust Pressure (bara)	1.04	1.04	1.04	1.04
Exhaust Flow (kg/hr)	503,640	1,026,000	294,876	394,200
Steam Mass Flow (kg/hr) @50 bar, 315°C	82,944	161,064	43,164	44,568
Heat Recoverable (kW)	65,000	126,667	32,000	34,944

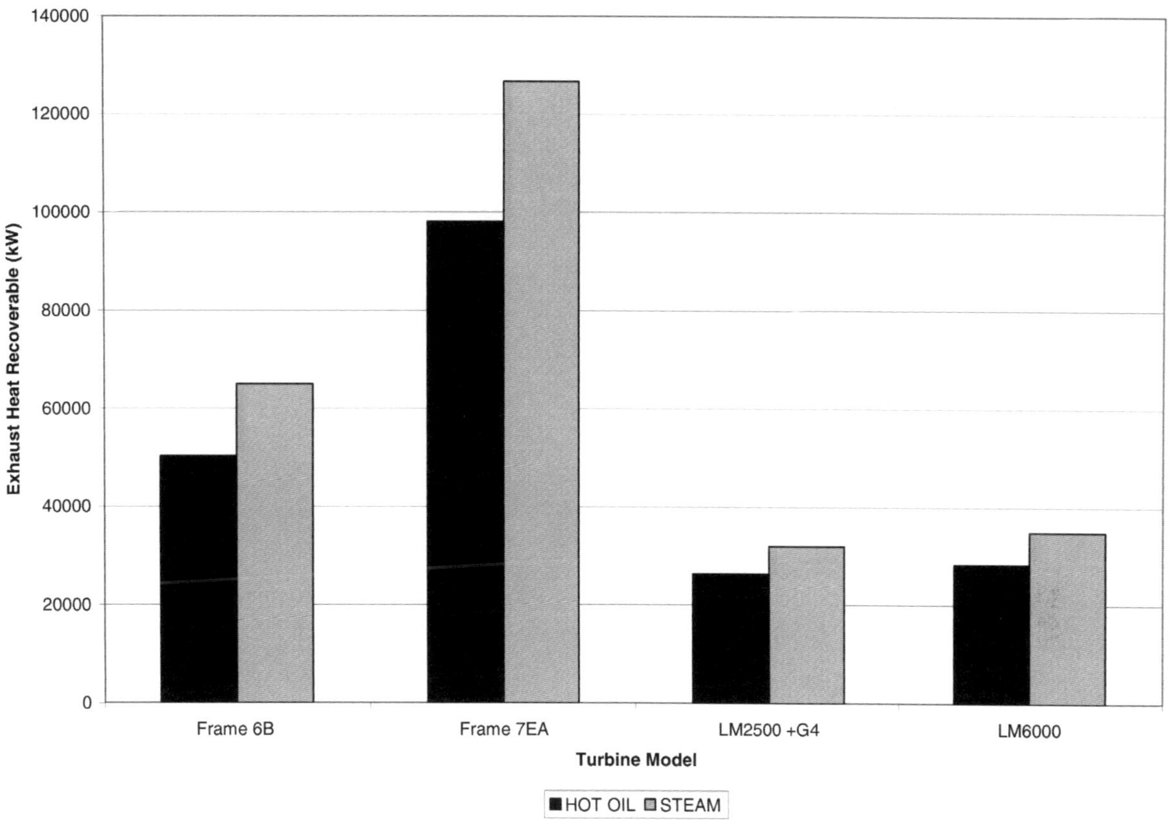

Figure 1.0: Heat Recoverable from Turbine Exhaust for Hot Oil and Steam Recovery Mediums

4. OVERALL PLANT THERMAL EFFICIENCY

Waste heat recovery increases gas turbine driver exhaust pressure loss, causing a slight increase in turbine heat-rate, decreasing available power. This pressure reduction is approximately 10 inches water for a typical WHRU coil arrangement (of which 6 inches account for normal losses in the stack and an additional 4 inches resulting from the waste heat recovery coil). For a Frame 6 gas turbine driver this equates to a loss of about 0.4 % available power.

The installation of a waste heat recovery unit will not significantly affect the overall fuel gas requirements. However elimination of the fired heaters decreases low-pressure fuel gas requirements. This net decrease in fuel gas consumption draw improves the overall plant thermal efficiency. Previous studies have shown a net increase of up to 1.4% in overall plant thermal efficiency is possible.

Overall plant thermal efficiency is defined as the ratio of the product higher heating value (HHV) to the plant feed gas higher heating value (HHV):

$$\eta_{TH} = \left(\frac{HHV_{product}}{HHV_{feed}} \right)$$

The overall plant thermal efficiency improvements for operating LNG facilities with a variety of duty/LNG production (MW/MTPA) ratios are shown in Figure 2.0.

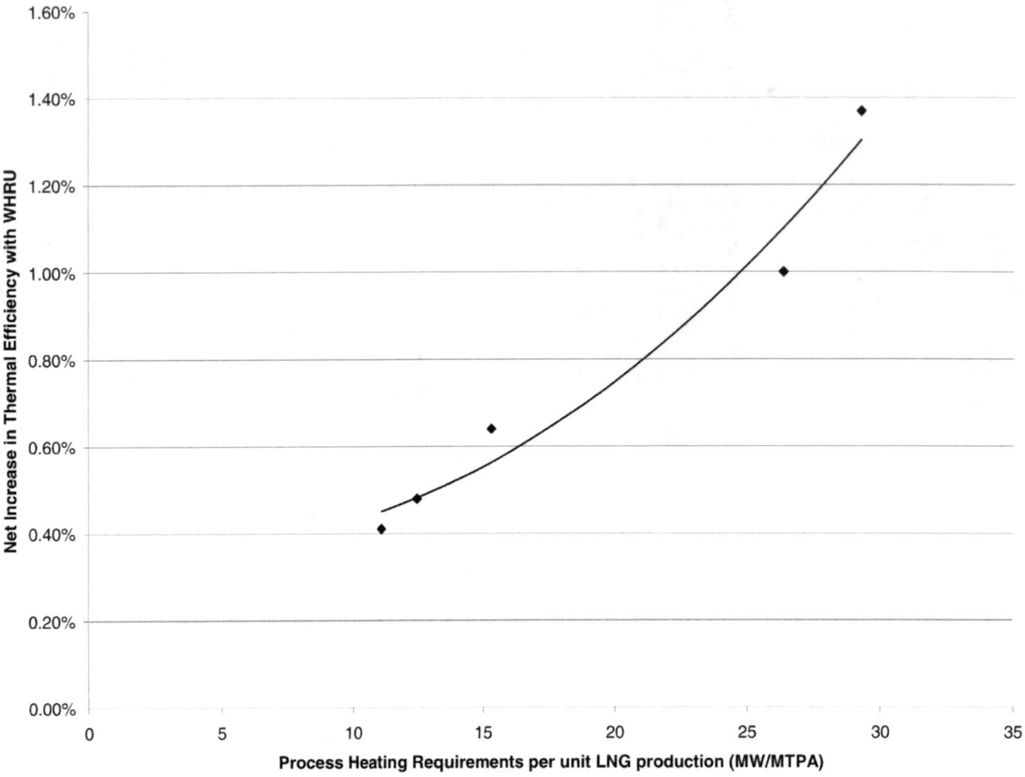

Figure 2.0: Overall Plant Thermal Efficiency improvements with waste heat recovery

Thermal efficiency is a function of many factors such as the feed gas inlet conditions including pressure, temperature, composition and ambient air conditions. Fuel gas can be obtained from both the front and back-end of an LNG facility. The withdrawal of fuel gas from the back-end of the LNG facility can provide a purge for unwanted inert gases, such as nitrogen. Inert gases can build up over time and consume unnecessary compression power, reducing overall plant thermal efficiency. As the fuel gas demand is reduced, so is the amount of nitrogen that can be rejected to the fuel gas system. Therefore plants with a high inert content in the feed gas, a Nitrogen Removal Unit may be required.

5. HEAT TRANSFER MEDIUM CONSIDERATIONS

5.1 HOT OIL

Use of synthetic heat transfer fluids is a common trend in the LNG plant. These are typically hydrocarbon based fluids with properties designed for long life, low degrees of fouling and low temperature pumpability. Modern synthetic heat transfer fluids are most often referred to simply as 'Hot Oil'. LNG plants require Hot Oils that can comfortably operate in a moderate temperature range; Hot Oil is typically supplied to users at approximately 200 – 250°C and returned at 110-130°C. Synthetic fluids resist the effects of oxidation many times better than mineral oils (manufacturers have quoted oxidation resistance up to 10 times that of comparable mineral oils). Less oxidation means less solids formation and much less fouling. Typical design parameters for hot oil include a minimum oil flow through the WHRU coils to keep the temperature below the maximum film and bulk temperature limits.

For a hot oil application, the temperature of the hot oil leaving each WHRU is controlled by dampers in the turbine exhaust stack; excess flue gas is diverted away from the hot oil coil through a bypass stack. Final temperature control is achieved by blending with an unheated bypassed hot oil stream.

In order to ensure availability of hot oil system at all times it is common to install an auxiliary/standby heater. A flow controller at the inlet of the WHRU allows the operator to balance hot oil flows between the WHRU and the auxiliary hot oil heater. Under normal operation, the auxiliary hot oil heater will be operating at minimum duty.

5.2 BOILER FEED WATER/STEAM

Steam as a heat transfer medium has a higher heat transfer coefficient compared to hot oil. As a result, steam requires a lesser surface area compared to hot oil for a given heat demand.

Steam is also fire resistant, so the safety issues associated with hot oil and direct heating are minimized. Steam has an unlimited fluid life, unlike hot oils and can be readily available.

The preparation and production of steam requires significant monitoring and treatment of the boiler feed water to remove minerals which could cause deposition and therefore fouling in heat exchangers and pipes. In addition steam systems with cogeneration require piping and equipment rated for high pressures. In addition there are high maintenance costs for the removal of scale, the repair of steam traps and leaks. The high costs associated with steam have to be balanced with the ability to recover more heat from gas turbine driver exhaust streams than hot oil.

5.3 REGENERATION GAS HEATING

There are specific applications where the circulated fluid can be directly heated in the gas turbine driver exhaust utilizing a WHRU coil. One common example is the heating of the regeneration gas used to dry the molecular sieves in the dehydration beds.

Direct heating does however come with some safety considerations. It is considered possible for the regeneration gas tubing welds to fail within the exhaust duct, resulting in the ignition of the regeneration gas. Such a problem can be resolved by locating all welds on heat recovery coils outside of the WHRU ducting.

The maximum allowable film temperature for regeneration gas is higher than that of hot oil; a typical value is around 400°C.

6. START-UP/CONTROLLABILITY

Increased heat integration makes the process more complex. Waste heat recovery provides a less flexible heat source than fired heaters, since waste heat is only available when the turbine is running.

Replacing fired heaters with waste heat recovery can complicate first time start-up. Most refrigerant gas turbines are not available early in start-up; only waste heat from power generation turbines is available. A start-up fired heater therefore may be beneficial to provide the heat necessary to dry out the dehydration system and start the acid gas removal unit. However during normal operation WHRUs require less maintenance than fired heaters.

The control scheme for a WHRU is often more complicated than that for a fired heater. If the same unit heats the regeneration gas and the hot oil, for example, the fact that the regeneration gas is an intermittent service can cause difficulties in controlling heat supplied to the hot oil.
Waste heat recovery schemes have found extensive application the in Oil and Gas industry, as such safety issues associated with their installation and operation have been well documented and managed. The source of waste heat and the choice of heat transfer medium will determine the potential risk of tube leak and ignition. The degree of risk and the level of safeguards should be addressed every time when waste heat recovery is considered.

Waste heat recovery for gas turbine driver exhaust is considered safer than fired heaters because of indirect heating. A WHRU can be considered as a shell and tube heat exchanger, and must be designed in accordance with API 660.

7. CONSTRUCTABILITY

A sketch of a simplified WHRU arrangement is shown in Figure 3.0.

Plot space must be made available for the sizeable ductwork required. The ductwork will likely be in the main process area, where the large turbines are located providing challenges in the plant layout design.

During start-up of the refrigerant gas turbine, it is possible that the mechanical coils installed in the exhaust ducting can be exposed to high temperatures prior to the introduction of heat transfer medium. Thermal cyclic fatigue can occur in intermittent services, for example, regeneration gas heating requirements are dictated by dehydration bed sequencing. The material of construction of the heat transfer coils for WHRUs must be chosen such that the coils can run dry, withstand thermal cyclic fatigue and avoid moisture and acid gas condensation.

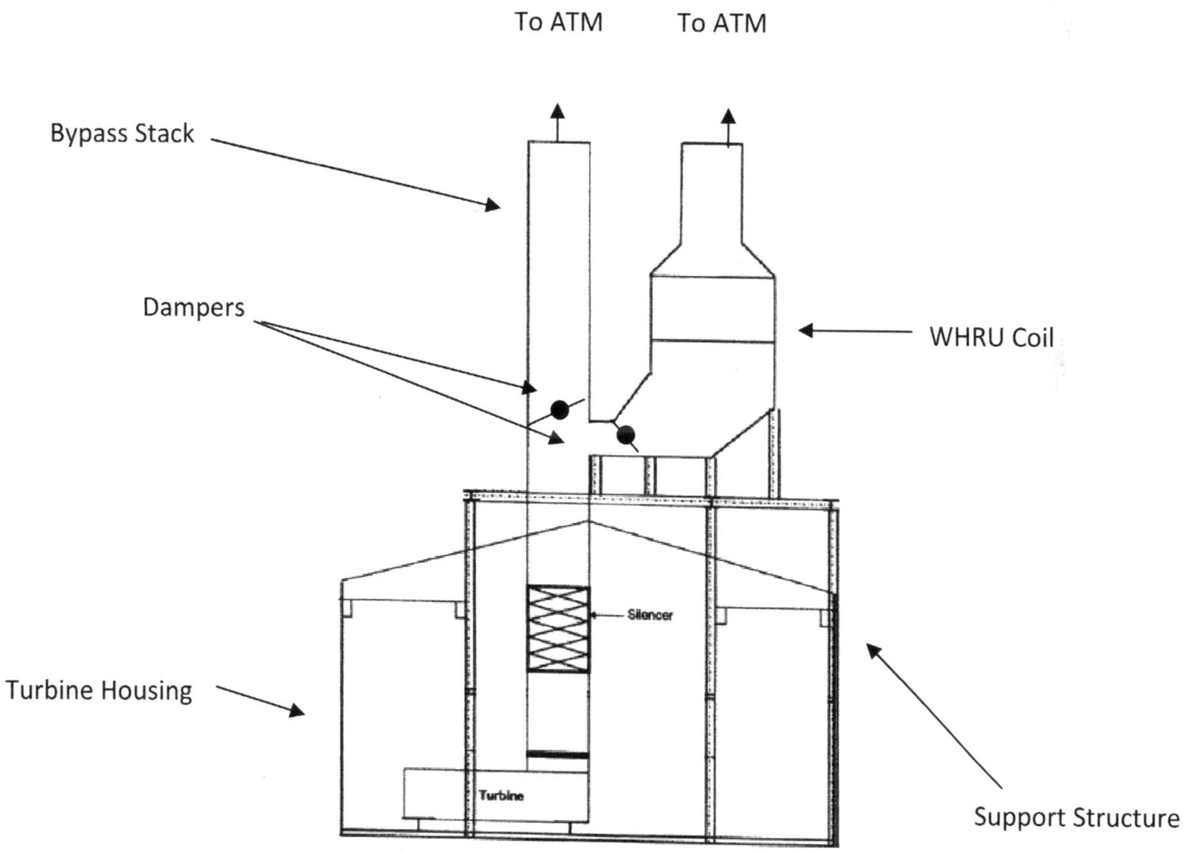

Figure 3.0: A Simplified WHRU Layout

It is possible to design the waste heat recovery unit to include an arrangement of multiple coils, containing different fluids, within one common exhaust. A combination that is generally used includes regeneration gas and hot oil or steam coils in the same gas turbine driver exhaust.

With multiple fluid services included in the same exhaust stack, there are design considerations on how to orientate the coils with respect to one another. The coil that is installed lower in the stack will be exposed to the hottest exhaust gas. Similarly coils further up the stack will be exposed to cooler exhaust gas; the lower coil having absorbed a portion of the exhaust heat before the exhaust comes into contact with the upper coil.

Regeneration gas is typically heated to a much higher temperature than hot oil. An exit temperature of approximately 315°C is generally required for dehydration bed regeneration. Regeneration gas entering the WHRU coil is typically 20°C. Such a large temperature gradient favors a design that locates the regeneration coils closer to the hotter gas or lower in the exhaust stack. On the other hand, hot oil entering the coil at 130°C is heated to approximately 250°C. Therefore, locating the hot oil higher up in the coil bundle is desirable. It can be seen that such an arrangement does not come without its problem. The regeneration gas heating duty is cyclical; hot regeneration gas only being required in the heating stage of the absorption/regeneration cycle of the dehydration beds. During the remaining stages of the heating/regeneration cycle the heat sink provided by the regeneration gas will not be available. As a result hot oil will be exposed to temperature swings. It is important to understand these swings when configuring coil arrangement.

8. FINANCIAL CONSIDERATIONS

An effective economic model will provide the owner/developer of an LNG plant with a guide to evaluate the factors that impact a financial decision as to whether to invest in a WHRU. This section outlines the inputs to such a model.

WHRU total installed costs can include the cost of ductwork, high temperature tubing, trim coolers, back-up fired heaters, plot space inside the battery limit and multiple structures to support the units on the exhaust stacks of multiple turbines. The economic viability will depend on the accrual of operating cost savings, and their ability to counteract the initial capital outlay. Savings can be in the shape of reduced fuel gas costs and reduced carbon dioxide emissions. Ultimately the impact of these savings will depend on the owner's measurement of the value of fuel gas that would otherwise be burnt in a fired heater. The owner could account for fuel gas usage as lost feed or lost product.

In order to determine the attractiveness of a WHRU, it is necessary to perform a comparison of the cash flows associated with waste heat recovery and those associated with a fired heater.

The cumulative cash flows (CCF) for fired heater and waste heat recovery unit can be represented by the model shown below:

$CCF_{\text{Fired heater or WHRU}}$ = -(Capital Cost,\$) - (Annual Fuel Gas Expenses, \$) – (Annual CO_2 taxes, \$)

A typical cash flow model for an LNG plant comparing various waste heat recovery options is presented in Figure 4.0

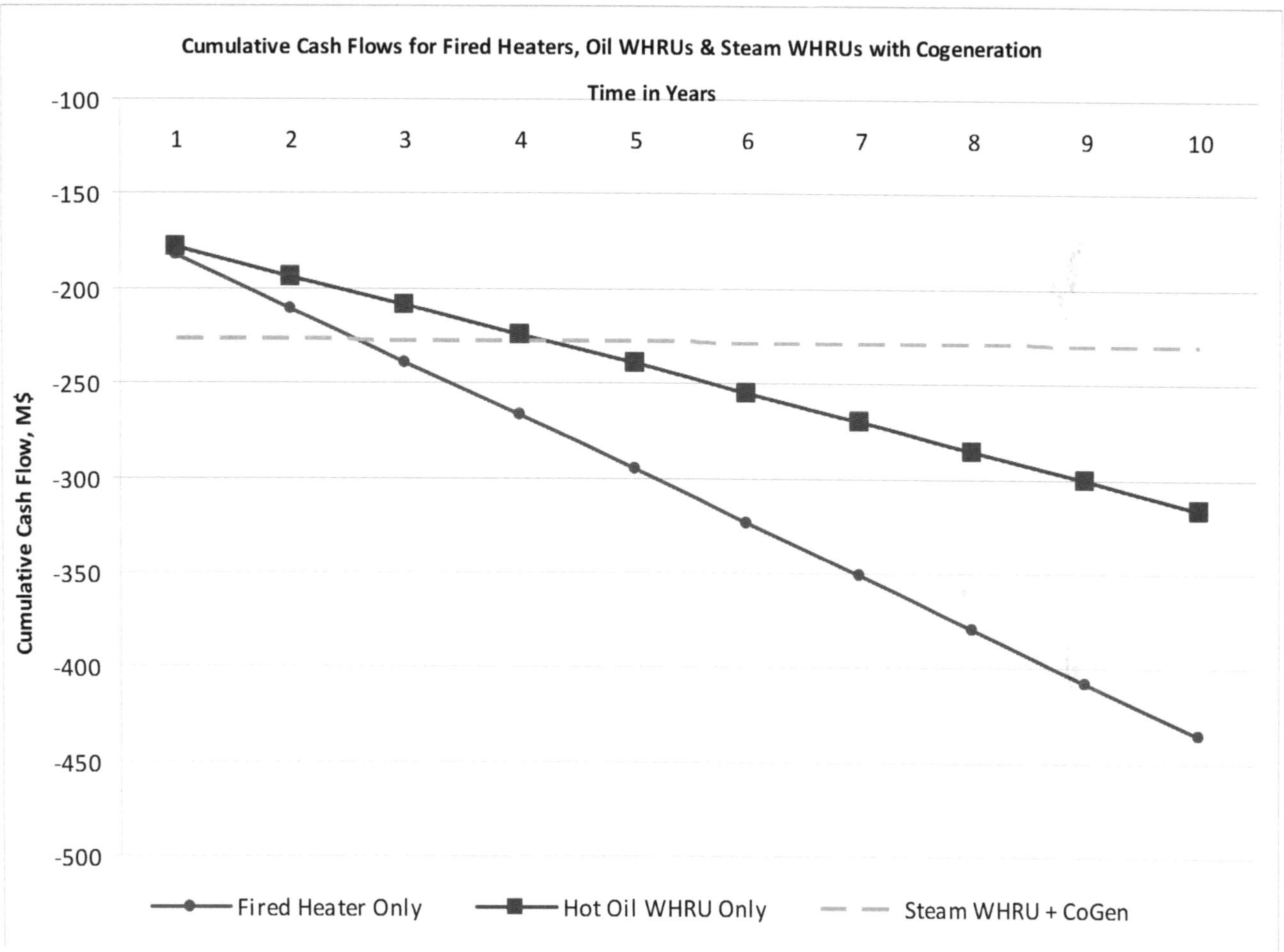

Figure 4.0: Cumulative Cash Flow Chart for Fuel Gas valued at 3$ / MMBTU

The input variables for the model are summarized in Table 2.0

Table 2.0: Total annual savings from Waste Heat Recovery

Description	Fired Heater Only	Hot Oil WHRU Only	Steam WHRU with Cogeneration
Total Fuel for process heat + electric power, MMBtu/hr	800	437	9
Assumed Fuel Cost $/MMBtu	3$	3$	3$
CO_2 tax ($/tonne)	20	20	20
Total Annual Cost for fuel and CO_2 ($ millions/year)	28 M$	15 M$	0.3 M$
Total Annual Savings from Base Case ($ millions/year)	0	13 M$	28 M$
Total Installed Cost ($ millions)	yy M$	1.05 * yy M$	1.45 * yy M$

A similar cash flow model can be developed for fuel gas valued as LNG product, say 8 $/MMBtu.

From figure 4.0 it can be inferred that eliminating fired heaters and providing all process heat from waste heat can provide significant reductions in fuel consumption and emissions with relatively little increase in capital cost and operating and maintenance complexity. The greater the process heat requirements the greater the savings.

Using steam for process heating and electric power generation offers the potential for even greater savings in fuel and emissions, although there could be a significant increase in capital cost and equipment complexity. A similar analysis would be conducted on a case-by-case basis to estimate expenses or cost savings associated with the different waste heat recovery options.

9. EXTERNALITIES

Owner priorities and environmental regulations could both influence the decision as to whether to install waste heat recovery. Future technological developments are likely to make waste heat recovery more attractive to the plant owner. Gas turbine driver suppliers may look at leveraging synergies between the waste heat recovery unit and the turbine, becoming the single point of responsibility for the entire unit.

Some LNG plant owners may have corporate policies that encourage the justification of additional expenditure in order to comply with environmental restrictions or statutes in the country in which they operate.

10. SUMMARY

This paper investigates the factors impacting the viability of WHRU in an LNG plant and identifies scenarios where WHRU equipment can add value to a plant.

11. REFERENCES

1. C. B. Meher-Homji, H. Patel, V. Rajkumar, D. Messersmith; Bechtel Corporation, J. Rockwell; ConocoPhillips Company ***"Thermo-Economic Analysis of Combined Cycle Based Liquefaction Plants"*** GASTECH 2011 Amsterdam, Netherlands 21-24 March 2011

2. Norrazak Hj. Ismail, MLNG Tiga Sdn Bhd, PETRONAS and Thomas Roy Stuart, Foster Wheeler Energy Ltd, Malaysia **"The Train 7 Fire at PETRONAS' LNG Complex, Bintulu, Malaysia"** LNG journal July/August 2005 page 12-14

3. Solutia Inc, audience guide slides **"Heat Transfer System Options and Considerations in Fluid Selection"** Conrad E. Gamble, MTS Fellow.

12. AUTHORS

Douglas Wiles

Douglas Wiles is a Process Engineer for Bechtel Houston, Texas. He has worked on multiple LNG FEED and EPC projects from conceptual design to commissioning and start-up.

Douglas has ten years of experience in LNG and Gas Processing. He holds an MEng degree in Chemical Engineering from the University of Nottingham, United Kingdom, and an MBA from the University of Texas, USA. He is a chartered engineer in the United Kingdom.

Stanley Kwan

Stanley Kwan is a Process Technology Principal Engineer of the LNG Technology Group for Bechtel Houston, Texas. He is responsible for providing project support of LNG Technology and conducting LNG Product Development Studies, and Technical Research.

Stanley has over thirty years of experience in refining and heavy oil upgrading technologies and process design for LNG, refinery and petrochemical projects. He holds a B.A.Sc. degree in Chemical Engineering from the University of Toronto, Ontario, Canada, and is a registered professional engineer in the state of Texas and the province of Ontario, Canada.

Vinod Rajkumar

Vinod Rajkumar is the Manager of the LNG Product Development Center – for Bechtel Houston, Texas. He has served in various roles in LNG Projects for the last 11 years. His LNG experience includes everything from conceptual design through commissioning, start-up and testing. He holds a Bachelor's of Science degree in BioChemical Engineering from National Institute of Technology, Jalandhar.

(25d) Demonstration Test Result of High Pressure Acid-Gas Capture Technology (HiPACT)

Koji Tanaka, Advanced Fuels Research & Development Division, JGC Corporation, Yokohama, Japan, Takehiro Komi, Domestic Project Division, INPEX Corporation, Niigata, Japan, Yasushi Fujimara, Research & Development Division, JGC Corporation, Ibaraki, Japan, Torsten Katz, Gas Treatment Process Technology, BASF SE, Ludwigshafen,, Germany, Oliver Spuhl, Sour Gas Treatment Process Engineering, BASF SE, Ludwigshafen, Germany, Erick Contreras, OASE Gas Treating Excellence, BASF East Asia Regional Headquarters Ltd., Hong Kong, Hong Kong and **Kumar Sugavanam**, Gas Treatment Process Technology, BASF Corporation, Houston, TX

The technology HiPACT (High Pressure Acid-gas Capture Technology), jointly developed by JGC and BASF, is an advanced CO2 capture technology to contribute to economical CCS implementation in natural gas processing plants. The solvent has high stability against thermal degradation, which enables high pressure and elevated temperature operation during solvent regeneration to reduce CO2 compression cost downstream. The solvent also has a higher CO2 absorption capacity than commercially available solvent technologies, which results in decreased solvent circulation rates and lower AGRU (Acid Gas Removal Unit) capital and operating costs.

The demonstration test was conducted in INPEX's Koshijihara natural gas plant in Nagaoka City. INPEX consistently improves the natural gas production efficiency to reduce environmental impact. As for the AGRU technology, DGA was initially selected, followed by aMDEA(TM) then finally by HiPACT for demonstration purposes. A commercial AGRU train was temporarily modified and aMDEA(TM) solvent was swapped with HiPACT. Treated gas from HiPACT was sent to the pipeline via dew point control system during the demonstration test.

HiPACT showed excellent performance in CO2 absorption; 20% reductions of solvent circulation rate and reboiler duty compared with aMDEA(TM) were achieved, keeping CO2 specification at absorber overhead unaffected. The temperature profile in the absorber indicated that the CO2 absorption kinetics was much higher compared to aMDEA(TM). HiPACT showed much sharper and narrower temperature bulge within the lower packing bed in contrast to aMDEA(TM)'s broader bulge. There was no negative indication for operation such as foaming. Through numerous data acquisition, the process modeling and simulation tools were updated.

(25e) Application of Exergy Analysis for LNG Plants and Receiving Terminals

Chen-Hwa Chiu, Chevron Energy Technology Company, Houston, TX and Lixin You, Chevron Technology Ventures LLC, Hosuton, TX

Natural gas liquefaction is an energy-intensive process. New world scale LNG plant usually has multiple trains with each train has a very high capacity (up to 8 million tons per annum of LNG). Increasing the plant efficiency can increase LNG plant output and economics, with reduced greenhouse gas emission.

On the other hand at LNG receiving terminals, LNG needs to be regasified to be put in pipelines to deliver to customers. LNG has a great cold energy potential. This cold energy can be used to generate power. The power generation scheme needs to be optimized to fully utilize the LNG cold energy potential.

Exergy analysis is based on the Second Law of Thermodynamics. It can identify the exergy loss distribution in the whole LNG plant or receiving terminals. Improving exergy efficiency for the unit operations with highest exergy loss will result in increased production, lower power consumption and better overall economics. For power generation using LNG cold energy, exergy analysis can identify the best power generation scheme for different LNG feed, composition and required delivery pressure.

This paper first review applications of exergy analysis in LNG plants and import terminals and then present major exergy analysis results and discoveries. Finally, recommendations for improving exergy efficiency in both LNG plants and receiving terminals are proposed.

(25f) Maximizing LNG Capacity for Liquefaction Processes Utilizing Electric Motors

Jonathan Berg, Air Products and Chemicals, Inc., Allentown, PA and Yu-Nan Liu, Air Products & Chemicals, Allentown, PA

In some locations currently being considered for LNG projects, relatively low cost electricity from the local power grid may be available. For these facilities, refrigeration compressors driven by electric motors drawing power from the grid may be attractive. The fuel requirements for these plants will be relatively low since there will be no need for fuel to drive refrigeration compressors or to generate electricity. This presentation will describe several natural gas liquefaction process configurations that are particularly amenable to LNG facilities without substantial fuel requirement. The applicability of the process configurations based on specific project requirements will also be discussed.

(37a) Impact of Low Levels of Ammonia in Syngas on the Fischer-Tropsch Synthesis Performance of Cobalt and Iron Catalysts in Fixed-Bed Operation

Heinz J. Robota, University of Dayton Research Institute, Dayton, OH, Jhoanna Alger, Energy and Environmental Engineering, University of Dayton Research Institute, Dayton, OH and P. Julius Pretorius, AITF, Edmonton, AB, Canada

The potential for liquid fuel production using the Fischer-Tropsch synthesis with biomass as the carbon source is confronted by a variety of practical issues. Among these is the ability to cost effectively reduce a variety of potential catalyst poisons to acceptably low levels within the capital constraints of a modest capacity biomass-based plant. As noted by Tsakoumis*et al.*in connection with the deactivation cobalt-based FT synthesis catalysts, "Little is known on the effect of nitrogen containing compounds and the available information is mainly found in patents".[1]We have undertaken an investigation into the effects of ammonia on both cobalt and iron FT synthesis catalysts operating under fixed-bed conditions at ammonia levels between 6 ppmv and 300 ppbv. Conversion performance after exposing the Fe catalyst to a steady 6 ppmv ammonia in the syngas remains unaffected relative to the performance in an ammonia-free gas. Conversion performance is stable over several hundred hours of operation without detectable deactivation. In contrast, two different cobalt catalysts both exhibit considerable sensitivity to the presence of ammonia in the syngas. The loss of conversion performance is a linear function of the total ammonia exposure until a N:Surface Cobalt ratio near 0.15 is reached, at which point half of the catalyst's activity is lost. The impact of each adsorbed N entity is quantitatively similar to the impact of S observed during H_2S exposure. Unlike the effect of S, the impact of N entities on cobalt is self limiting. Further exposure beyond N:Surface Cobalt = 0.15 results in no further change in performance. Altered performance persists unchanged for over 100 hours after removal of ammonia from the syngas stream. While the measured α value of the cobalt catalysts is unaffected, the relative selectivity to olefins and alcohols increases.

1) Tsakoumis, Nikolaos E.; Rønning, Magnus; Borg, Øyvind; Rytter, Erling; Holmen, Anders;*Deactivation of cobalt based Fischer-Tropsch Catalysts: A Review*; Catalysis Today 154 (2010) 162-182

(37b) The Synergism of Natural Gas Conversion with Biomass for Maximizing Biofuels Production

Daniel Resasco[1], Lance Lobban[2]and **R. G. Mallinson**[2], (1)School of Chemical, Biological and Materials Engineering and Carbon Nanotube Technology Center, University of Oklahoma, Norman, OK, (2)Chemical, Biological, and Materials Engineering, University of Oklahoma, Norman, OK

Hydrogen from natural gas maximizes the liquid fuel yield from thermochemical biofuel production pathways. This produces the greatest net fossil CO2 reduction compared with biomass produced hydrogen and renewable wind and solar based hydrogen production. This also appears to result in the lowest overall cost per unit of liquid fuel produced.

(37c) Utilization of Supercritical Fluids As Reaction Media for the Synthesis of Higher Alcohols From Syngas

Rui Xu, Sihe Zhang and Christopher B. Roberts, Department of Chemical Engineering, Auburn University, Auburn University, AL

Synthesis of higher alcohols, isobutanol in particular, from synthesis gas (syngas) has been attracting interest due to the potential for producing clean and high octane value fuels. Higher alcohols could be either used directly as fuels, as fuels additives for octane enhancement, or as feed stocks in the chemical industry. The abundance of various carbonaceous sources for producing syngas, including coal, natural gas and biomass, makes this process appealing as an alternative to non-renewable petroleum resources. However, due to the highly exothermic nature of the reactions and the underlying kinetic, this procedure suffers from poor productivity and selectivity towards higher alcohols preventing a large scale and economically feasible commercialized process.

For the last two decades, there has been intensive research centered on shifting the synthesis away from methanol and towards higher alcohols, including; i. the use of modified catalysts, ii. utilizing dual-bed reactors, or iii. slurry phase reactors. In this study, a supercritical medium (i.e. hexanes) was employed to investigate the effect of supercritical solvent on the formation of higher alcohols.

In this work, 0.5 wt% K promoted $CuCo/ZnO/Al_2O_3$was investigated as a catalyst for the synthesis of higher alcohols from syngas. The catalyst was prepared by coprecipitation methods and tested in a fixed bed reactor using a supercritical hexanes as the reaction medium. The results suggest that the presence of supercritical hexanes improved the heat transfer from the catalyst bed. It has also been observed that the formation of $CH_4$was significantly reduced under supercritical hexanes reaction conditions. The results of the catalytic investigations both with and without the supercritical hexane medium demonstrated that supercritical hexane has a significant effect on the selectivity and the productivity towards higher alcohols.

(37d) Cleaner Conversion of Coal to Liquid Fuels with Proven Gasification Technology:

Ravindra Agrawal, Process, KBR, Houston, TX

Gasification is typically conducted at high temperature and high pressure with aid of gasification agents such as oxygen/steam to generate syngas that is rich in CO and H_2. CO and H_2can be used as an intermediate for a wide variety of products including: chemicals, fuels, and metals. The high pressure reduces the volume of the syngas and makes it relatively simpler to remove potential pollutants. Traditionally sulfur compounds from coal are captured with lime and results in a large amount of solid and liquid waste that needs to be disposed off. In gasification process, sulfur compounds can be captured and discarded as sulfur rather than as calcium sulfate, thereby reducing the amount of solid waste to landfill. Also, the high pressure makes it easier to capture CO2 reducing the amount of greenhouse emissions. The high oil price in recent years has increased the demand for CO2 for EOR. This has resulted in a number of projects now seeking to capture CO2. This presentation will discuss some recent projects at KBR that involve proven gasification technologies. The advantages of integrating proven gasification technologies with technologies proven in oil and gas sector to produce liquid fuels from coal will be discussed.

(59a) Sodium Plus Sulfur Promoted Supported Iron Catalysts for the Selective Production of Lower Olefins From Synthesis Gas

Hirsa M. Torres Galvis, Johannes H. Bitter and Krijn P. de Jong, Inorganic Chemistry and Catalysis, Debye Institute for NanoMaterials Science, Utrecht University, Utrecht, Netherlands

The Fischer-Tropsch to olefins (FTO) process is a direct route for the conversion of synthesis gas into light olefins ($C_2^=$-$C_4^=$). The FTO process has to be carried out at high temperatures in order to drive product selectivity to short chain hydrocarbons. Under these conditions, the methanation reaction is kinetically favored resulting in a detrimental effect on product selectivity.

Many studies have been carried out to find a promoter or a combination of promoters that allow an important decrease in methane formation and that could improve the selectivity towards lower olefins simultaneously[1]. It has been reported that promoted bulk iron catalysts modified with sodium and/or sulfur can achieve methane selectivities as low as 10%C and C_2-C_4olefins selectivities of about 35%C[2]. However, up to now, there is no industrial application of these catalysts probably due to their low mechanical stability.

Bulk iron catalysts that are subjected to the very demanding conditions of FTO tend to disintegrate as a result of carbon lay-down[3]. An improvement in the mechanical stability of iron-based catalysts has been achieved by dispersing iron on an inert support that will provide mechanical anchoring for the promoted Fe nanoparticles while allowing the formation of the carbidic active phase[4].

Catalysts with high selectivity towards lower olefins, low methane production and excellent mechanical stability have been synthesized by incipient wetness impregnation of ammonium iron citrate or iron nitrate on alpha Al_2O_3. The sodium and sulfur promoters were incorporated by co-impregnation of aqueous solutions of sodium citrate (or nitrate, depending on the iron precursor) and iron or ammonium sulphate.

The supported and promoted catalysts were tested in the Fischer-Tropsch reaction at 1 bar, 350°C and $H_2/CO=1$ v/v. The sodium and sulfur promoted catalysts displayed a low methane selectivity (~15%C) while having a high selectivity towards lower olefins (~60%C). The promoted samples prepared with ammonium iron citrate showed a catalytic activity ten times higher than the catalysts prepared with iron nitrate. This behavior is attributed to a higher extent of aggregation of Fe nanoparticles on the samples synthesized with iron nitrate.

[1] Büssemeier, B. et*al.*, U.S. Patent No. 4564642 (1986)

[2] Crous, R.*et al.*, Int. Pat. Appl. No. WO 2010066386 A1(2010)

[3] Kalakkad, D. S.*et al.*, Appl. Catal. A 133 (1995) 335.

[4] Torres Galvis, H. M.*et al.*, Int. Pat. Appl. No. WO2011049456 A1(2011)

E-mail:h.m.torresgalvis@uu.nl. Financial support from NWO/ACTS-ASPECT is gratefully acknowledged

(59b) Hydroisomerization of n-Hexadecane Over Anion Modified Pt/HfO2 Catalysts

Muthu Kumaran Gnanamani, Gary Jacobs, Wilson D. Shafer, Dennis E. Sparks and Burtron H. Davis, Center for Applied Energy Research, University of Kentucky, Lexington, KY

Abstract

Nowadays, isomerization of n-paraffins plays an important role in the petroleum industry [1]. The most widely appliedisomerization catalysts include toxic and corrosive liquid acids, such as H_2SO_4 and HF, metal halides, and metal oxides promoted withCl or F. These reagents are highly corrosive and hence, environmentally hazardous." Many solid acids, such aszeolites, modified metal oxides, and clays, evade these difficulties, and offers to tune not only the number of acid sites but also their strength [2-5]. Among the solid acid studied, ZrO_2 modified by anions such astungstate, sulfate, or phosphate, has attracted considerable attention. ZrO_2promoted with sulfate groups have first been proposed by Holm and Bailey [6] which are active forisomerizationof light straight-chainalkanes at low temperatures (303-423 K)." Further, the stability of sulfatedzirconia catalysts was improved by adding noble metals and other transition metal oxides.

Hino and Arata [7,8], the pioneer in this field reported thatzirconia-tungstatecan be used as a strong solid acid catalysts. Althoughtungstated-zirconiacatalysts are markedly less active than sulfatedzirconia, they showed superior stability and selectivity towardsisomerization by promoted with platinum for largeralkanessuch as n-heptane[9,10]. The reason for the appearance of strong acidity intungstate-containing ZrO_2 remains unclear in many cases and continues to be discussed in literature [11]. Iglesiaet al. [11,12] proposed that platinum particles dissociate molecular hydrogen into hydrogen atoms which spill over to acid sites on the catalyst surface increases n-alkaneconversion rates and isomerization selectivities.

In the present work, the effect ofhafnia on acidic and hydroisomerizationactivity of n-hexadecane was studied using anions such astungstateand sulfate promoted with Pt. Diffuse FT-IR pyridine adsorption technique was employed to estimate the relative amount ofBr nsted and Lewis acids present in the catalysts. Activity and selectivity ofhafniacatalysts were compared withzirconia-tungstate catalysts.

References

1. R.A. Meyers, Handbook of Petroleum Refining Processes, McGraw-Hill, New York, 1996.

2. R.J. Taylor, R.H. Perry, Appl.Catal. A 119 (1994) 121.

3. S. Ernst, J.Weitkamp, J.A. Martens, P.A. Jacobs, Appl.Catal. A 48 (1989) 137.

4. J. Weitkamp, Appl.Catal. 8 (1983) 123.

5. R. Molina, S. Moreno, A.V.-Coelho, J.A. Martens, P.A. Jacobs, G.Poncelet, J.Catal. 148 (1994) 304.

6. V.C.F. Holm, and G.C. Bailey, (Phillips Petroleum Co.), US Patent 3,032,599 (1962)

7. M. Hino, and K.Arata, J. Chem. Soc., Chem.Commun., (1987) 1259.

8. M. Hino, and K.Arata, in Proc. 9^{th}Int.Congr.Catal.,(1988) 1727.

9. E. Iglesia, D.G. Barton, S.L. Soled, S.Miseo, J.E. Baumgartner, W.E. Gates, G.A. Fuentes, G.D.Meitzner, in Proc. 11^{th}Int.Congr.Catal.,(1996) 533.

10.S.L. Soled, E. Gates, E.Iglesia, U.S. Patent 5,422,327 (1995)

11.R.A.Boyse and E.I.Ko, J.Catal. 171 (1997) 191.

12.D.G. Barton, S.L. Soled, and E.Iglesia, Top.Catal. 6 (1998) 87.

(59c) Development of a Fischer-Tropsch Gasoline Process for the Steam Hydrogasification Technology

Yang Li, Chemical and Environmental Engineering, University of California, Riverside, Riverside, CA

Development of a Fischer-Tropsch Gasoline Process for the Steam Hydrogasification Technology

By

Yang Li, J. M. Norbeck, C. S. Park

Chemical and Environmental Engineering

Bourns College of Engineering

Center for Environmental Research and Technology

University of California, Riverside CA

92521

Abstract This paper will present preliminary results from a novel process being developed in our laboratory for the production of high quality F-T gasoline. The system is a combination of a conventional F-T catalyst combined with a zeolite-shell catalyst. This system is being designed for the Steam Hydrogasification process using various carbonaceous feedstocks. The chemical speciation and carbon distribution of the liquid fuel produced with the conventional F-T system will be compared to the new system using the zeolite-shell catalyst. It is found that the zeolite based system shifts the carbon distribution to the lower carbon number species (i.e., C6-C12 range) from the more typical diesel fuel fraction (C12-C20 range) found with FT catalyst systems. Chemical speciation results will be presented along with a assessment of the total carbon conversion expected for this process within the steam hydrogasification technology.

(59d) Direct Conversion of Gasified Biomass to Long Alcohols in Solid Hybrid-Nanoparticles At the Liquid-Liquid Interface of Water/Oil Emulsions

Jimmy Faria[1], Daniel E. Resasco[1], Ali A. Rownaghi[2]and Raymond L. Huhnke[2], (1)University of Oklahoma, Norman, OK, (2)Biosystems and Agricultural Engineering, Oklahoma State University, Stillwater, OK

We have developed a new reaction/separation technology based on a family of recoverable nanohybrid catalysts that simultaneously stabilize emulsions in biphasic systems[1]. These nanostructured solid particles exhibit a unique advantage in the streamlining of biomass syngas to fuels, where the presence of tars greatly complicates purification procedures. These novel catalyst/emulsifier hybrids can catalyze reactions with high "phase-selectivity" either in the aqueous or organic phases. The amphiphilic-catalysts are obtained by fusing carbon nanotubes to metal-oxide particles, which results in a "Janus" like nanoparticle that is able to stabilize water/oil emulsions by forming a rigid film at liquid-liquid interface of the droplets, increasing the apparent viscosity of the system[2]. In the aqueous phase, the tar molecules and small oxygenates can be extracted. Upon phase migration, these species react on the inorganic oxide, which may act both as the hydrophylic side of the emulsifier and as a condensation catalyst. Hence, it is able to catalyze condensation reactions in the aqueous phase, by which small oxygenates soluble in water, with low fuel value, condense via aldol-condensation, ketonization, or etherification. The resulting products are no longer water-soluble molecules and therefore migrate to the organic phase. The oxide used can vary in acid/base characteristics (MgO, and SiO_2).

Ruthenium clusters were anchored onto the hydrophobic carbon nanotubes of the nanohybrids to catalyze Fischer-Tropsch reactions to produce a significant fraction of long alcohols (C5-C15).

The direct conversion of the entire gasified biomass in a biphasic liquid system could greatly simplify the processing of the syngas obtained from biomass gasification, as it does not require the conditioning of the gas prior reaction.

[[1]] S. Crossley, J. Faria, M. Shen and D. E. Resasco,*Science*,**327**, 68 (2009).

[[2]] M. Shen, D. E. Resasco,*Langmuir*, 25, 10843 (2009).

(59e) Comprehensive Kinetic Model for Fischer-Tropsch Synthesis Over a Re-Promoted Co/Al_2O_3Catalyst

Branislav S. Todic[1], Tejas J. Bhatelia[1], Wenping Ma[2], Gary Jacobs[2], Burtron H. Davis[2]and Dragomir B. Bukur[1,3], (1)Chemical Engineering Program, Texas A& M University at Qatar, Doha, Qatar, (2)Center for Applied Energy Research, University of Kentucky, Lexington, KY, (3)Artie Mc Ferrin Department of Chemical Engineering, Texas A& M University, College Station, TX

Abstract

In spite of considerable effort to develop comprehensive kinetic models for Fischer-Tropsch synthesis (FTS), there is still a debate about the reaction mechanism and possible reasons for deviations from Anderson-Schulz-Flory (ASF) distribution. The goal of this work is to develop a comprehensive kinetic model for FTS on promoted cobalt-alumina catalyst and provide an insight into this topic.

Experiments were conducted over 25% Co/0.48%Re/Al_2O_3catalyst in 1L stirred tank slurry reactor over a range of operating conditions (i.e. temperature 220 °C, pressures 200 and 350 psig, H_2/CO feed ratio 1.4 and 2.1 and gas space velocity from 3.6 to 15 NL/g-cat/h). Rate equations were derived on the basis of Langmuir-Hinshelwood-Hougen-Watson approach using well known carbide mechanism for FTS[1]. In order to explain non-ASF behavior it was assumed that the desorption rate of an olefin precursor is a function of carbon number[2].

Kinetic parameters were estimated by minimizing multiple response objective function using Genetic algorithm and Levenberg-Marquardt method, as a global and local optimization tool, respectively. Model evaluation was based on physical meaningfulness of estimated parameters as well as on statistical relevance of the fit.

Model yielded physically meaningful values of parameters. Furthermore, it can be seen from Figure 1 (a-c) that it was also able to correctly predict deviations from the ASF distribution and olefin to paraffin ratio dependence on carbon number.

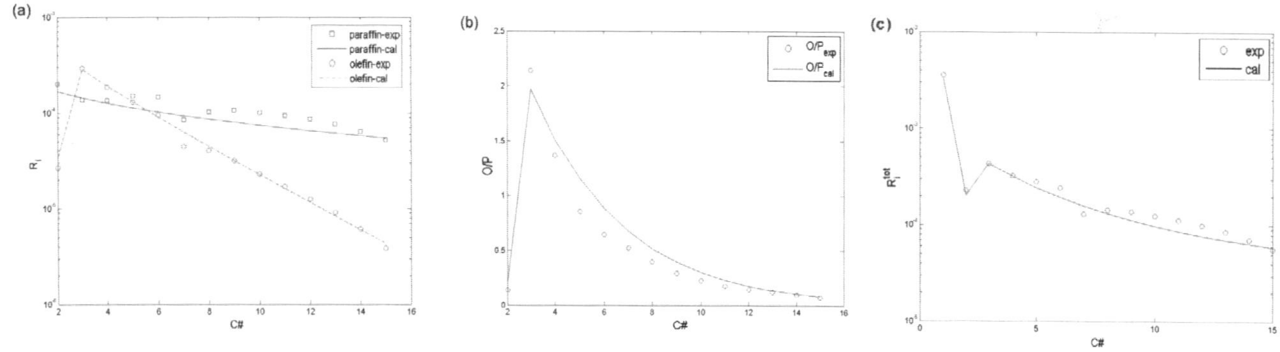

Figure 1 - Comparison of experimental data and model predictions for C_{1-15}hydrocarbons at T=220 °C,

P=203.3 psig, H_2/CO=2.1, SV=8.0 NL/g-cat/h (a) rates of formation for n-paraffin and 1-olefin; (b) O/P ratios (1-olefin/n-paraffin); (c) ASF plot (total rates vs. carbon number)

References

[1] LOX E. S., FROMENT G. F. (1993) Kinetics of the Fischer-Tropsch Reaction on a Precipitated Promoted Iron Catalyst. 2. Kinetic Modeling. Ind. Eng. Chem. Res.*32*,71-82.

[2] BOTES G. F. (2007) Proposal of a New Product Characterization Model for the Iron-Based Low-Temperature Fischer-Tropsch Synthesis.*Energy & Fuels*,*21*,1379-1389.

Acknowledgement

This paper was made possible by NPRP grant 08-173-2-050 from the Qatar National Research Fund (a member of Qatar Foundation). The statements made herein are solely the responsibility of the authors.

**Corresponding author: Dragomir Bukur, dragomir.bukur@qatar.tamu.edu*

(59f) Progress on a Comprehensive and Flexible Fischer-Tropsch Synthesis Reactor Model

Joseph W. Pratt, Energy Systems Engineering and Analysis, Sandia National Laboratories, Livermore, CA

Liquid fuels synthesized from renewable CO and H_2could displace petroleum-based fuels to provide clean, renewable energy for the future. Small scale (<20 bbl/day) reactors for synthetic liquid fuels production are an emerging development area that may enable mobile biomass-to-liquids plants suited for on-demand liquid fuel production from diverse, underutilized, local resources. A system compatible with these size requirements may require novel reactor configurations which cannot be efficiently explored without the aid of computer models. Such a model must be general enough to explore non-traditional designs while also being powerful enough to address the known issues of (1) non-uniform temperature distributions, (2) wax buildup, (3) conversion and selectivity optimization, and (4) lifecycle degradation.

Over the last 90 years, the determination of optimal Fischer-Tropsch synthesis reactor conditions has evolved starting with primarily experimentally determined inputs, to increased use of empirical modeling, to incorporation of more theory so that optimal conditions can be predicted and laboratory and pilot plant tests are no longer needed. However, although reactor size and capacity has increased over the years, the same general reactor design today is not much different than that used in the 1940's, especially in the case of fixed-bed reactors. This has led, rightly so, to models that explicitly or implicitly take advantage of this wealth of experience and knowledge with some degree of empiricism. Although they give very accurate results for modeling traditional reactors and/or a specific catalyst, the existing models are unable to examine the significant design changes and innovations that may be needed to solve the challenges described above.

The goal of this work is to create a Fischer-Tropsch synthesis model with the minimum amount of empiricism and assumptions allowed by the current state of the theory, enabling design of efficient, durable, and flexible small scale and/or novel reactors. Accomplishing this requires multiscale heat and mass transport models accounting for inter- and intra-particle gradients, coupled with generalized chemical kinetics models and dynamic (time-variant) degradation phenomena. There is no evidence in the open literature of a model of this type being successfully developed.

This presentation presents our work to-date on the physics component models and the combination of these components into a single comprehensive model, including implementation of schemes for efficient simulation across length scales varying by more than six orders of magnitude. The six component models are:

1. Two-phase interparticle (bulk) flow
2. Interparticle (bulk) heat transfer
3. Intraparticle mass transfer (within pore)
4. Intraparticle heat transfer (within solid and pore)
5. FTS chemical kinetics
6. Time-dependent catalyst deactivation by coke deposition, hydrothermal sintering, liquid/wax accumulation, and poisoning.

The successful, validated model of this type should enable accurate simulation of innovative and non-traditional reactors.

Maximizing Benefits in Selection of Gas Monetization Methodology: Fixed FLNG

Joseph H. Cho, Ph.D., P.E.
Vice President, LNG Plant Unit

Henry Ha
Process Engineer Lead

Wan-Jae Lee
Sr. Vice President, LNG Business Div.

SK Energy & Services
Seoul, Korea

AIChE Spring Meeting 2012
10th Topical Conference on Gas Utilization
Houston, Texas, April 1-5, 2012

ABSTRACT

When Shell announced FID of the Prelude FLNG project, even uncertainties have not been totally cleared up, it is true to get a bright picture in developing stranded wells, which cannot be economically commercialized with the conventional development methodology. However, there are key concerns affecting decision process of an FLNG project development. These may include reserve size with its characteristics (location, water depth, gas to liquid ratio, etc.), technical/commercial risks involved in entire ranges of the project development from evaluating to execution stage.

This paper presents selection criteria and process of natural gas development options in conjunction with technical & commercial issues and risks in developing either a conventional LNG plant or an FLNG option. Newly transformed Fixed FLNG options have been proposed for developing an LNG value chain in areas of environmentally sensitive, remote fields, extremely high labor cost countries, and labor forces shortage areas.

INTRODUCTION

The concept of FLNG, originally known as LNG FPSO (Floating Production and Storage and Offtaking), is to mount the treatment and liquefaction facilities on to a barge-like floating structure, so that processing and distribution can be undertaken at the offshore gas field rather than installing pipelines back to a shore based treatment, liquefaction and export facility. The concept avoids the environmental and social impacts associated with a pipeline to deliver the raw gas onshore from the offshore field, and the coastal modifications and land take that is required for a shore based facility.

FLNG is a niche technology suitable for developing stranded or associated gas fields with midscale production potential. It may be a better commercial solution than conventional LNG value chains, if the size of the field is between 1 and 3 trillion cubic foot (Tcf). Such production potential requires production capacity of 1.0-2.5 MTPA.

The FLNG era has been materialized by Shell's FID in May 2011. It is expected the first FLNG unit will produce LNG by 2017. Since there are many available candidates of wells to be developed with FLNG, the FLNG development concept will be adopted for gas monetization of small or stranded wells in near future.

The FLNG concept can also be transformed for gas monetization at remote areas (no infrastructure is available for construction of a conventional LNG Plant) or specific areas where labor costs are extremely high (or labor productivity is quite low because of both labor costs and labor union). The barge type FLNG can be moored at the fixed mooring facility near shore. This transformed FLNG called F-FLNG (Fixed mooring FLNG) can reduce total installed costs and construction period because the FFLNG can be built at a shipbuilding yard where the working conditions are better than an expected plant job site.

This paper highlights application of F-FLNG and its associated economic and technical benefits compared to both FLNG and conventional LNG plant.

TECHNICAL DRIVERS OF FLNG

FLNG Concept

The FLNG concept provides a technically innovative solution for developing small to medium stranded gas field in a cost-effective manner. This avoids the need for a conventional development scenario comprising off shore platforms, subsea pipelines, onshore liquefaction plant and export jetty. Conceptual outline of an FLNG is illustrated in Fig. 1.

Since an FLNG can be built at a shipbuilder's yard where working conditions are better, higher working productivity can be achieved than other remote areas where less infrastructure is available. This makes FLNG cost competitive against a conventional LNG plant if the development costs are to be compared based on a typical FLNG capacity ranges between 2-2.5 MTPA.

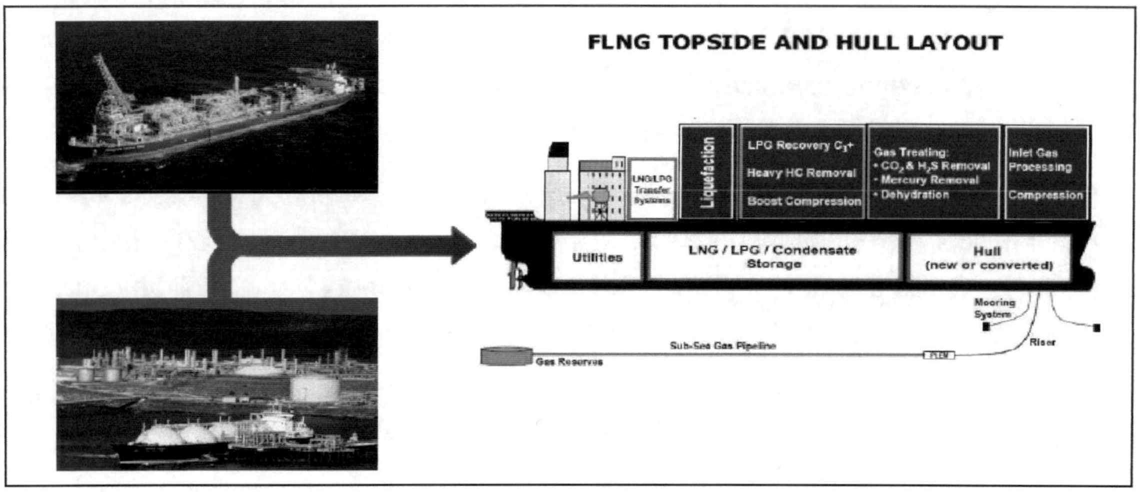

Fig. 1- Conceptual Outline of FLNG

Selection Criteria of FLNG

There are many screening processes and selection criteria before any project developer decides to adopt FLNG concept in developing a gas field. Simple selection criteria comprises of well location and its recoverable reserve size. If subsea pipeline and its associated production platform costs are considerably high, FLNG development can become favorable option. However, in depth feasibility study has to be performed to determine the best development option and an alternative, which identifies technical risks of each feasible development options, economic analysis and other issues. This includes environmental permit, operation permit, geopolitical issues, etc.

The following summarizes selection criteria for FLNG development option [1]:

- Technical - feasibility and safety;
- Environmental - impact during construction, operation and decommissioning, and assessment against compliance with an environmental regulation;
- Commercial - cost, schedule and business case;
- Organizational - do-ability and logistics; and
- Political - access to site and approvals process.

Commercial and Environmental Advantages of FLNG

With a reserve size of about 1 – 3 Tcf, long distance of subsea pipeline and its production platform make the conventional LNG development uneconomical. Elimination of a production platform and considerable distance of subsea pipeline reduces overall development capex and schedule, as such the project makes more economical and less complex.

For the F-FLNG development option, infrastructure that is required for conventional land based LNG plant, such as land acquisition, land development, an export pipeline, trestle and dredging, can be eliminated.

Environmental permit is relatively easier and faster than that of land based case. Opposition from residents, which can be frequently occurred in development of a land based conventional LNG plant, will be insignificant. Moreover, the facility can maintain security and safety against from sabotage at politically unstable areas.

Chemical injection lines contain storage fluid which will be displaced into the flowlines with Monoethylene glycol (MEG) for the MEG lines or compatible buffer fluids for the remaining chemical injection lines. These fluids will be displaced into the flowlines and routed back to the FLNG facility for handling and recycling or onshore disposal.

Technical Advantages of FLNG

FLNG can accommodate facilities for flow assurance line and well control. This includes, gas hydration control (MEG facility), formation water treatment, and umbilicals to control the wells and associated facilities.

In order to minimize the environmental impact, a closed loop freshwater cooling system can be employed. The cooling water can be taken from a depth of around 100 - 200 m, where the seawater temperature is stable and significant lower than that of subsurface. The cold seawater reduces cooling water volumes and increases LNG production.

Modularization with safety distances of each process can allow work safety and easy fabrication at a shipyard, resulting in expedite of overall construction schedule.

Consideration Points in FLNG Option

It is noted that the following should be taken into account in selection of FLNG for offshore gas fields:

- The production capacity of an FLNG facility is limited, due to the deck space required for process equipment, constraining throughput and economics, especially for projects with substantial gas assets. The FLNG sizes are mainly restricted by process configuration and compressor driver output. Two (or less than three) production modules are preferable in view of deck equipment arrangement. The gas turbine drivers currently available at the market have limits of power production. This is main reason of many studies have considered 2-2.5

MTPA FLNG. In order to increase LNG output beyond this capacity, different drivers rather than gas turbines may be considered.

- Expansion for FLNG is limited. It is only achievable by installation of an additional facility. However, expansion of onshore plants can be achieved with less capital costs as a brown field. Onshore expansion requires only incremental, as such saves materials and minimizes environmental impacts because new train uses common infrastructure and utilities of existing train(s).
- FLNG also has higher operating and maintenance costs compared to onshore plants, with very limited operational synergies available for multiple FLNG facilities compared to an onshore plant operating multiple gas trains [1].
- Risks associated with FLNG development have to be taken into consideration. Risks have been well identified and mitigation methodology can be found in the Reference 2.

CONCEPT OF F-FLNG

The basic concept of F-FLNG is almost identical to FLNG except that the F-FLNG is moored at the jetty without weathervaning facility (Turret). The feed gas is supplied by pipelines on a trestle jetty from near shore. Some of facilities, such as quarters, canteen, warehouse, administration offices can be reduced their capacity installed on F-FLNG and moved their capacities to near shore location.

LNG Storage containment system can also be located to onshore. The F-FLNG can then be barge type FLNG. However, the relocation of LNG storage tank to near shore may require vigorous studies in view of land acquisition, ground improvement, construction schedule, construction logistics, and overall costs. Under any situation, the barge requires buoyant system to support the topside gas processing and LNG production facilities.

Figure 2 illustrates F-FLNG compared to FLNG.

.

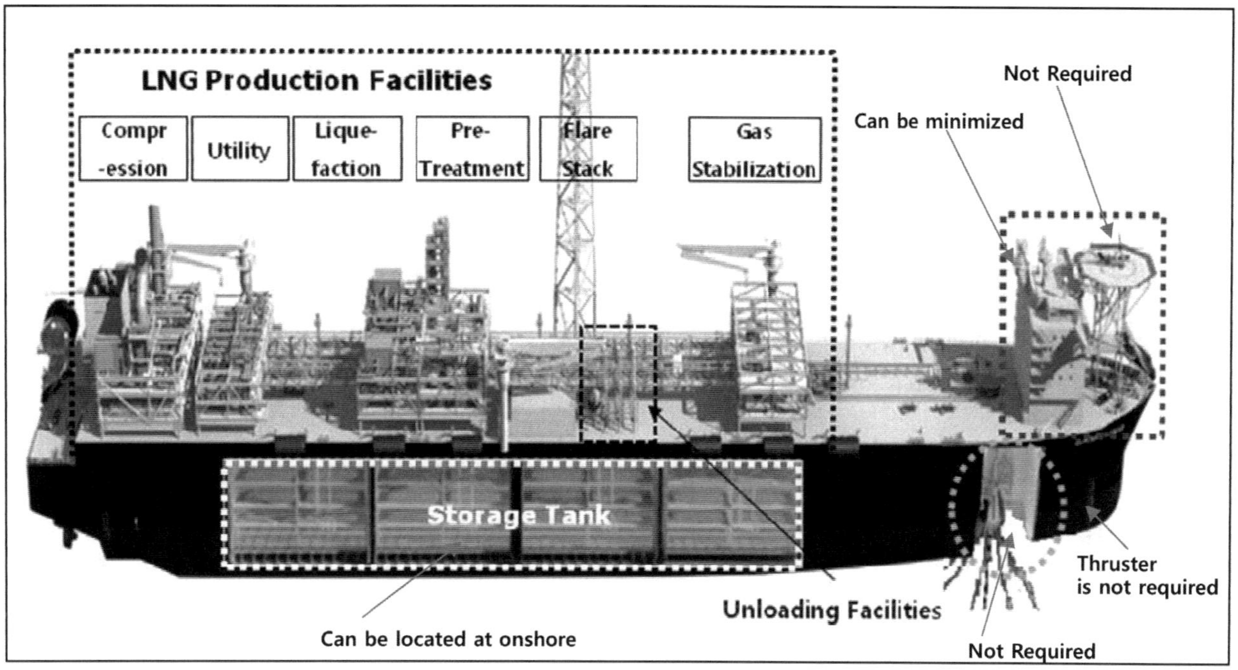

Fig. 2- Typical F-FLNG Outline

APPLICATION OF F-FLNG

Application and Advantages

If many projects are ongoing in any region, acquiring skilled labor is quite difficult, which results in an increase in labor costs. This may cause cost overrun and significant delay in execution of a project. F-FLNG can be applied to near shore or onshore gas monetization where infrastructure has not been well constructed or labor costs are high. Since the F-FLNG needs to be moored on the jetty facility, the sea conditions shall be calm or a manmade breakwater is required.

Because F-FLNG is moored and protected by the breakwater, it also reduces several concerns on even liquid distribution in the high towers, such as the main cryogenic heat exchanger, Natural Gas Liquid (NGL) extraction towers and acid gas removal unit, due to rolling.

F-FLNG needs to make sure available water depth is enough to allow berthing LNG carriers for LNG transfer from the F-FLNG. If it is not enough, there are two possible options may be considered:

- Extension of jetty until the enough water depth can be secured
- Dredging the ship lane so that the carriers can voyage without restriction

F-FLNG achieves high quality workmanship from the well controlled working conditions. The total construction schedule can be reduced compared to stick build conventional LNG plant.

F-FLNG can also apply to a place where getting LNG plant permit is difficult because of environmentally sensitiveness. Different variations of F-FLNG can be designed fit-for-purpose of the site conditions.

Side-by-side LNG transfer technology can be applied to LNG offloading from the F-FLNG because the F-FLNG is moored. The side-by-side technology using the flexible hose has been commercially proven by LNG transfer operation to Floating Storage and Regasification Units (FSRU) from LNG Carriers (LNGC) [3].

Type of Mooring

There are three types of jetty mooring as follows;

1) Jetty moored F-FLNG with side-by-side offloading (Fig.3)

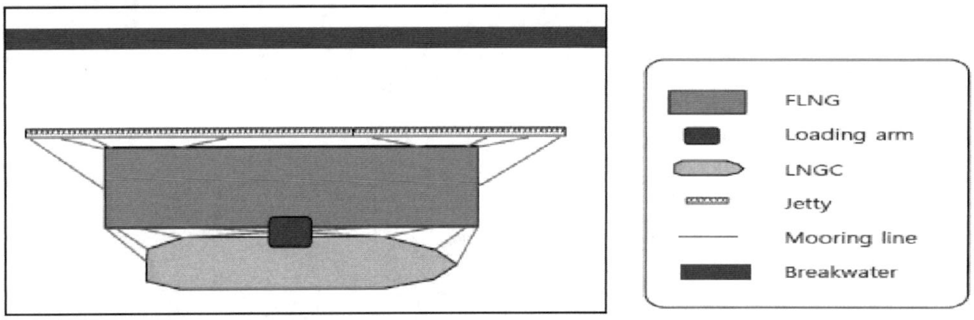

Fig. 3 - Jetty Moored F-FLNG

- No thruster is required
- LNG offload directly to LNGC (Loading arms installed at FLNG)
- Breakwater may be required
- Downtime may be less than turret moored pattern
- Feed gas to be supplied by piping along jetty
- Applicable to shallow water

2) Parallel berthing (Fig. 4)

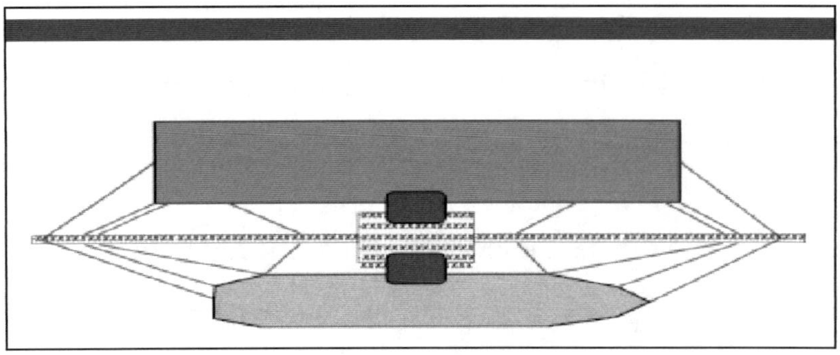

Fig. 4 - Parallel Berthing

- No thruster is required

- LNG offloaded through jetty (FLNG→ Jetty → LNGC)
- Breakwater may be required
- Less downtime (less relative motion)
- Feed gas to be supplied by piping along jetty
- Applicable to shallow water

3) Longitudinal berthing (Fig. 5)

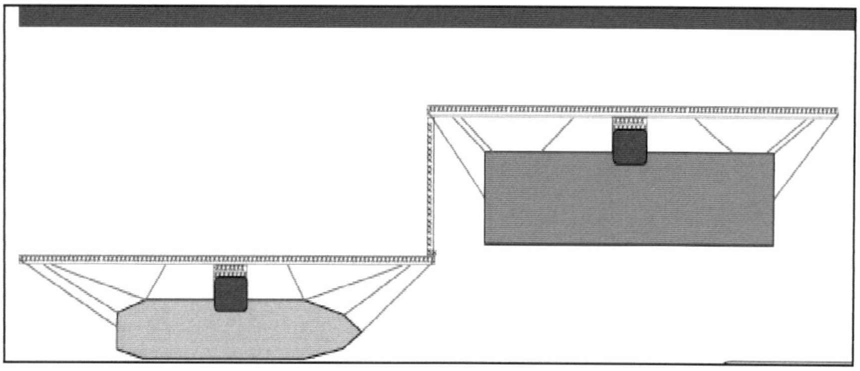

Fig. 5 – Longitudinal Berthing

- No thruster is required
- LNG offloaded through jetty (FLNG→ Jetty → LNGC)
- Longer jetty and breakwater are required
- Less downtime (less relative motion)
- Feed gas to be supplied by piping along jetty
- Applicable to shallow water

ECONO-TECHNICAL ADVANTAGES OF F-FLNG

F-FLNG Facilities

Since F-FLNG is moored and partial function of the following can be moved from FLNG to onshore, overall construction costs can be reduced:

- Quarters
- Warerhouse
- Administration building
- Heliport
- Azimuth Thruster
- Turret mooring system

One of the expensive equipment of the FLNG is a turret mooring system, which is defined as a mooring system where lines are connected to the turret which via bearings allows the vessel to rotate around the anchor legs. This turret can be mounted either internally or externally. An external turret is fixed, with appropriate reinforcements, to bow or stern of the ship. In the internal case the turret is placed within the hull, in a so called moon pool. The chain table, connecting the mooring lines to the turret, can be either above or below the waterline.

External Turret Mooring Systems provide an excellent solution for a wide range of FLNG and oil FPSO applications. These proven single-point mooring systems permit the vessel to freely "weathervane" 360 degrees, allowing normal operations in moderate to extreme sea conditions. External turret systems are less expensive than internal turret designs and can be delivered in a shorter period of time [4]. External turret mooring systems can be mounted at either the bow or stern of converted tankers or new-built vessels.

Internal permanent turret mooring systems provide FLNG/FPSO operators with maximum mooring and fluid transfer capabilities in remote, deepwater and/or harsh environmental conditions. For locations with severe cyclonic weather and harsh sea conditions, the internal turret mooring system provides full weathervaning and enhanced mooring load capacities that enable the FLNG/FPSO to remain permanently on location under the most demanding conditions [4].

The manufactures of the turret mooring system are also limited. Figure 6 presents Internal turret and External turret mooring system.

Fig. 6 – Internal and External Turret (courtesy of Offshore mooring.org)

Since the F-FLNG can be classified as non-vessel, the following safety facility can also be reduced:

- Life boat and Survival raft
- Heli-Deck

Economic Advantages

Majority of quarters can be located to onshore and minimum number of quarters will be equipped in the F-FLNG. Canteen is also minimized on the board. Instead, onshore cafeteria can serve meals for onshore support team and small onboard cafeteria can serve F-FLNG crews only.

An azimuth thruster is a 360 degrees rotatable thruster that is often used for main propulsion. The control system comes with an interface for the auto pilot system. An interface for dynamic positioning systems is optional when enhanced maneuverability is required in applications such as dynamic positioning, ship escort and ship docking services. However, this system can be eliminated because F-FLNG is moored and protected by the breakwater.

For the F-FLNG, mandatory safety facilities for a vessel can be reduced. Inspection and design requirements are less severe than those of vessel classification.

F-FLNG crews' wages and paid time off are different from FLNG. It is intuitively believed that F-FLNG operating costs are less than FLNG.

Construction costs of a breakwater is quite site specific depending on water wave direction and strength, water depth, and its required length is major factors in the cost estimate.

Following table 1 present relative costs comparison of FLNG and F-FLNG. The main factor impacting CAPEX of F-FLNG is seawater condition. If a breakwater is required, the F-FLNG project will increase by about 5%~10% of the CAPEX of the F-FLNG.

Table 2 shows the CAPEX distribution of FLNG and F-FLNG. The basis of CAPEX estimate is as following LNG production.

- Capacity : 2 MTPA
- Sea State : Benign
- Breakwater : Not required (for F-FLNG)
- LNG Storage : Equipped F-FLNG

Table 1 - Relative Costs Comparison between FLNG and F-FLNG (Unit, %)

Item	FLNG	F-FLNG	Remark
CAPEX	100	85~95	5~10% Difference by Breakwater
OPEX	100	75~85	

Table 2 - CAPEX Distribution of FLNG and F-FLNG (Unit, %)

Item	FLNG	F-FLNG	Remark
Pre-FFED/FEED/PMC	2.5~3.0	2.5~3.0	
Topside	60~65	65~70	Fire fighting and emergency gen. costs for F-FLNG can be reduced
Hull	25~30	25~30	For F-FLNG, sloshing is not considered in selecting cargo containment system. The thruster and survival craft is not required.
Turret	10~15	0	
Jetty	0	2.5~4.0	
Onshore Office	0	2.5~4.0	
Logistics Base	2.5~4.0	2.5~4.0	Additional warehouse for emergency supply is not required for F-FLNG
Total	100	100	

CONCLUSIONS

The following conclusions were obtained from this study:

- If many projects are ongoing in any region, acquiring skilled labor is quite difficult, which results in an increase in labor costs. The F-FLNG can be economically and effectively applied to near shore or onshore gas monetization where infrastructure has not been well constructed or labor costs are high.
- F-FLNG can also apply to a place where getting LNG plant permit is difficult because of environmentally sensitiveness.
- Because F-FLNG is moored, it also reduces several concerns on even liquid distribution in the high towers, such as the main cryogenic heat exchanger, NGL extraction towers and acid gas removal unit, due to seawater moving.
- Overall F-FLNG CAPEX and OPEX can save by 5~15% and 15~25% of the FLNG case, respectively.

REFERENCE

1. Shell, Project Description of Prelude FLNG Project, 2011.
2. Cho, J., Ha, H. and Kim D.Y.: “Critical FLNG Project Enablers: Identified Risks and Mitigation”, proceeding of AIChE 2011, Chicago, Illinois, March 13-16, 2011.
3. Bokhorst E.V.: “Progress in the Qualification of Components and Systems for LNG Transfer”, presented at LNG 16 conference, Oran, Algeria, 17-21 April 2010.
4. Web data: http://www.offshoremoorings.org/moorings/2006/groep4/, 2012.

Author Biography

First Author:
Name: Joseph H. Cho
Position: Vice President
Company: SK Energy & Services

Biography:
Joseph Cho is a Vice President in the LNG Biz Development Div. at SK E&S. He is a registered professional engineer and has been in the LNG industry for twenty six years. He was involved in process design of LNG liquefaction plants and import terminals for various regions as a senior engineer at top tier EPC companies. His scope of work includes managing LNG/Oil/Gas technology, feasibility and optimization studies, economic evaluations, computer simulations, process equipment specifications, control philosophy and review of output from general engineering. He graduated from the University of Oklahoma with a Ph.D. degree in Petroleum Engineering.

ADVANCED TECHNOLOGY HORIZON OF FLNG

Kiil Nam, Jinsang Park, T.J. Kim and Yoon Choon Kim, Hyundai Heavy Industries, Ulsan, Korea

Introduction

With depletion of conventional oil and gas fields, offshore installations have become complex and capital extensive, requiring development of new processes that should be able to operate satisfactorily and economically even for non-conventional environments. The FLNG (Floating Liquefied Natural Gas Plant) is introduced to develop stranded gas fields that have been considered as uneconomical in the past. The FLNG enables natural gas to be produced and liquefied on a floating vessel and offloaded directly onto a carrier without touching onshore facilities. However, there are many challenges that have interrupted the realization of the FLNG project. The installation needs gas purification and liquefaction processes which are completely new in the offshore industry though they are proven technology in the onshore LNG sector. During the conceptual design of such a new process, the selection of the best among the candidates is a formidable task due to the intrinsic features of offshore plants which are characterized by being uncertain, safety-critical, independent and motional.

The FLNG is uncertain for not only adopting onshore LNG process to fit with limited space but also achieving performance and high safety level as per offshore regulation. Safety-criticality stems from the hazards due to treating the flammable hydrocarbon materials in the limited and congested space. A list of common hazards in offshore plants is fire and explosion. The FLNG should additionally offer cryogenic spill protection. Being independent implies that the offshore installations should produce such required utility as electric power, cooling medium and instrument air for themselves. Moreover, they should be able to protect themselves and the people onboard without external helps. Motion comes from offshore environment. The design of the FLNG should consider the mechanical fatigue on structures and the dynamic impact on liquid motions in process facilities including sloshing effect in the tanks.

HHI (Hyundai Heavy Industries), the world's biggest shipbuilder and offshore facilities contractor developed its own Hyundai FLNG model in association with Linde AG, an international industrial gases and engineering company of Germany. The FLNG production capacity is 2.5 MTPA (Million Ton Per Annum) for generic design approach.

The paper presents advanced design features developed by HHI and Linde AG for materialization of FLNG as a market dynamic driver.

Design Approach of FLNG

Selection criteria for liquefaction process

The selection criteria for the liquefaction process are classified into two categories: safety and performance.

The safety requirements should align with the risk management standards and methodologies recommended by ISO 31000 and ISO 17776. The offshore requirements in material selection and layout should be considered in order to prevent the potential risk caused by loss of containment of flammable fluids. A safety and loss prevention management should implement a risk management process including safety studies for the identification of all health and safety related hazards, the assessment of risks to personnel, facility and environment and the reduction of risks by suitable control measures.

The requirements for performance are as follows;

- The process should be economical. The economic evaluations should be performed in a global sense, not in a local sense. The optimized design of a partial system does not necessarily guarantee the optimum of the whole process. As one of global economic performance parameters, the process availability should be considered in the economic evaluation. Complicated process configuration may give better energy efficiency and simultaneously less availability.
- The capacity and configuration of the liquefaction process should be inclusive of the flashed gas and BOG (Boil-Off Gas). The flashed gas occurs when the pressurized LNG is introduced into the cargo tanks while the BOG is generated due to the heat ingress into the cryogenic cargo containment system.

The candidate processes proposed for FLNG are between two fundamentally different processes; refrigerants and nitrogen expander types. The nitrogen expander process uses gaseous nitrogen as refrigerant. Most large LNG plants apply the refrigerants type, either as single or dual mixed refrigerant. For the capacity of less than 2 MTPA, the nitrogen expander process is preferred since the nitrogen expander process is simple and safe in small capacity. For the range capacity of 2~3 MTPA, the SMR (Single Mixed Refrigerant) is suitable and for the capacity of more than 3 MTPA, the DMR (Dual Mixed Refrigerant) is selected since the nitrogen expander and the SMR processes need the more trains due to the limitation in capacity per train.

Based on 2.5 MTPA, the SMR is a good selection due to higher efficiency and lower rotating equipment count than the nitrogen expander.

The results are given in Table 1, showing advantages of refrigerants type for the selection criteria.

Table 1 Comparison of Liquefaction Processes for FLNG

Criteria	N2 Expander	SMR
Energy Consumption	144%	100%
Safety	Inherent Safe	• Flammable Refrigerant • Safety gap required
No. of Refrigerant Train (Based on 25MW Gas Turbine)	5 trains	4 trains
Min. required rotating equipment count	• 1 N2 Compressor • 1 warm expender/compressor • 1 cold expender/compressor Total count : 15	• 1 refrigerant compressor Total count : 4
Plot space requirement	High	Medium

Operation	• Simple process • High No. of rotating equipment • One phase (Gas) flow of Refrigerant	• Relative simple process • Two phase flow of Refrigerant

Selection criteria for power generation system

The main power consumers of an FLNG are the liquefaction unit. The compressors of the refrigerant cycle of the liquefaction unit consume a huge amount of power, which requires a massive quantity of fuel gas. For the power generation of the compressors, various alternatives are considered, each of which is based on electrically driven types and mechanically driven types by steam turbines and gas turbines. The selection criteria are set up to answer the following questions.

- What kind of power generation systems is applicable? What are the basic characteristics of them, including efficiency, required fuel gas properties, physical dimensions, reliability, etc.?
- What configurations are conceivable based on the power generation systems?
- What configuration is economical? What is the implication of the economical evaluation?
- Which is the optimum solution?

The requirements for power generation system should be economical and flexible to fuel gas. The fuel gas consumption is not important in terms of the overall economic evaluation. The process availability should be considered in the overall economic evaluation. The power generation system is capable of utilizing the following gas as the fuel gas.

- Regeneration gas from the columns of pre-treatment Unit
- Non-condensable gas from liquefaction unit
- Flash gas from depressurized LNG
- LNG boil-off gas (BOG) from LNG storage tanks

The combination of the power generation systems gives five alternatives as shown in Table 2. These systems are commonly capable of utilizing gas fuels, which are abundant in the FLNG. The use of the fuel oil as the fuel source is not considered in the economic evaluation.

Table 2 Alternatives for Power Generation Systems

Alternatives	Compressor Drivers of Liquefaction Unit	Others Power Consumers
Alt. 1	Steam Turbine Generator	
Alt. 2	Gas Turbine Generator	
Alt. 3	Mechanically driven by GasTurbine	Gas Turbine Generators
Alt. 4	Mechanically driven by GasTurbine	Steam Turbine Generator
Alt. 5	Mechanically driven by SteamTurbine	Steam Turbine Generator

Table 3 shows efficiency, failure rate and repair time of the power generation systems. The efficiency of boiler and steam turbine is inferior by about 30 %. However, the failure rate of boiler and steam turbine is about 85% lower. Consequently, boiler and steam turbine is less efficient but more reliable and less maintenance-demanding.

Table 3 Comparison of Alternatives for Power Generation Systems

Power Generation System	Thermal Efficiency, %	Failure rate, per 10^6 hr	Repair time, hr
Boiler + Steam Turbine	32	106.5	39.5
Gas Turbine	40	756.8	23.7

The production availability is the measure of the production loss. It is caused by the failure, repair and maintenance of equipment. It is defined as the ratio of actual production rate to planned production rate over a specified period of time.

The production availability was estimated with five alternative configurations as shown Fig. 1. Alt. 3 and 4 show lower production availability since the mechanically driven type by a gas turbine not only has lower availability than that by a steam turbine but also cannot be considered as redundancy.

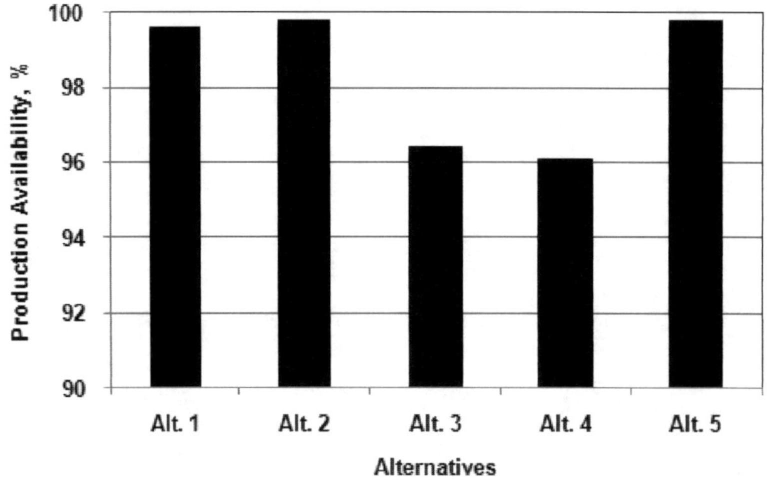

Fig. 1 Production Availability of Alternatives for Power Generation Systems

An increase in the production availability means a decrease in the production loss. The production availability entails the whole examination of the operability of the target plant since it describes how much the plant is capable of meeting the production demand. Fig. 2 represents the fuel gas cost including the production loss cost. The steam turbine driver and generator are high availability systems, while they have low efficiency. In addition, very large plants such as boilers, steam pipes and cooling systems are required with high seawater demand. The electrically driven compressors for the liquefaction unit should install a generator, motor, high voltage switchgear, etc. An increase in equipment count is accompanied by the overall cost and space escalation. The mechanically driven type is able to reduce the risk of the electrical system instability due to trip of large motors. Even though expected low availability of the gas turbine, Alt. 3 was applied due to good efficiency and the above disadvantages of other systems.

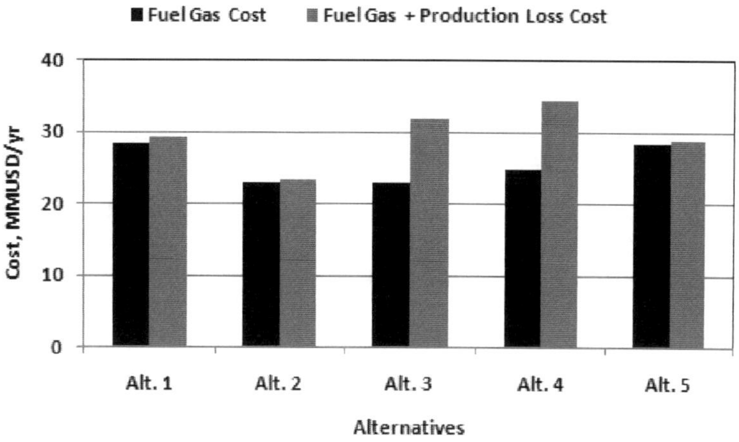

Fig. 2 Fuel Gas Cost including Production Loss Cost

Topside Process Design of FLNG

Fig. 3 shows a block diagram of the topside process for the FLNG. The topside processes are located on a deck platform 6 m above the hull deck. The storage tanks for LNG, LPG and Condensate are located below the hull deck.

Gas treatment process flow

Natural gas, as a multi-phase stream such as gaseous, condensate and water, is taken from the gas reservoir. The three-phase flow is fed to the Inlet facilities and is separated in them. The purpose of the gas pre-treatment units is to remove any of substances which become solid at liquefaction conditions. CO_2 and H_2S in the gas phase stream are normally removed by the amine absorption process. The dehydration system consists of three molecular sieve vessels for removal of water. The mercury removal is fixed bed adsorption unit.

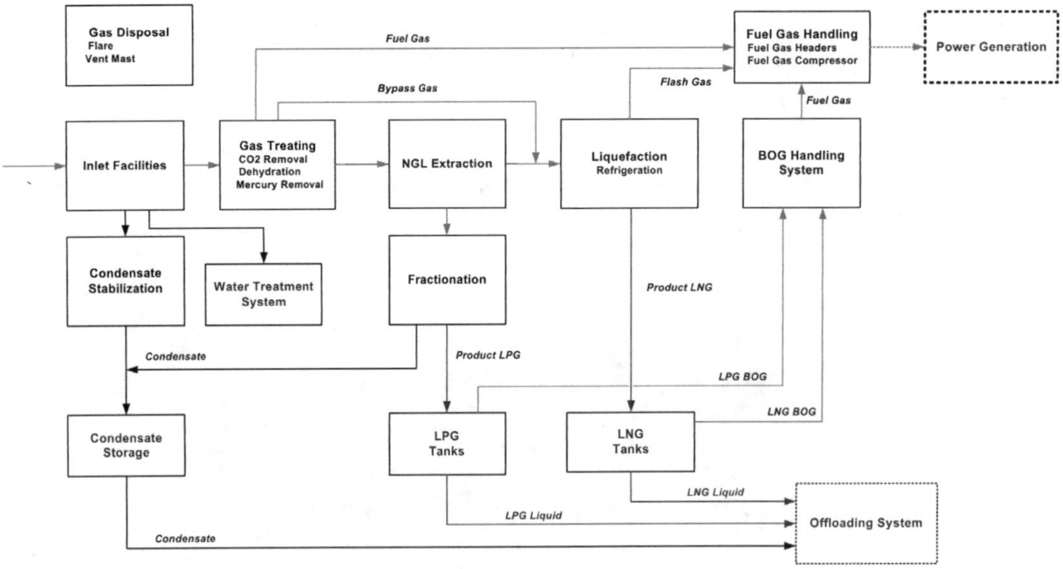

Fig. 3 Block Diagram of FLNG Topside Process

Liquefaction process flow

The liquefaction process applies a single mixed refrigerant (SMR) cycle which is separated into two refrigerant streams of different molecular weights by a two-stage refrigerant cycle compression and partial condensing downstream the compressor stages. The liquefaction section of the FLNG topside process applies LIMUM® process licensed by Linde AG to liquefy the treated gas. A schematic diagram of the process is shown in Fig. 4. The heavier hydrocarbon fraction (red line in Fig. 4) is used for pre-cooling of the natural gas, the lighter one (sky-blue line from separator downstream of HP MR compressor stage in Fig. 4) is first pre-cooled and then separated into two fractions from which the heavy one (yellow line in Fig. 4) is used for liquefaction and the light one for sub-cooling (sky-blue line in Fig. 4) of the natural gas.

In the liquefaction process, heavy hydrocarbons can solidify at LNG temperature (-162 ℃). Consequently, heavy hydrocarbons should be removed in the liquefaction unit to prevent the solidification. The natural gas is directed out of the main heat exchanger and into the heavy hydrocarbon recovery system, where the heavy liquids are removed. The vapor from the heavy hydrocarbon recovery system is returned to the main heat exchanger for liquefaction and sub-cooling.

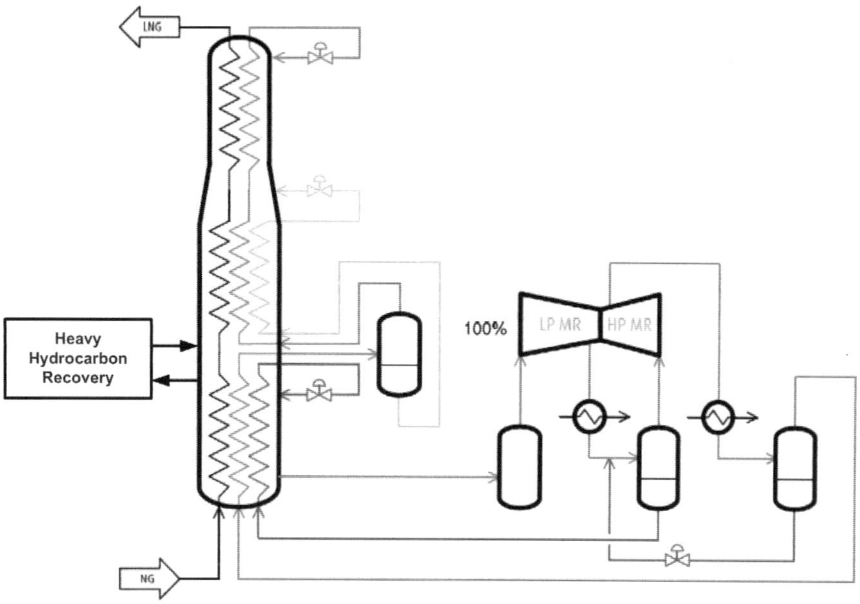

Fig. 4 Schematic Diagram of Liquefaction Unit (LIMUM® Process)

Refrigerant Loop

The liquefaction unit uses a closed loop refrigerant cycle system of compression, partial condensation, expansion, cooling and heating. The refrigeration loop contains two-stage compressors driven by gas turbines.

The low pressure refrigerant from the main heat exchanger is compressed to a midpoint pressure in the first stage of the refrigerant cycle compressor, and routed to the first inter-stage cooler. The first inter-stage drum separates the vapor and liquid. The liquid from the first inter-stage drum is the heavier hydrocarbon fraction used for pre-cooling of the natural gas. The vapor is sent to the second stage of the refrigerant cycle compressor. The refrigerant from the second stage of the refrigerant cycle compressor is cooled by the second inter-stage cooler. The liquid of the refrigerant is separated in the second inter-stage drum, exits the bottom of the drum and is routed after expansion to the first inter-stage drum. The vapor refrigerant enters the main heat exchanger where it is pre-cooled and partially condensed against the shell side refrigerant. Then it enters the refrigerant separator where it separates the light refrigerant as vapor and the heavy refrigerant as liquid. The heavy refrigerant is cooled in the main heat exchanger and is expanded by a Joules-Thompson valve. Thus the heavy refrigerant provides cooling duty for itself, the light refrigerant and liquefying the natural gas. The gaseous light refrigerant is routed to the main heat exchanger and is expanded by a Joules-Thompson valve into the shell side of the main heat exchanger. The low pressure refrigerant in the shell side migrates to the bottom of the main heat exchanger and vaporized. The refrigerant cycle is repeated.

Model test

The model test was performed to verify the sea keeping and offloading operability. The test was focused on the evaluation of the station keeping with the turret mooring arrangement and the offloading with the side-by-side arrangement.

The test demonstrated the safety for the turret mooring system and the side-by-side offloading system in the conditions of the wave, wind and current of 100 year return period and 1 year return period, respectively.

Fig. 5 Model Test of FLNG

Topside process layout

The primary consideration when laying out offshore facilities is their impact on people, the environment and asset integrity. The main criteria of the topside process layout are as follows;

- Compact, safe and cost efficient arrangement
- Use of optimized modules
- Fulfilling requirements for constructability and maintenance
- Integration of modules on main deck
- Minimize escalation of accidents between different units as far as possible
- High overall safety level
- Identify cost effective solutions, but keep conservative approach for the generic design
- Optimized regarding process and operation requirements

The plot plan was developed to identify the various areas and to obtain a safe layout and an optimum use of areas. Fig. 6 depicts the topside general arrangement of the FLNG.

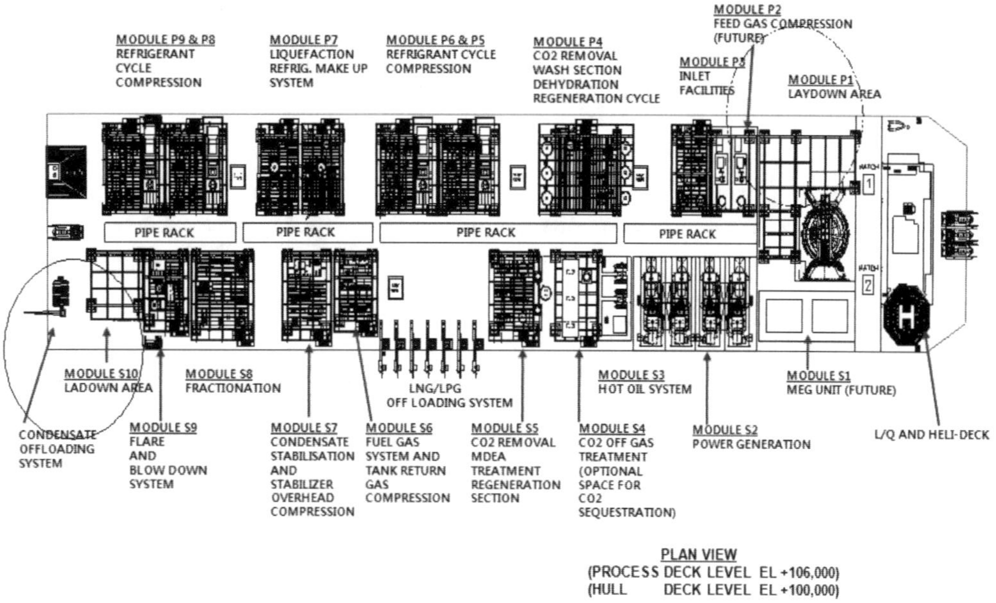

Fig. 6 Topside General Arrangement of Hyundai FLNG

The use of 3D model for each module ensures to maximize the degree of pre-fabrication and to minimize the amount on work in the integration yard. Fig. 7 shows the 3D Models for the refrigerant cycle and liquefaction unit, respectively. The refrigerant cycle module accommodates the compressors including inter-stage coolers, suction vessels and gas turbine. For the liquefaction unit module, the two modules are arranged side by side due to the weight limitation since for offshore facilities, the lift weight of the individual modules is critical.

Fig. 7 3D Models for Refrigerant Cycle and Liquefaction Unit

Conclusions

The paper assessed the feasibility of the design of the FLNG topside process. The selection criteria of FLNG topside process focused on making two key decisions regarding the liquefaction units and the power generation systems.

For 2.5 MTPA LNG production, the SMR process offers more advantages of less equipment count, smaller footprint and higher efficiency than the nitrogen expander type.

The compressors of the refrigerant cycle of the liquefaction unit consume a huge amount of power, which requires a massive quantity of fuel gas. For the power generation of the compressors, the electrically driven compressors causes a loss of efficiency and an increase in cost due to converting the power into electricity. The application of the steam turbine is uneconomical since very large plants such as boilers, steam pipes and cooling systems are required with high seawater demand. Consequently, the mechanically driven type by gas turbines was applied.

The model test was performed to verify the sea keeping and offloading operability. The test demonstrated the safety for the turret mooring system and the side-by-side offloading system in the conditions of the wave, wind and current of 100 year return period and 1 year return period, respectively.

The plot plan was developed to identify the various areas and to obtain a safe layout and an optimum use of areas. The living accommodation was at the opposite end of the FLNG to the flare, upwind of any possible gas release.

Further investigation and development will be performed for the robustness of the design and the operability of the FLNG.

References

1. Kiil Nam, Daejun Chang, Kwangpil Chang, Taejin Rhee, In-Beum Lee, (2011), "Methodology of life cycle cost with risk expenditure for offshore process at conceptual design stage," *Energy*, 36, pp. 1554-1563.
2. Kiil Nam, Taejin Rhee, Kwangpil Chang, Daejun Chang, (2006), "An analysis of gas turbine propulsion system for LNG carriers based on reliability and economic evaluation." *The Society of Naval Architects and Marine Engineers: Maritime Technology Conference & Expo and Ship Production Symposium,* Paper No. B1, Fort Lauderdale, U.S.A.
3. OREDA Participants (2002), "Offshore Reliability (OREDA) Handbook." 4th ed. Høvik, Det Norske Veritas (DNV).
4. Ayyub BM (2003), "Risk analysis in engineering and economics." London, CRC Press Company

(66c) Application of the Anti-Sloshing Blanket to LNG Carriers and FLNG

Sangeon Chun, Samsung Heavy Industry Co. Ltd., Gyeongsangnam-Do, South Korea

Sloshing is long problematic issue on achieving safety and reliability of LNG carriers and FLNG if the liquid level is below at a certain point in the containment system. Sloshing motion can be significantly reduced by covering the free-surface using a patented floater type blanket which is formed by connecting multiple floating units.

This paper presents effects of the anti-sloshing blanket on minimization of dynamic energy and removal of impact loads, which results in an increase in structure safety.

(66d) Cost Effectiveness of 2-Row Membrane System for FLNG

Herbert Moon, Daewoo Shipbuilding & Marine Engineering Co. Ltd., Seoul, South Korea

One of the most critical issues in engineering and design of FLNG is to reduce top side weight along with its load distribution to the sub-structure. Concentrated top side loads will result in an increase in hull structure costs. The developed 2-row containment system allows flexibility of top side load distribution. This system allows flexible modularization of top side equipment and reduces in hull structure cots as well. This system also provides great benefits in minimization of sloshing effects of the membrane storage FLNG. FLNG storage capacity can be increased with the 2-row system without sloshing problem. This paper presents the cost effectiveness of the 2-row containment system adopted to FLNG.

(67a) Gas Separation Membrane Tutorial

Lubo Zhou, UOP LLC, a Honeywell Company, Des Plaines, IL

Due to its low energy cost, membrane has been applied to different gas separation processes such as acid gas removal from natural gas, hydrogen purification, nitrogen production from air, and hydrocarbon recovery from refinery off gas, etc. In most recent years, membranes have also been proposed for flue gas CO2 capture. In this tutorial, the author will start the introduction of fundamentals of membrane for gas separation, followed by the different types of membranes and their applications. Different membrane process configurations will also be introduced at the tutorial. At the end, large scale commercial membrane units in the field of natural gas purification will be presented as the successful membrane application examples.

(67b) Cost-Effective Catalytic Membranes for H2 Purification

Bhanu Vardhan Reddy Kuncharam, Artie McFerrin Department of Chemical Engineering, Texas A& M University, College Station, TX and **Benjamin A. Wilhite**, Artie McFerrin Dept. of Chemical Engineering, Texas A& M University, College Station, TX

Conventional membrane systems for high purity hydrogen production from hydrocarbons typically employ permselective Pd or Pd-alloy materials in conjunction with hydrocarbon reforming. Palladium is expensive, prone to poisoning by carbonaceous materials and suffers embrittlement over multiple thermal cycles. In the light of these challenges our research group is exploring the use of catalytic membranes to achieve hydrogen separation without relying upon costly and delicate permselective materials.

In this presentation, we will review our work in the development of catalytic membranes for hydrogen extraction from fermentation liquors[1]and demonstrate the extension of this technique to hydrogen purification from reformate mixtures (H_2, CO, CO_2, H_2O) using a catalytic membrane active for water-gas-shift reaction. Design simulations of this membrane system predict H_2:CO permselectivities that are competitive to palladium membranes at a fraction of the materials cost.

[1] B. Kuncharam and B.A. Wilhite. Design of a Composite Catalytic Membrane for Permselective Extraction of Hydrogen from Ethanol via Reaction and Diffusion.*Journal of Membrane Science, in review*(2011).

(67c) Offshore Gas Treatment Technology for Natural Gas

Saadet Ulas Acikgoz[1], Shain Doong[2], Pengfei Chen[2]and Lubo Zhou[2], (1)UOP LLC, a Honeywell Company, Des Plaines, IL, (2)UOP LLC, a Honeywell Company, Des Plaines, IL

Offshore liquefaction of natural gas is expected to be the next technological breakthrough for capitalizing stranded natural gas resources. There are several technology options that can be used for acid gas removal or pretreatment of FLNG (Floating Liquified Natural Gas) plants. The traditional amine absorption processes have been widely used for the treatment of feed gas for onshore LNG plants.Due to limited space available on the offshore facility, the gas treatment units on FLNG plants have to be designed with smaller weight and footprint.

A hybrid system is a combination of membranes for bulk CO_2removal and amine units as polishing systems. Membranes have been proven in multiple offshore applications and suitable for offshore environment due to their compactness, minimal weight, footprint and insensitivity to sea motion. These hybrid systems can reduce the size and weight of the total pretreatment unit, minimize the effect of sea motion on the amine column performance and offer a cost effective solution.

() A Novel Catalyst for Partial Oxidation of Natural Gas

Nicholas Xu, Research Institute, Dason Technology, Chicago, IL

A Novel Catalyst for Partial Oxidation of Natural Gas

Nicholas Xu* *Dason Technology Inc., Naperville, 60565, USA*

Yu Wu *Chongqing Chemical Industry Research Institute, Chongqing, 400023, China*

*nick.xu@dasontechnology.com

Introduction

Many people have worked on CPOX research. The main barrier in the area was catalyst lifespan being insufficient to meet industrial requirements. We developed a novel catalyst for partial oxidation of natural gas (CPOX) to synthesis gas. The experiment was carried out in the lab and pilot plant at 10-20 bar pressure. The feed consisted of a mixture of natural gas and oxygen at a ratio of C:$O_2$of about 2:1. The natural gas conversion and CO selectivity are at 95% and H_2selectivity is about 90%. The catalyst was continuously running for about a year without deactivation.

Experimental

The experiments were carried out in an adiabatic stainless steel reactor with a Rh and Ni catalyst system. The lab reactor size is 20 mm in diameter and the pilot plant capacity is 1000 ton/yr. There is inert material at the bottom and the top of the catalyst.

The feed was heated to a range from 150°C to 300°C in an electric heater. There is a thermocouple at the top of the catalyst bed and another thermocouple at the bottom of the inert material below the catalyst bed. The experiments were performed between 10-20 bar pressure. A gas sample was taken from the reactor inlet and outlet. A GC was used for the gas sample analysis.

Results and discussion

Our catalyst was running for about a year under 10 – 20 bar condition without deactivation. The pilot plant results are in the figure below.

CPOX is compared to steam reforming and ATR technology in the following table.

Technology	Advantages	Disadvantages
Steam Reforming (SMR)	• Mature, proven • Most commonly use for large scale operations	• Large reactor size, highest CAPEX and OPEX • Reaction requires heat (endothermic) and steam • Large CO2 emissions from furnace • High CO2 in syngas

Autothermal Reforming (ATR)	• Smaller reactor size, lower CAPEX and OPEX than SMR • Less CO2 in syngas product than SMR • Part of NG is burned in the reactor to supply heat for reforming, internal burning has higher efficiency than SMR furnace • Requires 20% more oxygen than reaction required (CH4:O2=1:0.6) • Must add steam to reduce soot (CH4:H2O=1:0.4-0.7)	
Catalytic Partial Oxidation (CPOX)	• Smallest reactor size, lowest CAPEX and OPEX • Highest carbon efficiency • Idea oxygen consumption (20% less than ATR) • Lowest CO2 in syngas product • Don't need steam • Low catalyst lifespan before Dason's invention • Need oxygen but less than ATR	

CPOX carbon efficiency is at least 15% higher than ATR as demonstrated in this chart.

	CH_4	CO	CO_2	H_2	CH_4Conv.	CO sel	H_2Sel
CPOX	1.68	33.0	1.23	62.4	95.3	96.4	89.0
ATR	1.0	24.5	7.62	66.9	96.98	76.28	100.2

CPOX data above is our pilot plant typical data. The ATR data is from V. Palma et al. (2010) and is a calculated results using GasEq software at the equilibrium condition.

The carbon efficiency for CPOX is calculated as

32.2/(1.68+33.0+1.23) = 91.9%,

and for ATR is

24.5/(1+24.5+7.62) = 74%.

For a large plant, the feed stock consumption difference at about 18% makes a huge difference in operating costs.

Reference

1. D. A. Hickman and L. D. Schmidt, "Syngas Formation by Direct Catalytic Oxidation of Methane", Science 259, 343-346 (1993).

2. Rostrup-Nielsen, "Syngas in perspective", *Catalysis Today*71 (2002), pp. 243-247

(67d) An Innovative Mixed Zinc Oxide Based Sorbent Regenerable At Low Temperature for the Desulfurization of Syngas in IGCC or XTL Processes

Vincent Girard, Catalyse and Separation Division, IFP Energies Nouvelles, Solaize, France, Arnaud Baudot, Catalysis and Separation Division, IFP Lyon, 69360 Solaize, France, David Chiche, Catalyse and Separation division, IFP Energies Nouvelles, Solaize, France, Delphine Bazer-Bachi, GMD Division, IFP Energies Nouvelles, Solaize, France and Christophe Geantet, IRCELyon, Villeurbane, France

Integrated Gasification Combined Cycles (IGCC) and the second generation biofuels processes, and more generally Fischer-Tropsch based XTL processes, are attractive alternatives for future power production[1-3]. In order to prevent the corrosion of the industrial units, especially the combustion turbine blade used in IGCC processes or, in the others processes, to avoid the Fischer-Tropsch catalysts poisoning, deep desulfurization of syngas is achieved with metal oxides according to the following reaction :

$$MeO_{(s)} + H_2S_{(g)} \rightarrow MeS_{(s)} + H_2O_{(g)}$$

The large amounts of solid wastes produced are one of the major disadvantages of the in-situ desulfurization of syngas. The in-situ regeneration of the sulfided sorbent, back to the oxide, during the process could resolve this problem. From an industrial point of view, the oxidative regeneration is the simplest and most straightforward way to regenerate the sulfided sorbent, so it was mainly studied[4,5]. However, this reaction presents several drawbacks. The first one is its high exothermicity that can cause the sintering of the particles and the loss of reactivity due to the alteration of the textural properties[6,7]. Another drawback, which can be considered as the most important, is the formation of sulfates[5,8]. These refractory phases require high temperature to decompose, which causes important modification of textural properties and decrease of reactivity due to the sintering of particles. Moreover, the resulting temperature increase will impose the use of expensive alloys to make the reactors and will affect the thermal efficiency of the process. The formation of sulfate has to be avoided in order to maintain the sorbent durability and the temperature of the in-situ oxidative regeneration must be as low as possible, ideally the closest to the desulfurization step, to improve the global efficiency of this process.

The main goal of this work is to bring a new insight on the phenomena involved during the oxidative regeneration of the sulfided solids. The methodology developed is based on the confrontation between a global theoretical approach based on thermochemical data and an experimental study. This work is divided into two parts. The first one is the evaluation and the classification of the behaviors observed during both sulfidation and oxidative regeneration of simple oxides, in particular ZnO. The second one is the evaluation of the mixed oxides, selected thanks to the results of the first part and by the anticipation of potential synergetic effects. The synergetic effects are expected to avoid the formation of refractory sulfates and the resulting increase of temperature.

The first part led to a classification of the simple oxides based on the stability of their sulfide and sulfate phases. This classification led to a selection of different simple oxides whose mixture could bring synergetic effect. Finally, this methodology led to an innovative material which composition allows a fast regeneration, at a temperature 150°C lower than ZnO regeneration temperature and without sulfates formation.

Reference List

1. Daniel Vallentin, *Energy Policy*, 2008, **36**, 3198-3211.

2. Osamu Shinada, Akira Yamada, and Yoshinori Koyama, *Energy Conversion and Management*, 2006, **43**, 1221-1233.

3. Mark J. Prins, Krzysztof J. Ptasinski, and Frans J. J. G. Janssen, *Fuel Processing Technology*, 25-1-2005, **86**, 375-389.

4. R. V. Siriwardane and S. Woodruff, *Industrial & Engineering Chemistry Research*, 1995, **34**, 699-702.

5. W. F. Elseviers and H. Verelst, *Fuel*, 1999, **78**, 601-612.

6. T. Grindley and G. Steinfeld, *Proceedings of the Second Annuals Contractors Meeting on Contaminant Control in Hot Coal Derived Gas Stream.*, 1981.

7. J. Huang, J. Zhao, X. Wei, Y. Wang, and X. Bu, *Powder Technology*, 14-1-2008, **180**, 196-202.

8. R. V. Siriwardane and S. Woodruff, *Industrial & Engineering Chemistry Research*, 1-12-1997, **36**, 5277-5281.

() Musings on Management

Frank van Lier, The Lubrizol Corporation, Wickliffe, OH

It really is all about the people!

Hiring good people.

Retaining them.

Adjusting for different personal styles, cultures, encouraging diversity, embracing different perspectives.

Motivating people to work together in successful global teams.

Achieving top notch results that are better than any individual can achieve alone.

Maintaining the focus.

Getting the results.

Management is all about the excitement managers gets from having their people and teams grow and succeed.

(81a) Thermodynamic-Analysis-Based Design for Liquefied Natural Gas Receiving Terminal

Meiqian Wang, Jian Zhang and Qiang Xu, Dan F. Smith Department of Chemical Engineering, Lamar University, Beaumont, TX

Natural gas (NG) now has become the most promising alternate fuel energy. Recently, the estimation about large quantity of technically recoverable unconventional NG, such as shale gas, has been announced by the Energy Information Administration (EIA), which draws more attention on the development of natural gas. In the work, we developed a novel configuration for liquefied natural gas (LNG) receiving terminal, integrated with LNG regasification process, shale gas liquefaction and NGL recovery process. A great deal of cold energy from LNG regasification has been recovered through shale gas liquefaction and NGL fractionators' condensers.

The flexibility of the newly designed receiving terminal has also been addressed under different conditions, which include the changes of received LNG quantity and composition, transported shale gas quantity and composition, and heating value specified at local marketing place, etc. Moreover, heat duty integration has been involved in this design, where thermodynamic theory is employed to analyze and fundamentally evaluate the performance. All of the conditions together with thermodynamic and economic performance are integrated in a mixed-integer nonlinear programming (MINLP) model. This model is solved using optimization tool GAMS and validated via rigorous simulations in ASPEN Plus.

(81b) Design Approaches for Unconventional Gas Processing

John Harness, Technical Service, UOP, Des plaines, IL

Design Approaches for Unconventional Gas Processing

Membrane systems are well suited to remove CO2 and water to pipeline specifications although many people still consider membranes to be an "unconventional" technology. When applied to processing unconventional gas such as North American shale gas, membrane systems can offer distinct advantages.

This presentation will discuss the technology within membrane systems and present a case study from a gas plant that has been using membranes to process shale gas for the past 18 years. The presentation includes discussions on membrane theory and applications as well as discussion of a cost and benefit analysis comparing the use of membranes with Amine systems in unconventional shale gas applications.

UCSRP: A Flexible Sulfur Removal Process for Sweetening Natural Gas

Arun Basu, Howard Meyer, Dennis Leppin, Jim Zhou and Ajay Makkuni

Gas Technology Institute

ABSTRACT

With the recent activities in the area of unconventional natural gas (such as, Shale Gas) in the U.S., the need for a low-cost sulfur removal technology for processing relatively small as well as large CO_2-rich natural gas streams has increased significantly. In a conventional amine-based natural gas (NG) processing plant, the H_2S-rich acid gas stream from the stripper unit is treated in a Claus process for the production of elemental sulfur (the gas-phase reaction is $2H_2S + SO_2 = 3S + 2H_2O$). The Claus process has specific inherent limitations including (i) the need for a tail-gas clean-up unit (e.g., Shell Claus Off-gas Treatment: SCOT) to achieve the required H_2S recovery efficiency, (ii) inability to handle low sulfur production capacities (typically, for less than 20 tons/day sulfur), and (iii) the need for an additional CO_2-separation step prior to the Claus step if the gas stream from the Amine unit contains relatively high CO_2 (typically, if the CO_2 /H_2S ratio is greater than 2.3).

Gas Technology Institute (GTI) has been working with the University of California, Berkeley, for further development of their UCSRP (University of California Sulfur Recovery Process) technology. In this process, the Claus reaction is conducted in liquid phase at about 250-300 ^oF (and at any pressure) to produce liquid sulfur in the presence of an organic solvent and a homogeneous catalyst. A gas mixture consisting of a part of the H_2S-rich feed gas plus the unreacted H_2S stream from the UCSRP reactor is combusted with air in a furnace to form SO_2 (plus a small amount of sulfur) which is recovered in an absorber/stripper unit for recycle to the UCSRP reactor. As the overall sulfur recovery is 99.9+%, there is no need for a tail-gas clean-up unit. The UCSRP process can be scaled down to sulfur capacities significantly lower than 20 LT/day. For CO_2-rich feed streams from an amine plant, the gas can be processed directly in a UCSRP unit without a separate CO_2 removal unit. For direct processing of natural gas with relatively low CO_2 levels that already meet pipeline specifications, the UCSRP process will replace Amine/Claus-SCOT units; for higher CO_2 levels in feed natural gas, the exit gas from the UCSRP reactor will be treated to remove CO_2.without the complexity inherent when H_2S is also present in the stream.

For the processing of a H_2S-rich gas stream from an amine plant stripper with sulfur production at about 76 LT/day, Kellogg, Brown and Root Inc. had estimated the capital cost for an UCSRP plant to be about 60% and the operating costs to be about 68% of those for a three-stage Claus/SCOT plant.

INTRODUCTION

With the recent discoveries of relatively large quantities of unconventional natural gas (e.g., Shale Gas), the U.S. natural gas industry is interested in sulfur recovery processes that offer lower cost with high sulfur removal selectivity. Various amine-based technologies are commercially used to remove sulfur compounds from high-pressure natural gas feed as an acid gas which then requires a sulfur-recovery process. The modified Claus process coupled with a tail-gas treatment (TGT) process such as the SCOT (Shell Claus Off-gas Treating) process are typically used to recover elemental sulfur and produce a dischargeable plant tail-gas. Some plants produce sulfuric acid rather than elemental sulfur.

Gas Technology Institute (GTI) has been working with the University of California, Berkeley, for further development in a number of varying applications of their UCSRP (University of California Sulfur Recovery Process) technology for removal of sulfur compounds from various feedstock including natural gas and coal-derived synthesis gas.[1,2,3,4] In the UCSRP process, the Claus reaction ($2H_2S + SO_2 = 3S + 2H_2O$) is conducted in liquid phase (typically, using an organic solvent such as Diethylene Glycol Methyl Ether (DGM)) at about 250-300 °F (and at any pressure) to produce liquid sulfur. A homogeneous catalyst (e.g., 3- Pyridinemethanol), which does not dissolve in liquid sulfur, is added to the solvent in small quantities. At these conditions, as liquid sulfur is essentially immiscible in DGM solvent and over twice as dense, it is removed as a separate product from the reactor column. In this paper, we have (i) presented comparative capital and operating costs data (based on a prior study by Kellogg, Brown and Root Inc.[1]) for the UCSRP process to a conventional Claus/SCOT sulfur recovery system, and (ii) outlined key advantages of the UCSRP process for the sweetening of natural gas.

PROCESS OVERVIEW

Claus

The Claus process produces elemental sulfur by the partial oxidation of H_2S to SO_2 followed by the reaction of uncombusted H_2S with SO_2. The key reactions are:

$$2\,H_2S + 3O_2 = 2SO_2 + 2\,H_2O \qquad (1)$$
$$2\,H_2S + SO_2 = 3S + 2\,H_2O \qquad (2)$$

The overall reaction may be expressed as: $2\,H_2S + O_2 = 2S + 2\,H_2O$ (3)

In the original unmodified Claus process, the reaction was carried out in a single reactor (at about 400-650 °F) over a catalyst followed by a sulfur condenser.[2] With only one reactor, the sulfur recovery was limited to 75-90%. Most modern sulfur recovery plants use the modified Claus process, where the reaction is carried out in two steps. In the acid gas burner, 1/3 of the total H_2S in the feed gas is totally oxidized (using air) to SO_2. In the next step, the unburned H_2S and the SO_2 formed react in the reaction furnace without catalyst. This reaction proceeds rapidly in the high-temperature reaction furnace, and leads to a conversion of about 60% of the H_2S to sulfur vapor. The gas leaving the furnace is cooled to condense sulfur and then reheated for processing in a catalytic converter. Most modified Claus plants have two to four catalytic converters, each with its own condenser. Each succeeding converter operates at a lower temperature. The three-converter system typically provides a sulfur yield up to 98%. The gas leaving the final catalytic converter is typically sent to a Tail Gas Cleanup Unit (TGCU), such as a Shell SCOT unit. A schematic of the modified Claus process is shown in Figure 1.

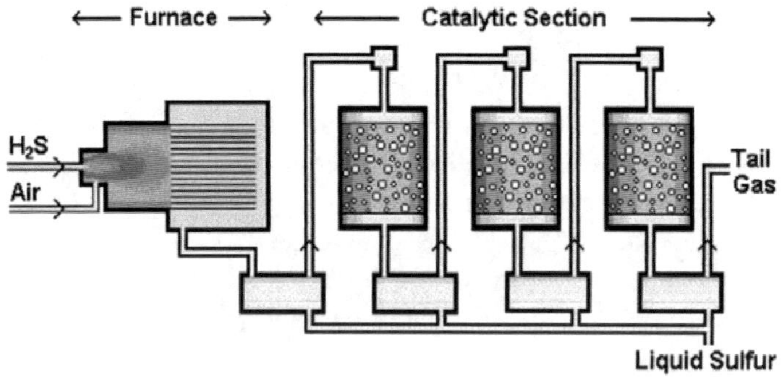

Figure 1: Schematic of the Modified Claus Process[5]

Several variations of the basic Claus process have been developed to handle acid gases containing different levels of H_2S. If the acid gas feed is relatively lean (less than 30 mol% H_2S on a dry basis), and if all of the acid gas is sent to the burner, the low level of H_2S would result in poor flame stability due to relatively low reaction furnace temperatures. Therefore, a fraction of the acid gas feed bypasses the burner to achieve suitable furnace temperatures necessary for flame stability. This configuration is referred to as the "Split-flow Claus". When the acid gas feed is sufficiently rich in H_2S (i.e., greater than 30-35 mol%), all of the acid gas is processed in the burner. This process configuration is known as the "Straight-through Claus".

A few key inherent problems of the modified Claus process are:

- As the furnace burns very fuel-rich, chances of soot formation increase significantly due to the conversion of hydrocarbons to COS and CS_2; this requires precise control of air flow to the furnace.
- Higher levels of CO_2 in the feed natural gas would significantly increase overall capital and operating costs of sulfur removal. As an example, if the CO_2/H_2S ratio is more than about 2.3, the off-gas from an amine stripper would have to be treated in an expensive CO_2 separation unit before the off-gas can be processed in a Claus plant.

SCOT

A schematic of a typical SCOT unit is shown in Figure 2.[2] The process was developed by Shell, and introduced in the early seventies as a process option for improving the efficiency of a Claus unit. By using the SCOT process, the sulfur recovery of a Claus plant can be increased to about 95-99.9%.

The SCOT process has two primary sections: (1) a hydrogenation/hydrolysis section, followed by water quench and (2) a selective amine gas treating section. In the first section, the tail gas from the Claus plant is heated to about 480-570 °F and then processed with H_2 (or a reducing gas) over a cobalt molybdate catalyst for converting all sulfur compounds (including SO_2,S, COS, CS_2) to H_2S. The effluent from the reactor is cooled to about 350 °F in a waste-heat exchanger (WHE) and the WHE effluent is further cooled by direct water quench. In the second section, the H_2S product stream is processed in an amine-based (e.g., MDEA or DIPA) absorber to recover most of the H_2S. The absorber overhead is usually incinerated (if the H_2S level is higher than 100 ppmv; using catalytic or thermal processes) to convert residual H_2S to SO_2 before discharge to the atmosphere. The H_2S-rich amine solution is processed in a regenerator to recover most of the H_2S which is recycled to the front end of the modified Claus unit. A schematic of the SCOT process is shown in Figure 2.

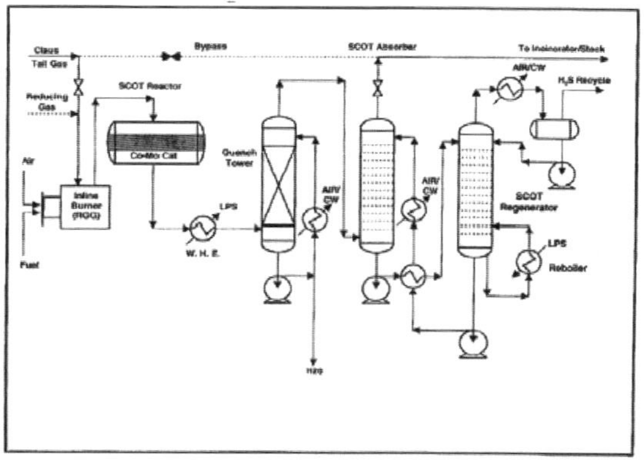

Figure 2: Schematic of the SCOT Process[2]

Although an integrated Claus/SCOT type process can meet current environmental regulations for sulfur emissions in the U.S., a SCOT-type tail-gas treatment process is quite expensive (typically costing as much as the modified Claus plant).[1]

UCSRP as a Replacement of the Claus/SCOT Processes

In the UCSRP concept (shown schematically in Figure 3), an organic solvent (e.g., DGM) containing a small quantity of a homogeneous catalyst is used to perform the Claus reaction in liquid phase at about 250-300 °F; this ensures that the reactor temperature is above the melting point (247 °F) of sulfur and below the polymerization temperature (310 °F) of elemental sulfur. The H_2S-rich acid gas is fed to the UCSRP reactor along with recirculated solvent plus SO_2 generated from combusting the reactor off-gases and part of the H_2S-rich feed with air in a furnace. As the liquid sulfur product is essentially immiscible and insoluble in DGM solvent and over twice the density as DGM, it is removed as a separate product from the reactor column thus shifting the chemical equilibrium toward sulfur formation.

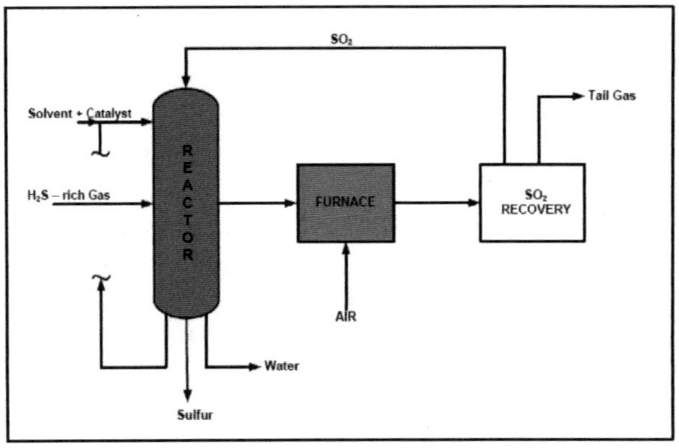

Figure 3: Schematic of the UCSRP Process

GTI is actively working with Prof. Scott Lynn (University of California, Berkeley), the inventor of the UCSRP process, to further improve the UCSRP technology, especially for high pressure (up to about 800 psig; the UCSRP-HP option) applications including sulfur removal from coal-derived syngas.[3,4] In the context of NG sweetening, GTI has demonstrated:

- The production of high-quality sulfur product (99.2% pure elemental sulfur) without any secondary treatment.
- Specific catalyst and solvent systems are stable under operating conditions based on a 1,000 hour test.
- Carbon steel is an acceptable material of construction for the UCSRP reactor

A few key advantages of the UCSRP process over the Claus process are: (i) the control of process stoichiometry is significantly simplified as the reactor is located before the furnace, (ii) slightly air-lean combustion precludes SO_3 and NO_x formation, (iii) hydrocarbons are burned primarily to CO_2 and water, (iv) relatively higher CO_2/H_2S ratios in the feed acid gas are tolerated, and (v) overall sulfur recovery is improved significantly.

- Cost Advantages of the UCSRP Process Over and Claus/SCOT: Scoping Economics Study by Kellogg Brown & Root, Inc.[1,2]

For sweetening of natural gas, KBR Inc. had conducted a scoping economics study to compare cost advantages of the UCSRP concept over the Claus/SCOT option in treating H_2S-rich gas from a conventional amine plant for the production of 76 long tons/day of sulfur.+1,2 A process flow diagram of the UCSRP process, as defined in the KBR study, is shown in Figure 4.

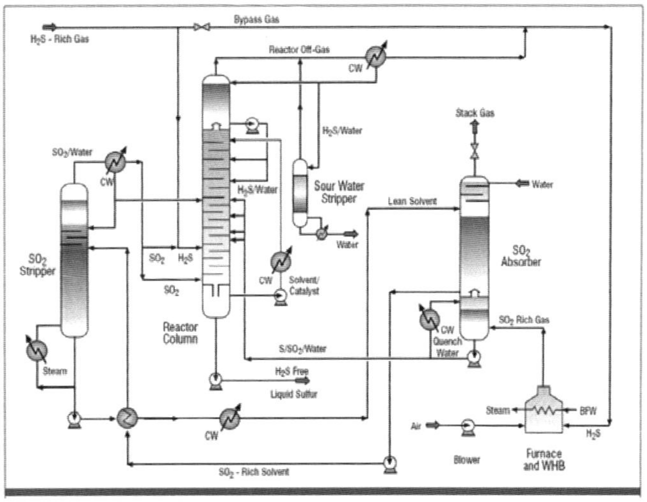

Figure 4: The KBR Study--Process Flow Diagram for the UCSRP Sulfur Recovery Process (for H_2S-rich Gas from an Amine Plant)[1]

As shown in Figure 4, the H_2S-rich gas stream enters the UCSRP unit at about 120 °F and 1.7 bar, typical of a Claus plant feed.[1] The reactor resembles a gas absorption column. A small fraction of the SO_2-rich gas stream enters the reactor column below the bottom bubble-cap tray. The remainder of the SO_2-rich gas together with about 75% of the H_2S-rich gas stream enters a tray or two higher and H_2S is in substantial excess at all points above. The remainder of the inlet H_2S bypasses the reactor and is sent to the furnace. The reactor is kept nearly isothermal by injecting water into the solvent flowing between trays to absorb the heat of reaction (from the Claus reaction) by vaporizing. The sulfur forms a second liquid phase. The solvent, which is recycled in a circulation loop from the bottom to the upper part of the reactor column, is saturated with elemental sulfur. The overhead vapor from the reactor is scrubbed with a water reflux to remove solvent vapor. A part of the water separated from the reactor off-gas is pumped back to the reactor as reflux. The net condensed water is sent to a sour-water stripper where dissolved H_2S and CO_2 are stripped overhead. The unreacted H_2S leaves the condenser and is mixed with the bypassed acid-gas feed and enters the furnace where it is burned with slightly sub-stoichiometric air to form SO_2. The hot combustion gas, containing SO_2, S, CO_2, H_2O and N_2 is cooled in a waste-heat boiler and is then fed to the SO_2 absorber where lean solvent from the SO_2 stripper is fed near the top. At the bottom of the absorber a water quench cools the combustion gas from the furnace to near ambient temperature, condensing a large fraction of the water vapors plus the sulfur vapor. The net dilute slurry of colloidal sulfur plus small amounts of higher sulfoxy acids is sent to the reactor column. The SO_2-rich solvent from the absorber is pumped to the SO_2 stripper to recover SO_2 for recycle to the UCSRP reactor.

As shown in Table 1, for the KBR study, two different H_2S/CO_2 feed gas ratios were evaluated:

Table 1: Feed Gas Composition (Design Basis: 76 LTPD Sulfur)[2]

	67/32 H_2S/CO_2	30/70 H_2S/CO_2
CH_4, mol%	0.2	0.2
H_2S, mol%	62.6	27.8
CO_2, mol%	30.2	64.9
H_2O, mol%	7	7
Feed Rate, kmol/hr	151.8	350.7
Claus Plant Configuration	Straight through flow	Split flow

Economic Results

- According to the KBR study[2], the relative capital cost for the UCSRP option is about 61% of a conventional Claus/SCOT unit (Figure 5) for a feed gas with 67% H_2S and 33% CO_2; for a feed gas with 30% H_2S and 70% CO_2, the estimate for the UCSRP option was about 60% of the Claus/SCOT plant.

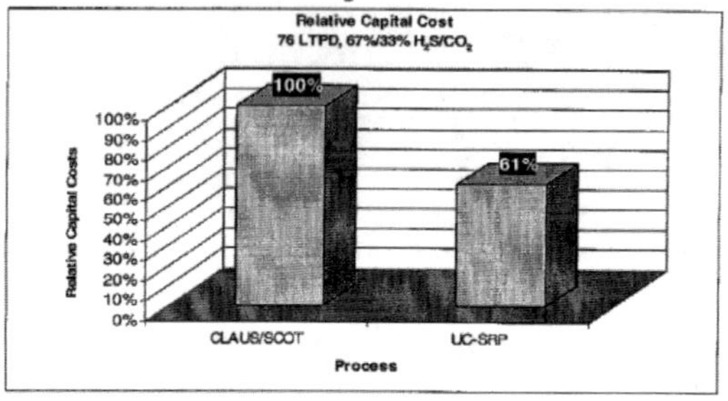

Figure 5: Relative Capital Cost – Claus/SCOT vs. UCSRP (Feed Gas with 67% H_2S, 33% CO_2)[2]

The relative Operating plus Maintenance costs for the UCSRP option were estimated at about (1) 68% of the Claus/SCOT design 67% H_2S/33% CO_2 Case, and (2) 67% of the Claus/SCOT design for the 30% H_2S/70% CO_2 Case.[2]

- Cost Comparison for a 935 LTD Sulfur Recovery plant[2]

As noted in Reference 2, another consulting firm performed an economic study for estimating relative costs for a 935 LTD sulfur recovery plant using the conventional Claus/SCOT and the UCSRP process. The relative capital cost for the UCSRP option was estimated to be about 62% of a Claus/SCOT system.

UCSRP as a Replacement of the Claus/SCOT Processes for a CO_2-rich Acid Gas Stream

If CO_2 level in a sour natural gas feedstock exceeds pipeline specifications (typically, less than 2 vol.%), an amine-type gas treating process is typically used to remove both H_2S and CO_2. In general, if CO_2/H_2S molar ratio is less than about 2.3, the off-gas from an amine stripper can be treated directly in a Claus/SCOT plant.[2] However, if this ratio is greater than 2.3, it may be necessary to carry out an expensive separation of the CO_2 prior to the Claus step. On the contrary, for an UCSRP unit, such a CO_2-rich off-gas from an "amine stripper" can be processed directly.

UCSRP as a Replacement of the Amine/Claus/SCOT Processes for a Natural Gas Stream

For direct sweetening of high-pressure natural gas feedstock, the high-pressure version of the UCSRP process would replace conventional amine plants integrated with a Claus/SCOT unit. A schematic of a two-stage UCSRP-HP concept, currently being evaluated at GTI for lab-scale demonstration, for high-pressure operations is shown in Figure 6; in this example, coal-derived syngas at about 800 psig is being processed to recover sulfur.[4] In this concept, the first UCSRP stage would be operated under "slightly excess SO_2" mode at the reactor outlet while the second stage would be operated under an "slightly excess H_2S mode". Typical sulfur balances for this example (in Figure 6) are shown in Table 2. In this example, the total SO_2 in the solvent plus syngas at the Reactor-1 outlet is about 9.7% of the SO_2 feed to this reactor; the H_2S content in the total solvent plus syngas flow at the Reactor-2 outlet is about 8.4% of H_2S in the syngas feed to the second reactor. For this example, the total syngas feed to the UCSRP reactors contains about 6,200 ppmv H_2S; about 1.34 vol.% of the feed bypasses the first reactor for processing in the second reactor. Based on GTI's design estimates, the exit syngas (sent to a CO_2 removal step) from the high-pressure separator (where Stream-11 is the inlet stream) would contain about 8 ppmv of H_2S plus COS.

Table 2: Stream Data for the Example Shown in Figure 6 (Basis: 1,000 lb-moles of total Syngas to the UCSRP Plant)[4]

Stream #	Stream Description	Lb-mole
1	Syngas to R-1[i]	986.6
2	Syngas to R-2[i]	13.4[ii]
3	SO_2 to R-1	3.1
4	Sulfur product	~ 6.2
5	Sulfur feed to Furnace	3.1
6	SO_2 from Furnace	3.1
7	SO_2 in solvent from R-1	0.26
8	SO_2 in syngas from R-1	0.04
9	H_2S in solvent from R-2	0.004
10	O_2 to Furnace	3.1
11	H_2S in syngas from R-2	0.003

i. Water vapor content: 1 mol%
ii. H_2S level: 0.083 lb-mole

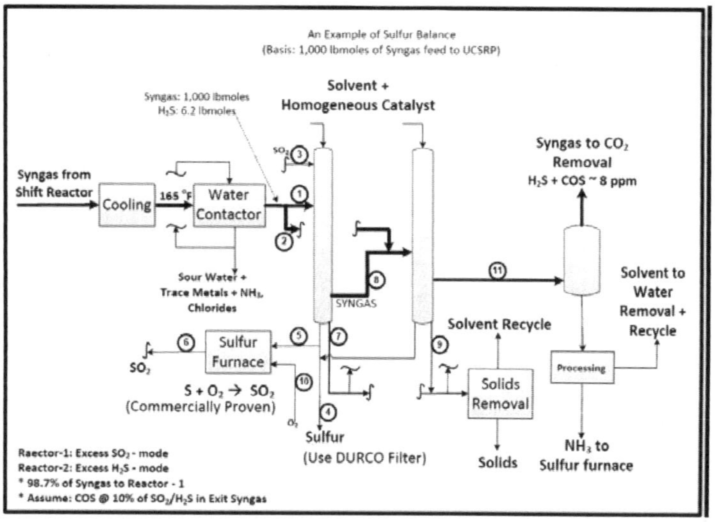

Figure 6: GTI's Two-stage UCSRP-HP Process Concept[4]

For direct sweetening of natural gas, an UCSRP-HP unit can replace an integrated Amine/Claus-SCOT plant. As shown conceptually in Figure 7, the USCRP-HP based plant designs would be relatively much simpler in handling various natural gas feedstocks with wide variations in (i) H_2S and CO_2 levels, and (ii) total gas flow rates.

As shown, if the CO_2/H_2S molar ratio is relatively high (~ greater than 2.3) in the H_2S-rich gas from an amine plant, we would need to first separate CO_2 before the gas can be processed in an integrated Claus/SCOT facility. For smaller plants (e.g., sulfur capacity less than 20 LT/days), we would need a "Liquid Redox" type process (and not a Claus/SCOT unit) to treat H_2S-rich gas (with CO_2/H_2S ratio of less than 2.3) from an amine plant. For the UCSRP-HP option, we can quite conveniently scale down to smaller sulfur capacities.[2]

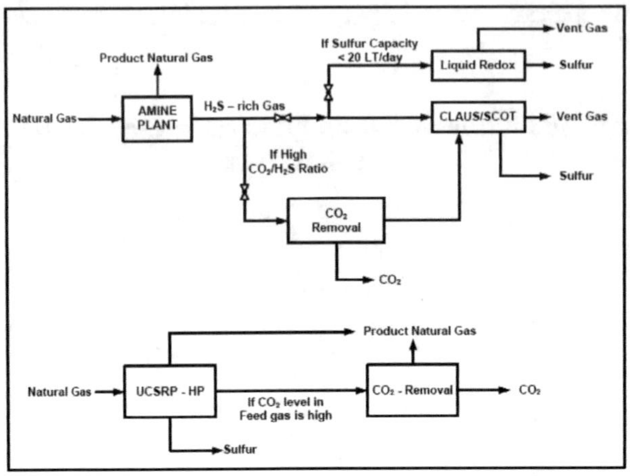

Figure 7: Relative Simplicity of an UCSRP-HP Process in Sweetening CO_2-rich Natural Gas

CONCLUSIONS

In summary, our key conclusions are:

- In comparison to the conventional Claus/SCOT process, UCSRP can reduce CAPEX by about 35-40% and OPEX by about 30-35%.
- In comparison to conventional Amine/Claus/SCOT for sweetening of natural gas feedstock, UCSRP-HP will reduce CAPEX/OPEX even more, especially for CO_2-rich NG streams.

References

1. Scot Lynn et al., "New Approach to Sulfur Removal Could Reduce Costs," GasTIPS, Winter 2002
2. Scot Lynn et al., "The University of California Berkeley's Sulfur Recovery Process: Claus Revisited," Paper presented at the 9th Gas Research Institute Sulfur Recovery Conference, October 24-27, 1999, San Antonio, Texas
3. S. J. Zhou, et al., "Economic Evaluation of the UCSRP-HP Process in IGCC Applications", Paper presented at the Clearwater Coal Conference, June 6, 2011.
4. Dennis Leppin, et al., "Economic Evaluation of the UCSRP-HP Process in IGCC Applications", SULPHUR 2011 International Conference & Exhibition, Houston, USA, November 6-10, 2011.
5. http://www.nelliott.demon.co.uk/company/claus.html

(89a) Increasing Propane Desuperheater and Condenser Reliability in PT Badak NGL

Mohammad Arief Setiawan, Technical, PT Badak NGL, Bontang, Indonesia

PT Badak NGL is the operator for natural gas liquefaction at Badak LNG Plant, Bontang, Indonesia. At present PT Badak NGL has 8 Trains. The mode for liquefaction is using the APCI process license. At the first step, Feed Gas is pre-cooled with propane refrigerant, followed by liquefaction process by using Multi Component Refrigerant (MCR) as cooler. The MCR is also using propane as its cooler. The propane refrigerant is using three level refrigerant pressures for different level of cooling.

The propane refrigeration system use sea water for condensing the propane after compression. To fulfill this, PT Badak NGL operates 2 parallel propane desuperheaters and 6 parallel propane condensers in each Train. The exchanger acts to condensed propane refrigerant prior being used as cooler media in refrigeration system. The units are all shell and tube exchangers, having sea water stream in the tube side.

The propane desuperheaters and condensers in Train G were frequently facing leakages problem. The last 2 years record showed at least 9 and 5 times repairs have been conducted in the propane desuperheaters and condensers respectively, while other Train's repair frequency only reached below 4 times. The problem has caused significant repair cost and reduction of propane inventory. Our analysis revealed that the sea water stream in the tube side has lower velocity than the minimum required specification of 6 ft/s. This has increased the tendency of fouling in the tube surface. Eventually the fouling effect will promote under deposit corrosion phenomenon which will lead to leakage.

Several options have been proposed for the trouble shooting actions, either by increasing the cooling water supply, installing special tube insert, or by modifying the service mode of the exchangers. PT Badak is planning on implementing the last option to solve the problem since it proposes the least cost of all options. To implement this program, the number of operating propane condensers will be reduced from 6 of 6 service mode to 5 of 6 service mode. This action is expected can increase the sea water velocity thus reducing under deposit corrosion and leakages phenomenon.

(89b) Absorption Refrigeration for Auxiliary Applications in LNG Plant

Jordan Belue, Energy Technology Company, Chevron, Houston, TX, Robert Havlik, Tengizchevroil, LLP, Farnborough, United Kingdom and Stanley Huang, Process Technology, Chevron Energy Technology Company, Houston, TX

ABSORPTION REFRIGERATION FOR AUXILIARY APPLICATIONS IN LNG PLANT

Jordan Belue[a], Robert Havlik[b], Stanley Huang[a],

a: Chevron Energy Technology Company, Houston, Texas

b: Tengizchevoil, LLP.

ABSTRACT

A Liquefied Natural Gas (LNG) plant demands a large amount of mechanical power for generating refrigeration. Over decades of development, the art of designing the prime mover trains utilizing steam or gas turbine drivers, combined with helper motors or synchronous generators, have been pushing the technological limits and approaching perfection in new constructions. This is witnessed by a rapid growth of global LNG capacity in the first decade of the 21stcentury. The largest gas turbines (GT) used are Frame 9 machines. Furthermore, various approaches to enhance the rotating equipment train and gas turbine output have been investigated and implemented to push the capacity boundaries. For example, inlet conditioning for power augementation utilizing fogging, inlet chilling and evaporative cooling of GT inlet air have been installed in various occasions.

In addition to gas turbine and prime mover capacity enhancements, the other frontier is improvement of thermal efficiency of the prime movers and power plant either to free up more gas towards LNG Production or utilize in a combined heat and power conceptual design to capture waste heat and integrate back into the LNG Process Plant. The most obvious approach is implementing waste heat recovery unit (WHRU) on the gas turbines. The recovered heat can be directed to various heat users in the plant which is known as Cogeneration designs. In most LNG plants, the recoverable waste heat is more than the demand of the process plant thermal users. Therefore, in some cases, the recovered heat is used for power generation, which is known as Combined Cycle (CC) designs. Relatively few cases were demonstrated to use recovered waste heat to enhance plant capacity, although a well-known steam-injection process is commonly practiced in the power industry.

In this presentation, comparison will be made between a selected air chilling and an alternative approach, called absorption chilling cycle, which directly converts heat to refrigeration. Absorption refrigeration has been used in refinery and other industries, but not LNG. Emphasis will be placed on dampening the impacts of seasonal temperature variations, instead of straightforward capacity enhancement, which requires enlargement of all supporting systems and underlying structures.

Work Output of Multicomponent LNG Mixtures in Two-Phase Expanders

- Dr. Linda St-Cyr, Ebara International Corporation, Sparks NV
  Presentation ID# 244808

- Abstract: A two-phase expander recovers more work from a multicomponent LNG mixture than a single phase expander. The calculation of the recoverable work involves the complex thermodynamics of two-phase mixtures, and is best done utilizing property tables from the National Institute of Standards and Technology (NIST). The NIST tables are in turn based on equations developed by the European Gas Research Group (GERG-2008). Building on the work done by these distinguished groups, we are able to document the contributions to the enthalpy integral from the PV term versus the internal energy.

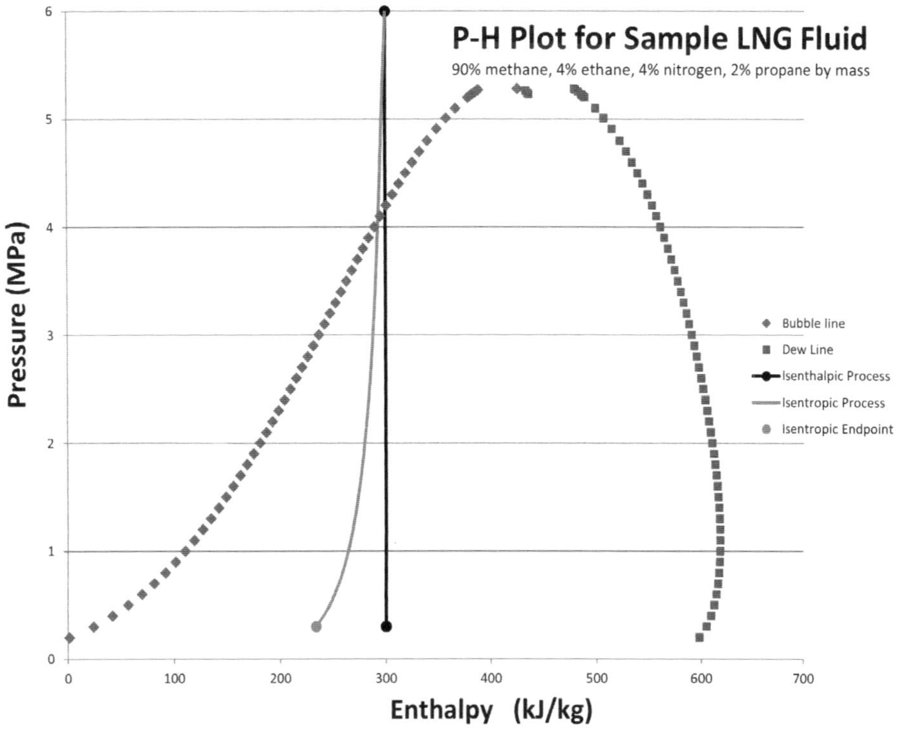

Expander Process vs Throttling Process

- The 2-phase region for a typical LNG fluid consisting of 90% methane, 4% ethane, 4% nitrogen and 2% propane is shown on this P-H diagram using values generated by the NIST REFPROP routine. The blue diamonds represent bubble points and the red squares represent dew points. Inside the 2-phase dome both liquid and vapor are present in equilibrium.
- An isentropic expansion process (green curve) and an isenthalpic expansion process (black line) show 2 possible fluid paths starting at h = 300 kJ/kg and P = 6 MPa and ending at P = 0.3 MPa.
- The isenthalpic line (black) represents the effect of using a throttling valve to reduce LNG pressure. This constant enthalpy process is known as a Joule-Thomson process.
- The isentropic curve (green) represents the most efficient expansion from the given starting position to an end state at 0.3 MPa. This trace follows a constant entropy curve and represents a 100% efficient (ideal) process. Expanders approximate this curve, usually at about 85% efficiency.
- For a given pressure, the farther to the left that a 2-phase point sits, the greater the percentage of liquid vs vapor. This Joule Thomson process ends at a liquid faction of 0.43 , while the isentropic process ends at a liquid faction of 0.57, which is 30% more liquid.
- Single-phase expanders operate only up to the beginning of the 2-phase region. After that, a Joule-Thomson valve is needed to continue the expansion to the desired ending pressure. For this example, the endpoint for the single-phase expander would lie at only 7% of the distance from the black line to the green curve.
- The endpoint for a 2-phase expander @ 85% efficiency would lie at 85% of the distance from the black line to the green curve. The expansion curve for an 85% efficient process can be approximated by a curve that lies at 85% of the distance from the black line to the green curve for each pressure from the beginning at 6MPa to the end at 0.3 MPa.

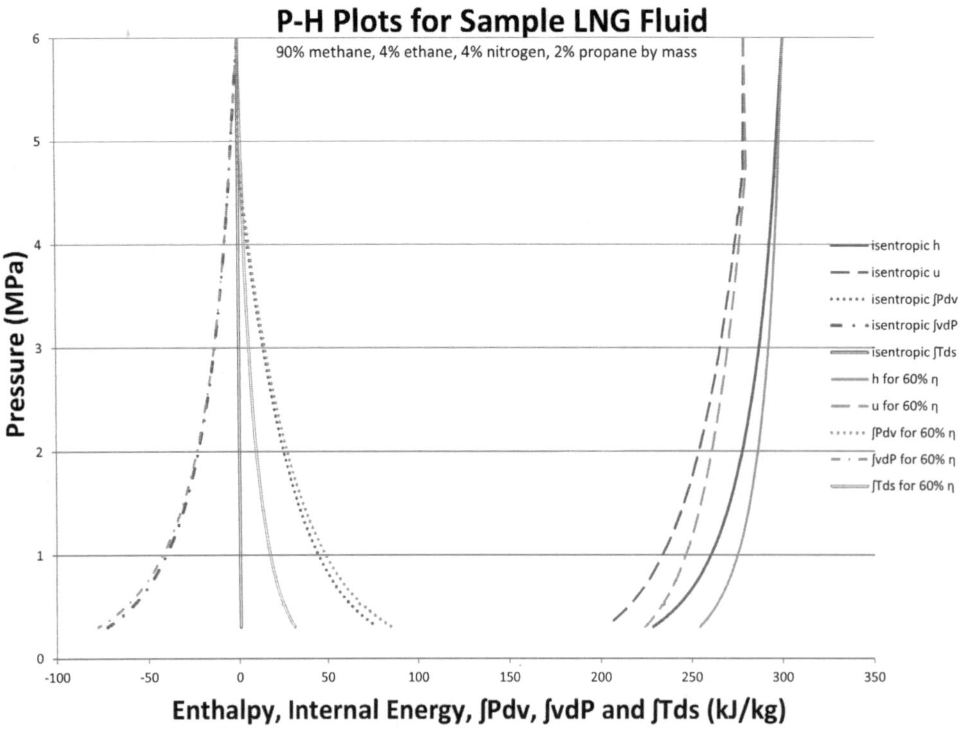

Enthalpy, Internal Energy, $\int Pdv$, $\int vdP$ and $\int Tds$

- The previous plot (made using NIST REFPROP) shows the process curves for enthalpy (h), internal energy (u), $\int Pdv$, $\int vdP$ and $\int Tds$ for a 100% efficient process (blue) and for a 60% efficient process (orange). 60% was chosen so that there was enough of a difference to be clearly visible on the plot.
- Enthalpy per unit mass is defined as $h = Pv + u$, thus $\Delta h = \int Pdv + \int vdP + \Delta u$. From the plots and the tabulated data, it can be verified that at 100% efficiency, $\Delta h = \int vdP$ and $\Delta u = -\int Pdv$. An interesting thing to note is that the traces for Δu (long dashes) and $\int Pdv$ (dotted) both show a distinct kink at about P = 4.7 MPa that the other curves don't have. Indeed, for a closed volume, the reversible Pdv work done by a fluid decreases its internal energy u by exactly the amount $\int Pdv$, so they are directly related. Similarly, Δh (solid line) and $\int vdP$ (dot-dash) are also directly related, since $\int vdP$ is the reversible work done by a steady-flow system.
- For a process that is not 100% efficient, we can write the Gibbs relation $\Delta h = \int vdP + \int Tds$. For this relation applied to an expansion, we note that $\Delta h < 0$, $\int vdP < 0$ and $\int Tds > 0$, the latter being another way to state that this process is *not* reversible. So the magnitude of $\int vdP$ must be greater than it was for an isentropic process by an amount that makes up for the irreversibilities in $\int Tds$ (the double-line plot). Thus we see that $\int vdP$ gives a measure of the available work only in the case of an isentropic process.

Numerical Experimentation with NIST REFROP

• By doing numerical experiments using properties from the NIST REFPROP tables, it is possible to clarify the thermodynamics of mixtures and 2-phase systems.

• The NIST tables for LNG components are based on equations developed by the European Gas Research Group (GERG-2008).

National Institute of Standards and Technology

Dr. Eric W. Lemmon
Thermophsical Properties Division
Theory and Modeling of Fluids Group
325 Broadway
Boulder, CO 80305, U.S.A.
Phone: +1-303-497-7939
Web: http://www.nist.gov/cstl/properties/fluids_modeling/index.cfm

Dr. Mark O. McLinden
Thermophsical Properties Division
Experimental Properties of Fluids Group
325 Broadway
Boulder, CO 80305, U.S.A.
Phone: +1-303-497-3580
Web: http://FluidProperties.NIST.Gov

Dr. Marcia L. Huber
Thermophsical Properties Division
Theory and Modeling of Fluids Group
325 Broadway
Boulder, CO 80305, U.S.A.
Phone: +1-303-497-5252
Web: http://www.nist.gov/cstl/properties/fluids_modeling/index.cfm

(89d) Two-Phase Pump Expanders for Liquefied Propane Gas

Hans E. Kimmel, Ebara International Corporation, Sparks, NV

Thermodynamic power cycles at cryogenic temperatures require new and different designs and materials than there are in conventional power plants operating at high temperature.

The working fluids in cryogenic power cycles are liquefied gases like ammonia, nitrogen and hydrocarbon gases. Most of these cryogenic working fluids are flammable or corrosive or both.

The development of a particular design of a two phase pump expander for liquefied propane gas was necessary to meet the requirements for a reliable and safe operation of the rotating machinery within an efficient cryogenic power cycle. The applications and the design of two-phase pump expanders for liquefied propane gas as working fluid are presented and the performance characteristics discussed.

(89f) Cryogenic Expander Application in Propane Liquefaction Process

Michael Cords, Ebara International Corporation, Sparks, NV

The use of cryogenic liquid expanders in the propane liquefaction process can increase the amount of liquefied product by reducing the boil-off losses. Current practice is to use a Joule-Thomson valve to reduce the high pressure condensate. Losses associated by throttling across a J-T valve result in additional heat causing unwanted boil-off. A cryogenic expander reduces the enthalpy of the liquid during expansion which sub-cools the liquid, significantly reducing the boil-off losses, and can export this extracted energy as electrical energy.

The thermodynamic properties of propane are well suited to exploit the benefits of a cryogenic expander. The technology for propane cryogenic expanders exists and has been proven in LNG liquefaction, both in the LNG stream and the Mixed Refrigerant cycle. Mixed Refrigerant is comprised of mostly propane, making the technological transition from MR expander to propane expander negligible and low risk. The benefit of increased production already proven in LNG liquefaction can be applied to propane liquefaction resulting in overall process efficiency and reduced costs.

(95a) Switching Between Hydrogenolysis and Bifunctional Hydrocracking on An Unsulfided Co/MoO3/SiO2-Al2O3 Catalyst

Heinz J. Robota, University of Dayton Research Institute, Dayton, OH and Jhoanna Alger, Energy and Environmental Engineering, University of Dayton Research Institute, Dayton, OH

In previous work using hexadecane as a feed molecule, we found that under typical catalytic hydrocracking conditions, an unsulfided $Co/MoO_3/SiO_2$-Al_2O_3catalyst exhibited primarily hydrogenolysis. Two types of hydrogenolysis were observed. One type involved terminal cleavage of methyl groups, in which the feed n-C_{16}underwent progressive cleavage to first n C_{15}, then, to a lesser extent to n-$C_{14}$and so on. This resulted in high quantities of product methane. The second type of hydrogenolysis reaction resulted in the cleavage of interior C-C bonds with near equal probability. This second pathway yielded ethane in molar quantities similar to other cracked products in the $C_3$to C_{13}range. Further, the overall yield of isomerized cracked products was quite low, typically less than 10 mole percent. However, when feeding a mixture of n-C_{26}/n-$C_{28}$to the same catalyst under similar conditions, methane yields were low, ethane was barely detectable, and isomer content of the cracked products was in excess of 80 mole percent. These latter observations are consistent with a bifunctional hydrocracking mechanism. Further, a comparison of the product selectivities as a function of conversion in these two cases are consistent with hydrogenolysis in the n-C_{16}case and bifunctional hydrocracking in the n-C_{26}/n-C_{28}case. In an effort to understand the reasons for this feed dependent reaction behavior, we have investigated the cracking of n-C_{18}under similar conditions. We have found that we are able to switch between hydrogenolysis and bifunctional hydrocracking by altering reaction conditions so as to either favor or suppress the relative olefin content of the reacting C_{18}. While holding the feed rate of hydrocarbon constant, hydrogenolysis is nearly exclusively observed at 300° C when $H_2$is fed at a molar ratio of 10:1 H_2:n-C_{18}. However, at the same absolute $H_2$and n-C_{18}feed rate, but with a supplementary flow of Ar at twice the H_2feed rate, nearly exclusively bifunctional hydrocracking behavior is observed. In contrast, when only $H_2$is fed at a 30:1 H_2:n C_{18}rate, an even higher rate of exclusively hydrogenolysis is observed. Similarly, we found that when n-$C_{16}$is fed with the supplementary Ar flow, it also exhibits predominantly bifunctional hydrocracking behavior. We believe that when the olefin content of the hydrocarbon reactant is sufficiently high, bifunctional hydrocracking predominates. Hence, since the equilibrium olefin content of a C_{26}/C_{28}feed is considerably higher than in a C_{16}feed, at equivalent molar flux, bifunctional behavior dominates in the C_{26}/C_{28}case. For this reason, when seeking to process a typical Fischer-Tropsch wax, where the dominant molecular sizes are in the $C_{26}$and larger range, quite typical bifunctional hydrocracking is exhibited by this catalyst.

(95b) The Effect of Chain Length of Feed on Hydrocracking Over Pt/SiO_2-Al_2O_3Catalyst

Jungshik Kang, Wenping Ma, Wilson D. Shafer, Keogh Robert, Gary Jacobs and Burtron H. Davis, University of Kentucky, Center for Applied Energy Research, Lexington, KY

The hydrocracking of hydrocarbons with carbon number $C_{16}$and $C_{28}$and their mixture was studied over Pt promoted amorphous silica-alumina catalyst. At various reaction conditions we determined the conversion, product selectivity, and the degrees of isomerization on the catalyst of each carbon number. We found out that gas phase hydrocarbons (at the reaction conditions) passed through the catalyst bed much faster than liquid phase hydrocarbons so that essentially only the liquid hydrocarbons were converted. We found that at a low hydrocracking conversion range of wax, all the liquid phase reactants show about the same hydrocracking reactivity. So that we can conclude that the hydrocracking reaction depends not only on the carbon chain length but the phase of reactant at the reaction conditions if predominantly in the gas phase it is not or only slightly converted. Therefore the dependence of conversion on carbon number is directly related to the liquid/gas phase equilibrium.

(95c) Middle Distillates Production Via Fischer Tropsch Synthesis with Product Upgrading Under Supercritical Phase Conditions

Sihe Zhang, Rui Xu, J. Ed Durham and Christopher B. Roberts, Department of Chemical Engineering, Auburn University, Auburn University, AL

Fischer-Tropsch (FT) synthesis involves surface catalyzed polymerization reactions that inherently yield a high selectivity towards methane and a broad product distribution of gaseous light products, middle distillate hydrocarbons (olefins + paraffins), and heavy wax, as well as $CO_2$and water. Subsequent product upgrading via catalytic oligomerization and hydrocracking/isomerization can be employed to convert light olefin FT hydrocarbons into middle distillate range liquids, while also converting the long-chain hydrocarbons of heavy wax into fuel range products. This results in a reduction in olefin selectivity, generation of aromatics, suppression of heavy hydrocarbons selectivity and an enhancement of cold-flow properties.

In an effort to further direct the FT product distribution towards middle distillate transportation fuel range products, we have undertaken a series of experiments where we have sequentially incorporated catalytic oligomerization and hydrocracking/isomerization stages subsequent to FT in single pass operation. A multiple fixed bed reactor system was designed and utilized in this investigation. A precipitated iron-based FT catalyst was applied due to its low price and high tendency to yield olefins, which can be upgraded in the oligomerization, hydrogenation and isomerization reactions within the sequential reaction beds. The reaction temperatures were maintained differently in each bed to provide optimal reaction performance, while the system pressure was kept uniform in all three reaction beds.

Moreover, we have utilized the enhanced heat transfer properties of a supercritical fluid reaction medium in the FT and subsequent upgrading stages in order to reduce the selectivity towards $CH_4$and CO_2. The use of the supercritical solvent provides prolonged catalyst life and activity maintenance. Distinguishable enhancement of propagation probability was observed under the supercritical phase operation. In particular, this study involves a systematic investigation into the influence of the supercritical fluid reaction medium on the overall product distribution and selectivity towards fuel range products.

(95d) Oxygenate/Hydrocarbon Distribution Compared Across Different Group I Promoters for Iron Catalysts

Wilson D. Shafer[1], Robert A. Keogh[2]and Burtron H. Davis[1], (1)University of Kentucky, Center for Applied Energy Research, Lexington, KY, (2)Center for Applied Energy Research, University of Kentucky, Lexington, KY

Oxygenate and Hydrocarbon Distribution Compared Across Different Group I Promoters for Iron Catalysts.

Wilson D. Shafer, Robert Keogh, Burtron H. Davis, Center for Applied Energy Research, University of Kentucky, Lexington, Ky

Abstract

A main catalyst used in FT currently is iron; this is driven by cost and the ability of Fe to play a role in water-gas-shift alongside the production of liquid fuels. Iron catalysts can be dispersed amongst a support and promoted with potassium or copper. The activity of iron catalysts promoted with potassium display an increase in the FTS and WGS activity [1]. Specific testing was done on different K promoter loading to find that greater than 5%molar K/Fe loading did not increase the CO conversion [2]. Low atomic loadings of potassium have been shown to increase the FTS conversion by playing a role in increasing the CO dissociation but slightly decreasing the H_2dissociation [3]. If this is the case for potassium; the other alkali metals, containing the same chemical characteristics should play a similar role.

A factor in describing the effects seen is the basicity of the promoters described by Roper et al [4], using a fluidized bed reactor, where the alkaline basicity was inversely correlated to the amount methane produced. In addition, 2002 specific work was done with slurry phase operations investigating the role in other row I alkali metals in comparison to potassium in iron catalysts [5]. This specific work was done looking into the WGS and FT activity for a slurry reactor system with similar alkaline promoter loadings and specific reactor conditions. The results based on the previous work, display varying results for each promoter across the same Fe catalyst. It was found that of the entire row I alkali metals, potassium promotes both FTS and WGS the best, in fact Li, Cs, and Rb promoted catalysts seem to show lower activity when compared to the unpromoted Fe catalyst.

This specific analysis work was to investigate the role I metal promoters by looking at the specific hydrocarbon/oxygenate distribution. The addition of each row I metal exclusively adds to the mole per carbon amount of olefin and oxygenated material over the unpromoted Fe catalyst. If, as previously mentioned, the addition of K as a promoter assists in the dissociation of CO and in turn suppression of H_2dissociation, this then could be seen in the overall product distribution for each row I promoter. The amounts of oxygenates and olefins increase linearly with each promoter added, where Li and the unpromoted produced the smallest amounts and Cs, Rb promoted Fe catalysts displayed the most. This distribution work can then compared to the alkali promoted Fe catalysts work done in 2002, which described basicity as a factor, others include atomic size and the ability to suppress $H_2$and assist in CO dissociation, to better explain the metals activity and why K seems to be the best row I promoter.

References

1.D.B. Bukur, D. Mukesh, S.A. Patel, Ind. Eng. Chem. Res. 29 (1990) 194.

2. M. Luo, R. J. O'Brien, S. Bao, B. H. Davis; Appl. Catal. A: General 239 (2003) 111-120

3.Gaube, J.; Klien, H.-F.; J. of Mol. Catal. A 283 (**2008**) 60

4. M. R"per, in: W. Keim, (Ed.), Catalysis in C1 Chemistry, D. Reidel Publishing Company, Dordrecht, 1983, pp. 41"88.

5.W. Ngantsoue-Hoc, Y. Zhang, R.J. O'Brien, M. Luo, B. H. Davis; Appl. Catal. A: General 236 (2002) 77-89

(106a) Synthesis of PVA Based Membranes Modified by Benzimidazole Group for Direct Methanol Fuel Cells

Pratima Gajbhiye, Chemical Engineering, Indian Institute of Technology, Kanpur, State of Uttar Pradesh, India, Jayant K. Singh, Chemical Engineering, Indian Institute of Technology, Kanpur, Ka, India and Anil Kumar, Department of Chemical Engineering, Indian Institute of Technology Kanpur, Kanpur, India

Abstract

In recent years, there has been a great effort to look for alternate membrane material and Polybenzimidazole (PBI) membranes have been synthesized where the conducting moiety of this polymer is benzimidazolium groups. The monomers used to synthesize these polymers are bis conjugated amines and diacids are expensive. In this work, we have synthesized grafted polymers of PVA as membrane materials where the grafts are either Poly Acrylic Acid, Poly maleic anhydride (PMA) or Poly Methacrylic acid (PMAA) and these materials were denoted by PVA-g-PAA, PVA-g-PMA, and PVA-g-PMAA. We then synthesized C1, C3 & C4 compounds (formula given at the bottom of Table S1) having bis conjugated amines whereas compound C2 having only one conjugated amine group. One of the steps in synthesizing C1 compound is nitration of phenyl ring, where there is a possibility of formation of two isomers out of which only one of them has strictly bis conjugated amines and can be polymerized with Isophthalic acid giving the formation of PBI. The other byproduct formed is an impurity and is minimized by carrying out the nitration step at low temperature but this cannot be completely removed from the product which is a mixture. Table S1 presents the summary of our experimental results on characterization of membrane, In the C1 to C4 compounds and in some cases (membranes 9, 10, 11), Sulfocuccinic acid were added to the casting polymer solution as additional crosslinking agent in order to achieve higher conductivity. We added the polymer, thus formed, to the casting solution, we first compared the results of the Fuel cell formed by adding C1 (C3 and C4) compound alone and heating at 180°C for some time. Table S1 has two sets of results, one on characterization (columns 1, 2, 3,4,5,6, and 7) of membranes while the other on Fuel cell (columns 8 to 11). For the latter columns 10 and 11 gave results on characterization on Fuel cell while columns 8 and 9 gave results when the Fuel cell was run on control mode for higher efficiency.

Experiments on membrane pore sizes are given in column 1 which lies between 8.88 to 9.52nm and they all are ultra filtration type. The values of the pore size are mostly governed by the nature of the graft chains, e.g. membranes 1,3,4,7 and 12 consist of PAA grafts and were found to have similar pore sizes. Similarly membranes 2,5,8,9 were of the similar pore size because the PVA have the same grafts of PMA. The next important thing is to determine the wall charges of these membranes and they were estimated by experimentally determining the rejection of Chromic (VI) acid solution through these membranes. Subsequently, a space charge model was proposed for the flow of the solution through their pores and a semi analytical solution was derived for the overall problem of separation of chromic acid solution. We observed that the high surface charge was because of the benzimidazolium groups present on the surface and they ranged from minimum of 465 to maximum of 525mV. The thermal degradation (TGA) experiments were carried out which showed that, through the modification of the polymer by benzimidazole groups, lead to stable membrane and the degradation temperature varies from 150°C for 10% weight loss and for approximately 80% weight reduction, a minimum of 300°C to 700°C.

The twelve membranes reported in Table S1 were first characterized completely in terms of pore size, wall charge, conductivity, IEC and thermal degradation. In fuel cell applications, higher temperatures give large current density and it is observed that some of the membranes (rows 1, 2, 5, 6, 7, 8 and 12) could easily be operated upto 150°C without the membrane degrading. However in our case, we insisted that the membrane of direct methanol fuel cell were operated at 90°C for which most of the fuel cell results for Nafion 117 are available. The catalysts used on the anode and cathode side were Pt-Ru and Pt loaded on carbon respectively, which were synthesized by following procedure given in literature. We find that in each case, the power density and the current density were higher and for some of our results (membranes 1,6,12) were atleast two to three times higher than that of Nafion 117 as seen in column 10 and 11.

It was recognized that the membrane material has hydrophilic PVA matrix and hydrophobic benzimidazolium groups at the surface formed by reaction of the C1 to C4 compounds with grafts (see Fig.S-1). When the reaction occurs (confirmed by IR and NMR), the hydrophobic benzimidazolium are formed which tends to separate as confirmed by the SEM results. In view of this we assumed that the pores of the membrane would have a thin hydrophobic layer with positive charge walls. This is confirmed by the contact angle measurements where the minimum contact angle is 51° and varied upto 89°. Even though most of the polymer matrix is PVA with its grafts, when the ions move through the pores (giving rise to current in Fuel cell), the pore walls get highly charged (as measured by column 2 of Table S1). Some of the membranes have Sulfo succinic acid SSA as part of the matrix (e.g. the membrane 9 to 11) which would tend to give sulfonate groups inside the PVA matrix as well as at the phase boundaries, this way giving rise to bipolar nature of the membrane (benzimidazolium group being strictly cationic in nature). In addition to this, some of the compounds such as C1 and C2 compounds have bonded sulfonate groups which would give the positive and negative charge separated by the molecular distance, once again producing a bipolar nature to the membrane. In remainder of the cases the charges would be induced on the phase boundaries as seen in Fig.S-1. The conductivity observed in our membranes is very high and could be attributed to the special bipolar structure of our polymers.

Table S1. Fuel cell and Membrane Characterization

Compound	Membrane	1	2	3	4	5	6	7	8	9	10	11

			MWCO (pore size, nm)	Wall Charge	Conductivity (S/cm)	IEC	Thermal Degradation temp (°C) 10% - 80%		Water Uptake	Contact Angle	Average Voltage Gain (V)	Average time cycle (min)	Power Density (mW/ cm)	Current Density (mA/ cm)
C1	1	PVA-g-PAA-C1	9.1	488	0.04	2.7	199.732	268.70	7.5	72	0.523	55.75	50.47	74.88
	2	PVA-g-PMA-C1	9.35	465	1.23×10^{-2}	1.8	142.659	447.6	10.691	74.917	0.568	42.896	0.305	3.81
	3	PVA-g-PAA-modified	9.52	488	1.51×10^{-2}	2.12	115.123	706.53 (28.93%)	13.665	75.581	0.778	69.85	37.37	64.44
C2	4	PVA-g-PAA-C2	9.52	525	6.59×10^{-2}	2.38	119.558	691.93	13.4	51.2	0.704	39.83	23.0020	50.55
	5	PVA-g-PMA-C2	9.36	519	1.04×10^{-2}	2.17	181.59	562.348	14.9	50.24	0.77	161.93	34.72	62.11
	6	PVA-g-PMAA-C2	8.99	482	1.16×10^{-2}	2.18	201.055	472.568	13.8	51.8	0.6066	147.81	76.72	92.33
C3	7	PVA-g-PAA-C3	9.0	512	2.51×10^{-2}	2.23	204.908	453.03	8.624	70.368	0.6883	82.55	36.397	75
	8	PVA-g-PMA-C3	9.5	510	3.51×10^{-2}	2.27	166.819	505.293	13.513	62.944	0.726	37.557	36.11	64.381
	9	PVA-g-PAA-SSA-C3	9.36	497	1.71×10^{-2}	2.51	154.914	813.769 (24.32%)	10.5	54.8	0.8096	99.423	23.45	57.778
	10	PVA-g-PMA-SSA-C3	9.83	480	2.28×10^{-2}	2.46	115.12	428.713	8.5	76.2	0.663	224.22	28.561	58.358
	11	PVA-g-PMAA-SSA-C3	8.88	525	3.66×10^{-2}	1.85	139.772	681.715 (30.78%)	7.2	89.2	0.77	18.36	20.832	48.11
C4	12	PVA-g-PAA-C4	9.5	495	2.56×10^{-2}	1.62	189.87	451.86	8.927	59.132	0.704	74.19	41.1	71
	13	Nafion 117			0.1								21.01	48.33

Compound C1, [Bis-3-Amino-4-{3-(triethylammoniumsulfonato) phenylamino}-phenyl sulfone] hydrochloride

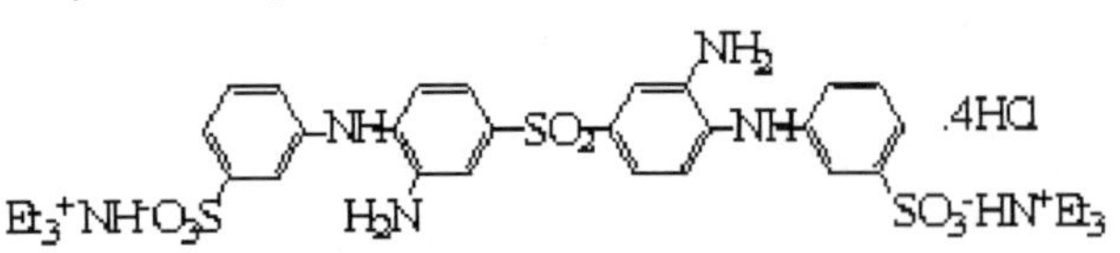

Compound C2, 3-Amino -4-[3-(triethyl ammonium sulfonato) phenyl amino] phenylene hydrochloride

THE GRAPHIC ON THIS PAGE WAS NOT AVAILABLE ON THE E-MEDIA VERSION OF THIS PAPER.

Compound C3, Bis -4 [1, 2 diamino phenyl]-phenyl sulfone) hydrochloride

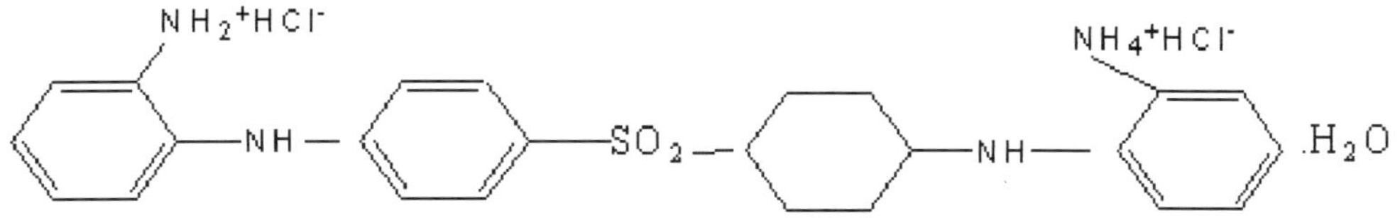

Compound C4, 1, 4 [1, 2 diamino phenyl] Phenylene hydrochloride

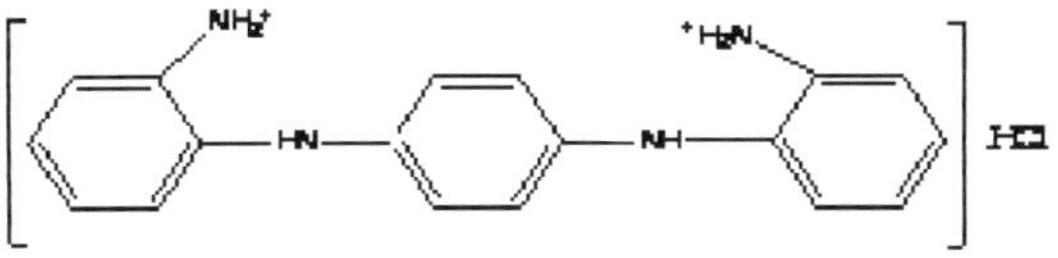

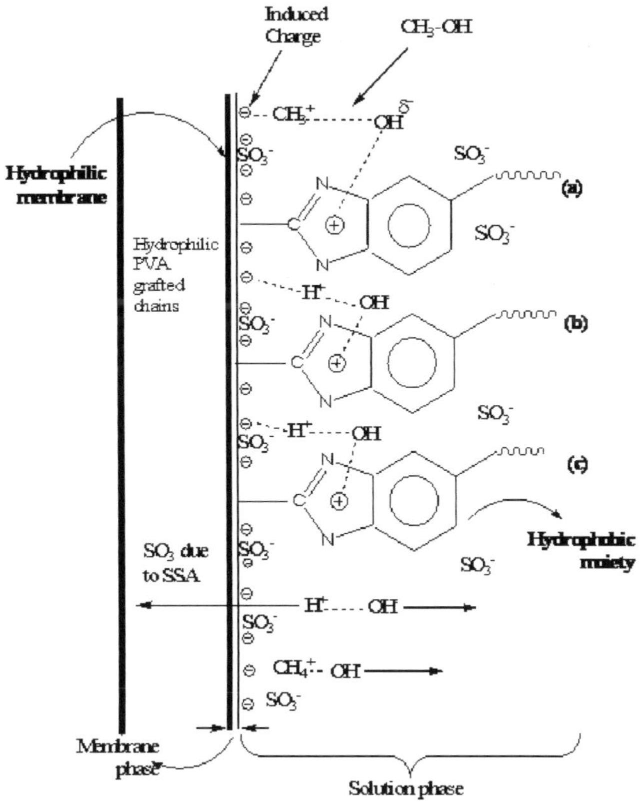

Figure S-1. The Bipolar nature of the membrane surface

(106b) Operation and Controls of Operating Direct Methanol Fuel Cells with PVA Based Membranes Modified by Benzimidazolium Groups

Pratima Gajbhiye, Chemical Engineering, Indian Institute of Technology, Kanpur, State of Uttar Pradesh, India, Pranjal Joshi, Chemical Engg, IIT Kanpur, Kanpur, India, Jayant K. Singh, Chemical Engineering, Indian Institute of Technology, Kanpur, Ka, India and A. Kumar, Department of Chemical Engineering, Indian Institute of Technology Kanpur, Kanpur, India

Operation and Control of Direct Methanol Fuel cell with PVA based membranes modified with benzimidazolium groups

Pratima Gajbhiye, Pranjal Joshi, Jayant K.Singh, Anil Kumar

Department of Chemical Engineering, Indian Institute of Technology, Kanpur 208016, India

Abstract

In recent years, in an effort to find alternate membrane material (other than Nafion 117), Polybenzimidazole (PBI) membranes have been synthesized where the conducting moiety of this polymer is benzimidazolium groups. The monomers used to synthesize these polymers are bis conjugated amines and diacids are expensive. In this work, we have synthesized grafted polymers of PVA as membrane materials where the grafts are either Poly Acrylic Acid, Poly maleic anhydride (PMA) or Poly Methacrylic acid (PMAA) and these materials were denoted by PVA-g-PAA, PVA-g-PMA, and PVA-g-PMAA. We then synthesized C1, C3 & C4 compounds (formula given at the bottom of Table S1) having bis conjugated amines whereas compound C2 having only one conjugated amine group. One of the steps in synthesizing C1 compound is nitration of phenyl ring, where there is a possibility of formation of two isomers out of which only one of them has strictly bis conjugated amines and can be polymerized with Isophthalic acid giving the formation of PBI. The other byproduct formed is an impurity and is minimized by carrying out the nitration step at low temperature but this cannot be completely removed from the product which is a mixture. Table S1 presents the summary of our experimental results on characterization of membrane, In the C1 to C4 compounds and in some cases (membranes 9, 10, 11), Sulfocuccinic acid were added to the casting polymer solution as additional crosslinking agent in order to achieve higher conductivity. We added the polymer, thus formed, to the casting solution, we first compared the results of the Fuel cell formed by adding C1 (C3 and C4) compound alone and heating at 180°C for some time. Table S1 has two sets of results, one on characterization (columns 1, 2, 3,4,5,6, and 7) of membranes while the other on Fuel cell (columns 8 to 11). For the latter columns 10 and 11 gave results on characterization on Fuel cell while columns 8 and 9 gave results when the Fuel cell was run on control mode for higher efficiency.

Experiments on membrane pore sizes are given in column 1 which lies between 8.88 to 9.52nm and they all are ultra filtration type. The values of the pore size are mostly governed by the nature of the graft chains, e.g. membranes 1,3,4,7 and 12 consist of PAA grafts and were found to have similar pore sizes. Similarly membranes 2,5,8,9 were of the similar pore size because the PVA have the same grafts of PMA. The next important thing is to determine the wall charges of these membranes and they were estimated by experimentally determining the rejection of Chromic (VI) acid solution through these membranes. Subsequently, a space charge model was proposed for the flow of the solution through their pores and a semi analytical solution was derived for the overall problem of separation of chromic acid solution. We observed that the high surface charge was because of the benzimidazolium groups present on the surface and they ranged from minimum of 465 to maximum of 525mV. The thermal degradation (TGA) experiments were carried out which showed that, through the modification of the polymer by benzimidazole groups, lead to stable membrane and the degradation temperature varies from 150°C for 10% weight loss and for approximately 80% weight reduction, a minimum of 300°C to 700°C.

The twelve membranes reported in Table S1 were first characterized completely in terms of pore size, wall charge, conductivity, IEC and thermal degradation. In fuel cell applications, higher temperatures give large current density and it is observed that some of the membranes (rows 1, 2, 5, 6, 7, 8 and 12) could easily be operated upto 150°C without the membrane degrading. However in our case, we insisted that the membrane of direct methanol fuel cell were operated at 90°C for which most of the fuel cell results for Nafion 117 are available. The catalysts used on the anode and cathode side were Pt-Ru and Pt loaded on carbon respectively, which were synthesized by following procedure given in literature. We find that in each case, the power density and the current density were higher and for some of our results (membranes 1,6,12) were atleast two to three times higher than that of Nafion 117 as seen in column 10 and 11. In normal Fuel cell operations, there is decay in the cathode catalyst which is further increased by the methanol crossover. The voltage of the Fuel cell increases, the average volatage gained is reported in column 8 and it is found that the voltage gain depends on the nature of the membrane which is not the same for all of them. In column 9, we have measured the average time cycle and we found that some of the membranes (membranes 6, 7, 9, 10) gave extremely good results.

It was recognized that the membrane material has hydrophilic PVA matrix and hydrophobic benzimidazolium groups at the surface formed by reaction of the C1 to C4 compounds with grafts (see Fig.S-1). When the reaction occurs (confirmed by IR and NMR), the hydrophobic benzimidazolium are formed which tends to separate as confirmed by the SEM results. In view of this we assumed that the pores of the membrane would have a thin hydrophobic layer with positive charge walls. This is confirmed by the contact angle measurements where the minimum contact angle is 51° and varied upto 89°. Even though most of the polymer matrix is PVA with its grafts, when the ions move through the pores (giving rise to current in Fuel cell), the pore walls get highly charged (as measured by column 2 of Table S1). Some of the membranes have Sulfo succinic acid SSA as part of the matrix (e.g. the membrane 9 to 11) which would tend to give sulfonate groups inside the PVA matrix as well as at the phase boundaries, this way giving rise to bipolar nature of the membrane (benzimidazolium group being strictly cationic in nature). In addition to this, some of the compounds such as C1 and C2 compounds have bonded sulfonate groups which would give the positive and negative charge separated by the molecular distance, once again producing a bipolar nature to the membrane. In remainder of the cases the charges would be induced on the phase boundaries as seen in Fig.S-1. The conductivity observed in our membranes is very high and could be attributed to the special bipolar structure of our polymers.

Table S1. Fuel cell and Membrane Characterization

Compound	**Membrane**		**1**	**2**	**3**	**4**	**5**		**6**	**7**	**8**	**9**	**10**	**11**
			MWCO (pore size, nm)	**Wall Charge**	**Conductivity (S/cm)**	**IEC**	**Thermal Degradation temp (°C) 10% - 80%**		**Water Uptake**	**Contact Angle**	**Average Voltage Gain (V)**	**Average time cycle (min)**	**Power Density (mW/ cm)**	**Current Density (mA/ cm)**
C1	1	PVA-g-PAA-C1	9.1	488	0.04	2.7	199.732	268.70	7.5	72	0.523	55.75	50.47	74.88
	2	PVA-g-PMA-C1	9.35	465	1.23×10^{-2}	1.8	142.659	447.6	10.691	74.917	0.568	42.896	0.305	3.81
	3	PVA-g-PAA-modified	9.52	488	1.51×10^{-2}	2.12	115.123	706.53 (28.93%)	13.665	75.581	0.778	69.85	37.37	64.44
C2	4	PVA-g-PAA-C2	9.52	525	6.59×10^{-2}	2.38	119.558	691.93	13.4	51.2	0.704	39.83	23.0020	50.55
	5	PVA-g-PMA-C2	9.36	519	1.04×10^{-2}	2.17	181.59	562.348	14.9	50.24	0.77	161.93	34.72	62.11
	6	PVA-g-PMAA-C2	8.99	482	1.16×10^{-2}	2.18	201.055	472.568	13.8	51.8	0.6066	147.81	76.72	92.33
C3	7	PVA-g-PAA-C3	9.0	512	2.51×10^{-2}	2.23	204.908	453.03	8.624	70.368	0.6883	82.55	36.397	75
	8	PVA-g-PMA-C3	9.5	510	3.51×10^{-2}	2.27	166.819	505.293	13.513	62.944	0.726	37.557	36.11	64.381
	9	PVA-g-PAA-SSA-C3	9.36	497	1.71×10^{-2}	2.51	154.914	813.769 (24.32%)	10.5	54.8	0.8096	99.423	23.45	57.778
	10	PVA-g-PMA-SSA-C3	9.83	480	2.28×10^{-2}	2.46	115.12	428.713	8.5	76.2	0.663	224.22	28.561	58.358
	11	PVA-g-PMAA-	8.88	525	3.66×10^{-2}	1.85	139.772	681.715	7.2	89.2	0.77	18.36	20.832	48.11

		SSA-C3						(30.78%)						
C4	12	PVA-g-PAA-C4	9.5	495	2.56×10^{-2}	1.62	189.87	451.86	8.927	59.132	0.704	74.19	41.1	71
	13	Nafion 117			0.1								21.01	48.33

THE GRAPHIC ON THIS PAGE WAS NOT AVAILABLE ON THE E-MEDIA VERSION OF THIS PAPER.

Compound C1, [Bis-3-Amino-4-{3-(triethylammoniumsulfonato) phenylamino}-phenyl sulfone] hydrochloride

THE GRAPHIC ON THIS PAGE WAS NOT AVAILABLE ON THE E-MEDIA VERSION OF THIS PAPER.

Compound C3, Bis -4 [1, 2 diamino phenyl]-phenyl sulfone) hydrochloride

THE GRAPHIC ON THIS PAGE WAS NOT AVAILABLE ON THE E-MEDIA VERSION OF THIS PAPER.

Compound C2, 3-Amino -4-[3-(triethyl ammonium sulfonato) phenyl amino] phenylene hydrochloride

THE GRAPHIC ON THIS PAGE WAS NOT AVAILABLE ON THE E-MEDIA VERSION OF THIS PAPER.

Compound C4, 1, 4 [1, 2 diamino phenyl] Phenylene hydrochloride

THE GRAPHIC ON THIS PAGE WAS NOT AVAILABLE ON THE E-MEDIA VERSION OF THIS PAPER.

Figure S-1. The bipolar nature of PVA-g-PAA-BASPAPS membrane which reduces the methanol crossover due to the molecular dissociation. In (a), the imidazolium group is shown to separate out from the PVA moiety of the chain into a separate phase. In (b), the water is shown to complex with the SO and the imidazolium group. As a result of this, interaction in (d) and (e), the water and the methanol is shown to split into anion and cation and the anion permeates through the membrane giving bipolar effect

(106e) Numerical Investigation in Heat Conduction Effect on Heat Exchange for Mini Cross-Flow Heat Exchanger (MCHE)

Hongyue Duan[1], Wei Mu[1], **Xigang Yuan**[1] and Lingai Luo[2], (1)State Key Laboratory of Chemical Engineering, Tianjin University, Tianjin, China, (2)LOCIE Chambery, Savoie University, Savoie, France

In a traditional heat exchanger, the effect of heat transfer within the wall material driven by the temperature gradient in axial direction of the fluid flow is usually very small because the area of the cross section of the tube wall is much smaller than that of the tube for heat exchange. However, in a mini cross-flow heat exchanger(MCHE), where the heat conducting medium is typically a cubic-shaped block of metal and the hot and cold streams are put into heat exchange via two sets of channels in the block, the heat conduction in the wall material in the direction parallel to the fluid flow may have a significant effect on the performance of the heat exchanger.

In this paper, a 2-channel-layer MCHE was simulated and a periodic boundary condition was applied so that the results could be applicable to entire MCHE. The proposed model and the mesh grid size were proved by the good agreement of the simulated overall heat transfer coefficient with that found in the experimental reported in the literature[1]. Then the effect of heat conduction on heat transfer performance in a MCHE was studied numerically. The simulation results suggest that the performance of the MCHE can be improved by increasing the wall thickness between the adjoining channels in the same layer or by decreasing the wall thickness between the adjacent layers. The heat conduction effect can also be weakened by optimizing the material conductivity or by increasing the diameter of the channels. New correlations were developed based on the simulated results to predict the value of the overall heat transfer coefficient as a function of the operating parameter such as Re in the channel and the design parameters of the MCHE such as the diameter of the channel, wall thickness and thermal conductivity.

[1]Luo L A, Fan Y L, Zhang W W*et al.*Integration of constructal distributors to a mini crossflow heat exchanger and their assembly configuration optimization[J]. Chemical Engineering Science, 2007, 62 (13): 3605-3619.

(106g) The Effects of Out Gassing Through Cylindrical Walls on the Transport Coefficients

Luma. H. Yaseen[1],**Nada B. Nakkash**[2]and M. N. Latif[1], (1)Al-Nahrain University, Baghdad, Iraq, (2)Chemical Engineering, Al-Nahrain University, Baghdad, Iraq

Abstract

This work concerns with studying the influence of blowing on friction factor and mass transfer coefficient when the flow is normal to a tube bank with outgassing porous cylinders. The friction factor and the mass transfer coefficients are measured for a range of Reynold's number and blowing rate and expressed in terms of various dimensionless groups.

The test rig consists of a rectangular duct (20.3 × 11.3 cm), five rows of three porous tubes in staggered arrangements fixed across the duct with 1.5pitches (both longitudinal and transversal). The outside and the inside diameters of tubes were 25.4 and 21.4 mm, respectively.

The friction factor measurements was studied for Reynold's number range of

$3000 < Re < 17{,}000$and blowing rate up to 4×10^{-3}. The results show that the blowing intothe boundary layer for the cylinders increases the overall friction factor from 10%to 50% depending on the blowing rate. The mass transfer coefficient was measured for Reynold's number range of $5000 < Re < 16{,}000$ and blowing rate of $10^{-5} < m/G_\infty < 10^{-3}$.

A new definition of the blowing rate was considered and expressed by the mass transfer driving force. Two successful formulas were obtained to account the outgassing effect on the mass transfer coefficient. The mass transfer coefficient for the outgassing boundary layer was obtained first by determining the non-blowing coefficient and multiplying it with the correction factor .

The first model was modified to account the density variation associated with the boundary layer change:

$0.51\ Re^{-0.5} \rho_s$

$St_m Sc^{2/3} =$ -------------- (------)$^{0.4}$

$(1 + \beta_m)^{4.7} \rho_\infty$

With the coefficient of determination R2 = 0.93.

The second model depends on the flat plate correction factor:

$Ln(1+ \beta_m)$

$St_m Sc^{2/3} = 0.5\ Re^{-0.5}$(---------------)$^{9.94}$

β_m

With the coefficient of determination R2 = 0.92.

These two models are found to be in agreement with the previous literature and a reduction of about 80% in the mass transfer coefficient has been obtained due to the outgassing.

(106h) Down the Hubbert Curve

Ian Sutton, Sutton Technical Books, Houston, TX

Down the Hubbert Curve

In March 1956 M. King Hubbert, who was at that time Chief Geologist with the Shell Development Company published a seminal paper with the title "Nuclear Energy and the Fossil Fuels" (Hubbert 1956). In this lengthy and authoritative paper, which was presented before the American Petroleum Institute in San Antonio, Texas, Dr. Hubbert showed how key fossil fuel resources in the United States, particularly coal and oil, were of finite size and that their output would reach a maximum, and then decline. Dr. King's forecasts for the production of oil in the United States were accurate; he predicted that oil production would peak in the early 1970s, and that is what has happened. (The widely-used term "Peak Oil" is frequently accredited to Dr. Colin Campbell (2001)).

In the fifty years since the publication of Dr. King's paper great advances have been made in finding and extracting oil from what were considered to be marginal fields. But Dr. King's overall conclusion is still valid. Production of oil in the United States remains below the 1970's peak. Although Dr. Hubbert's analysis was confined to the United States, subsequent analysts applied his insights to other parts of the world. For example, the book *Twilight in the Desert: The Coming Saudi Oil Shock and the World Economy* (Simmons 2005) by the late Matthew Simmons achieved wide recognition by showing that the concept of peak oil most likely applies to reserves in Saudi Arabia, and other nations such as Iran and Kuwait.

The purpose of this paper is to examine how oil production and consumption have changed in the last five years, and to examine whether or not the concept of Peak Oil remains valid. Key to this discussion is an understanding of the concept of Energy Returned on Energy Invested (EROEI). It is concluded that the concept of Peak Oil, as first postulated by Dr. King, remains valid and current, and the production of oil peaked in the period 2005-2011 and is now set to move down the right hand side of the Hubbert Curve.

The title of this session is *Challenges in Instituting Energy Savings Project in the Present Market Conditions*. If the production of oil and has indeed already passed a peak then the challenge is not to do with "instituting" energy savings; those savings are going to occur whether we like it or not. The challenge will be to determine how the process industries can restructure themselves given that the energy resources available (both as direct energy and as petrochemical feedstock) will be in decline and that there will be intense demand for those same resources from other areas of the economy.

(106j) Statistical Decision Making for In-Service Inspection of Underground Waste Storage Tanks

Emily Mitchell, Emory University, Atlanta, GA and **Stephen P. Harris**, Savannah River National Laboratory, Aiken, SC

Pitting is a localized form of corrosion that occurs on passivated metal surfaces exposed to aggressive chemical environments. The pitting process involves the breakdown of the passive oxide film and subsequent anodic dissolution of the metal at a local site, while the remainder of the surface remains protected by the film.

Deep pits may result in lowering the maximum level of waste that made be stored in the tank and eventually lead to loss of confinement. Pit depths, established by ultrasonic testing, are subject to variables that cause uncertainty, including the repeatability of metric location, time between measurements, different transducers, and different inspectors.

The generalized extreme value (GEV) distribution is fit to the data gathered from ultrasonic testing and statistical projections are made for the full tank wall interior. A comparison is made with results from other well known statistical distributions. The estimated parameters are used in statistical tolerance limits for bounding pit depth in a waste tank.

The tolerance limits give an indication of extreme values of pit depth for the current condition of the tank walls. The upper tolerance limit for pit depth will then be compared to action levels for pitting. If the upper tolerance limit for pit depth exceeds the action level, more frequent examinations or increased tank surface area inspections are conducted.

(106k) Thermal Inbreathing Requirements of Low Pressure Storage Tanks At Elevated Temperatures

Donald E. Brooks, Pressure Relief Analysis, Lloyd's Register Energy Americas, Inc., Houston, TX

ISO 28300 (API 2000) "Petroleum, petrochemical and natural gas industries - Venting of atmospheric and low pressure storage tanks" provides guidance on determining the required thermal inbreathing flow rates to protect atmospheric and low pressure storage tanks against vacuum due to weather changes (e.g. a decrease in atmospheric temperature)." For heated, un-insulated tanks with vapor space temperatures above 58.9 °C (120 "F), ISO 28300 recommends an engineering review of thermal inbreathing requirements.

This paper describes one approach to evaluating thermal inbreathing requirements for un-insulated storage tanks with vapor space temperatures exceeding 58.8 °C (120 "F)." The tank is assumed to be exposed to the cooling effects of a sudden rain. A simplified unsteady state heat transfer analysis is performed to determine the tank wall and vapor space temperatures versus time. Initial gas mass and pressure are assumed to remain constant in the tank. Gas volume in the tank is calculated as a function of vapor space temperature using the ideal gas law. The rate of volume change is calculated for each time increment. The rate of volume change goes through a maximum value. This maximum value is selected as the thermal inbreathing requirement of the tank.

The calculation methodology described in this paper can be programmed in a spreadsheet for ease of use.

Dynamic Operational Risk Assessment with Bayesian Networks

Shubharthi Barua, Xiaodan Gao, M. Sam Mannan
Mary Kay O'Connor Process Safety Center
Texas A&M University, Jack E. Brown Engineering Bldg., 3122 TAMU, Room 244, College Station, TX 77843
shubharthi.barua@tamu.edu, xiaodan_gao08@mail.che.tamu.edu, mannan@tamu.edu

Extended Abstract Prepared for the Poster Presentation at the
8th Global Congress on Process Safety
Houston, TX
April 1 – 4, 2012

UNPUBLISHED

8TH GLOBAL CONGRESS ON PROCESS SAFETY

Dynamic Operational Risk Assessment with Bayesian Networks

Shubharthi Barua, Xiaodan Gao, M. Sam Mannan

Mary Kay O'Connor Process Safety Center
Texas A&M University, Jack E. Brown Engineering Bldg., 3122 TAMU Room 244, College Station, TX-77843
shubharthi.barua@tamu.edu, xiaodan_gao08@mail.che.tamu.edu, mannan@tamu.edu

Keywords: risk assessment, Bayesian network, dynamic Bayesian network, fault tree, event tree

Abstract

Oil/gas and petrochemical plants are complicated and dynamic in nature. Dynamic characteristics include ageing of equipment/components, season changes, stochastic processes, operator response times, inspection and testing time intervals, and timing of safety system operations, all of which are time dependent criteria that can influence dynamic processes. Conventional risk assessment methodologies can quantify dynamic changes in processes with limited capacity. Several methods are proposed for dynamic risk assessment such as C-DET, D-DET, DORA, etc., but all of them have deficiencies of some kind. Bayesian networks, a graphical inference technique, have some advantageous characteristics for risk analysis over other techniques. This research focuses on providing a framework for dynamic risk analysis based on Bayesian networks. A case study will be provided to illustrate this model.

1.0 Introduction

Oil/gas, chemical, and petrochemical process plants consist of numerous equipment and unit operations, thousands of control loops, and exhibit dynamic behavior. Different hazards and several types of risk are associated with these process facilities. For example, large process areas have the potential to stock a large amount of hazardous chemicals, which can result in the process area being highly congested. It is reported that process risk, i.e., risk in process facilities due to fires and explosions, greatly contributes to the total risk of offshore installations. Also, reactivity, toxic, and mechanical hazards have a significant effect on the total risk of these facilities. Therefore, it is very important to identify hazards, perform risk assessments, and take proper initiatives to minimize/remove hazards and risks; else a catastrophic accident may result.

From case histories, it has been observed that catastrophic accidents have a significant effect on people, environment, and society. Catastrophic accidents such as the Flixborough disaster, the Bhopal incident, and the Piper Alpha disaster caused fatalities and unbearable economic loss. In the past few years, several disastrous accidents in refineries, power plants, and offshore platforms have significantly affected people's perception, and consequently raised concern to emphasize process safety. A series of explosions in 2005 resulting from insufficient process safety and lack of risk reduction measures at the BP Texas City refinery accident resulted in 15 fatalities, 180 injuries, and $1.5 billion in losses [1]. The Kleen Energy power plant explosion in

Connecticut in 2010 is considered to be one of the worst industrial explosions in the USA [2]. In 2010, a fire and explosion resulting from a blowout at the Macondo well resulted in 11 deaths and 17 injuries. Also the continuous spill from the wellhead for 87 days had disastrous effects on the environment and wildlife surrounding the Gulf of Mexico. It has been reported that most of the accidents could be prevented or their severity could have been minimized by effective risk assessment and process safety management. But existing risk assessment methodologies cannot properly quantify various time-dependent effects. Therefore, the objective of this research is to develop a risk assessment methodology with the capability to capture variations due to different time-dependent effects.

2.0 Literature Review

N. Siu [3] summarized available methodologies for risk assessment. For example, the fault tree and event tree methods are two most common methodologies for risk assessment. But both techniques have limited ability to address different time-dependent effects such as equipment or component ageing, stochastic processes, operator response times, and inspection and testing time intervals. N. Siu [3] also described other methodology limitations while applying dynamic system analyses. Therefore, for dynamic systems, conventional techniques are not able to provide accurate results because of their inadequate ability to describe the variation of operational risks as time-dependent deviations, or the changes occurring in the process. Continuous event trees [4], discrete event trees [5], and Markov theory [6] methods have been proposed for dynamic reliability and safety analysis. But all of the proposed techniques have some deficiencies. For example, the Markov theory based model does not consider the effect of inspection on system-state transitions. X. Yang et al. [1] proposed the Dynamic Operational Risk Assessment (DORA) technique, which is able to take inspection and testing time intervals and repair time into account for evaluating operational failure.

Recently, Bayesian networks are receiving wide acceptance in reliability assessment, fault diagnosis, safety, risk, and maintenance analysis. Several authors have [7,8] compared some commonly used methods for accident analysis with Bayesian networks, and described several advantages of Bayesian networks over these methods. N. Khakzad et al. [2] illustrated parallelism between Bayesian networks and fault trees, and discussed the potential of Bayesian networks as a tool for process safety and accident analysis. H. Pasman and W. Rogers [9] described an approach to incorporate Bayesian networks into layer of protection analysis to make it more effective. These studies demonstrate the potential advantages of Bayesian networks application in risk assessment. This research provides a methodology for dynamic risk assessment based on Bayesian networks.

3.0 Methodology and Expected Result

3.01 Bayesian Networks

A Bayesian network, which is a probabilistic graphical model, describes causal influence relations among variables. A set of random variables and their conditional dependencies are represented by Bayesian networks via a directed acyclic graph (DAG). It has the ability to represent dependency among events clearly, accommodate multi mode and continuous random variables, and incorporate information i.e. generic, system specific and expert judgment to support optimum decision making.

A simple Bayesian network is shown in Figure 1. In a binary network, variables are represented by nodes and causal relationships among different nodes are shown by an arc. Conditional probability tables or defined probabilistic relations among nodes represent how one variable is linked with one or many variables. The nodes that influence another variable and have unconditional probability are called parent or root nodes. Nodes that are conditionally dependent on their direct parents are called intermediate nodes. The top node is defined as a leaf node, which is similar to the top event of a fault tree and initiating event for an event tree.

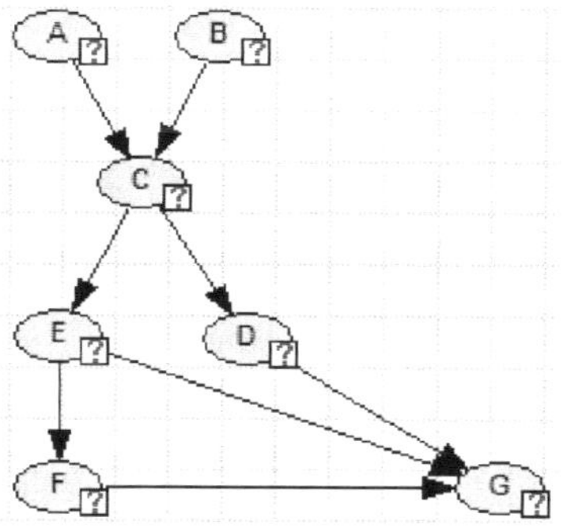

Figure 1: A simple Bayesian network

Let $N = (G, P)$ be a Bayesian network, where

$G = (V, E)$ is a directed acyclic graph; V (random variables) represents nodes; and E represents edges between pairs of nodes of DAG.

P represents probability distribution over V and $V = \{X_1, X_2, \dots, X_n\}$ is either discrete or continuous random variables [10]. These random variables are assigned to the nodes and the edges.

Bayesian networks can be represented by the joint probability distribution P(V);

$$P(V) = \sum_{X \in V} P\{X|pa(X)\} = P(X_1, X_2, \dots, X_N) = \sum_{i=1}^{n} P\{X_i|pa(X_i)\}$$

Here $pa(X_i)$ = parent nodes of X_i.

A main advantage of the application of Bayesian networks in risk analysis is the ability to update prior data using Bayes' theorem by incorporating new information as demonstrated by N. Khakzad et al. [2]. Also, Bayesian networks have an advantage of handling different types of uncertainty.

3.02 Dynamic Bayesian Networks

A general Bayesian network is a static model, i.e., the joint probability distribution is usually a representation of a fixed point or an interval of time [11]. A dynamic Bayesian network is a Bayesian network that has the ability to describe the evolution of joint probability distribution over time. Dean and Kanazawa [12] proposed modeling time in a discrete manner to represent the progression of time in a dynamic Bayesian network. In a dynamic Bayesian network, arcs links node/nodes from previous time slice to the node/nodes of the next time slice to represent temporal dependencies among them.

3.03 Methodology

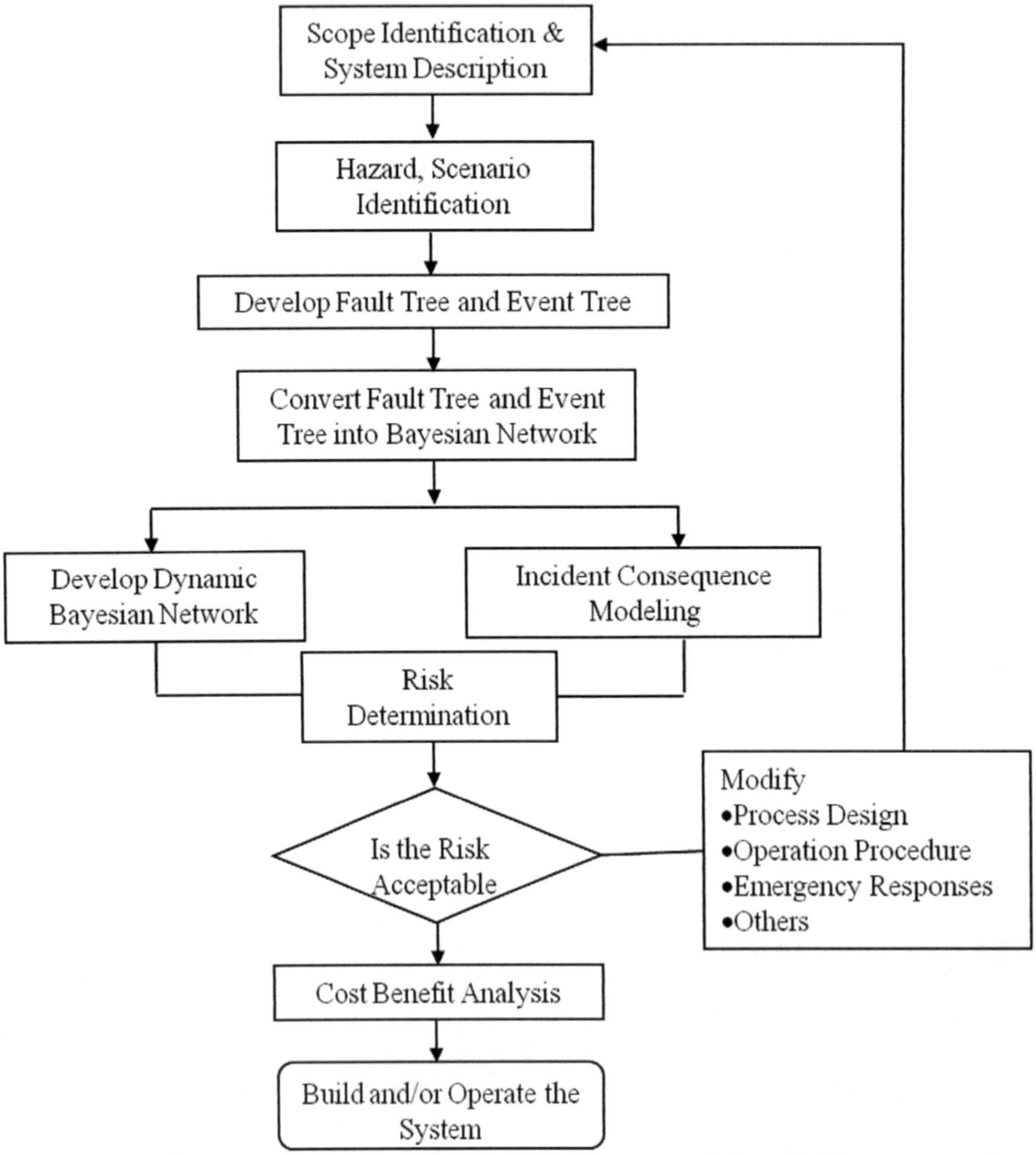

Figure 2: Framework for the dynamic Bayesian network based risk assessment

The objective of this research is to propose a dynamic risk assessment method based on dynamic Bayesian networks. According to the framework, a fault tree and event tree are developed for a system. The objective for following step is to convert the developed fault and event tree into a Bayesian network. Bobbio et al. [13] provided a methodology for conversion of fault trees into Bayesian networks. N. Khakzad et al. [2] showed equivalency between fault trees and Bayesian networks, and described the advantages of Bayesian networks for process safety analysis. Bearfield and March [14] proposed another methodology for generalizing event trees into Bayesian networks. The method has the ability to generalize simple event trees into Bayesian networks in a relatively easy manner, but the method becomes complicated for large event trees. Therefore, in the first phase, a methodology is proposed to convert fault trees and event trees into a static Bayesian network. In the second phase, a dynamic Bayesian network is developed from the static Bayesian network. The dynamic Bayesian network provides probability of a system at different time-slices for different states of the system. The probabilities are then combined with consequence modeling levels to calculate risk. GeNIe, which was developed at the Decision Systems Laboratory at the University of Pittsburgh, is used to create the model [15].

4.0 Conclusions

This article briefly describes a framework based on Bayesian network for dynamic risk assessment for the oil/gas, chemical, and petrochemical industries. This research illustrates the capability of Bayesian networks to represent all component interactions in complex systems, and to update all prior system probabilities in the presence of new information. The advantages of modeling flexibility and efficient computational mechanisms using dynamic Bayesian networks are also discussed.

4. References

[1] Yang, X. and Mannan, M.S., "The development and application of dynamic operational risk assessment in oil/gas and chemical process industry", Reliability Engineering and System Safety 95 (2010) 806-815

[2] Khakzad, N., Khan, F. and Amyotte,P., "Safety analysis in process facilities: Comparison of fault tree and Bayesian network approaches", Reliability Engineering and System Safety 96(2011) 925-932

[3] Siu,N. "Risk assessment for dynamic systems: an overview", Reliability Engineering and System Safety 43 (1994) 43-73

[4] Devooght, J. and Smidts,C. "Probabilistic reactor dynamics-I: the theory of continuous event trees", Nuclear Science and Engineering 1992; 111:229-40

[5] Amendola, A. "Accident sequence dynamic simulation versus event trees", Reliability Engineering and System Safety 1988; 22:3-25

[6] Howard, R. "Dynamic probability systems", Vols. I and II, New York, 1971

[7] Sklet, S. "Comparison of some selected methods for accident investigation", Journal of Hazardous Materials 2004; 111:29-37

[8] Nivolianitou, Z.S., Leopoulos, V.N. and Konstantinidou, M., "Comparison of techniques for accident scenario analysis in hazardous systems", Journal of Loss Prevention in the Process Industries 2004; 17"467-75

[9] Pasman, H. and Rogers, W.; "BBN, a tool to make LOPA more effective, QRA more transparent and flexible, and therefore to make safety more definable", 14th Annual Sysposium, Mary Kay O'Connor Process Safety Center, College Station, Texas, October 2011

[10] Donohue, S.K. and Dugan, J.B.; "Modeling the "Good enough to release" decision usin V&V preference structure and Bayesian belief networks", Reliability and Maintainability Symposium, 2003 Annual, 2003

[11] McNaught, K.R. and Zagorecki, A. "Using Dynamic Bayesian networks for prognostic modeling to inform maintenance decision making", IEEM 2009, Hond Kong, January 2010

[12] Dean, T. and Kanazawa, K., "A model for reasoning about persistence and causation", Computational Intelligence 1989; 5:142-150

[13] Bobbio A, Portinale L, Minichino M and Ciancamerla E., "Improving the analysis of dependable systems by mapping fault trees into Bayesian networks", Reliability Engineering and System Safety 71(3): 249-260

[14] Bearfield G and Marsh W, "Generalising event trees using Bayesian networks with a case study of train derailment", Available online at http://www.agenarisk.com/resources/white_papers/Generalising_Event_Trees.pdf , last modified: 23 February 2010, accessed: 7 January 2012

[15] Decision Systems Laboratory, University of Pittsburgh, GeNIe (Graphical Network Interface and SMILE (Structural Modeling, Inference, and Learning Enginer), Version 2.0 software, 2010. Available online at http://genie.sis.pitt.edu

(137b) Comparing the Performance of Compact Heat Exchange Reactor to CSTR for Fischer Tropsch Synthesis

Hani Gadalla[1], Zhijun Jia[1], Steven J. Vallee[1], Burtron H. Davis[2]and Dennis Sparks[2], (1)Chart Energy & Chemicals, La Crosse, WI, (2)Center for Applied Energy Research, University of Kentucky, Lexington, KY

A commercial Fischer Tropsch catalyst was studied in two Compact Heat Exchange Reactors (CHER) with different channel sizes, and in a continuous stir tank reactor (CSTR). The CHERs were constructed by Chart Energy and Chemicals with one reactor having a 5mm channels and the other having a 6mm channels, while maintaining the same catalyst volumeThe catalyst was crushed and sieved retaining particles ranging in size from 75-125μm. Testing was performed at the University of Kentucky Center for Applied Energy Research in Lexington, KY. The CO conversion, selectivity and hydrocarbon productivity were measured for each reactor over a wide range of operating conditions, as a function of temperature, pressure, H2:CO feed ratio, and flow rate. Both reactor types showed excellent temperature control, but the performances varied at different operating conditions. The larger channel CHER had a higher conversion than the 5mm channel, potentially due to slightly higher reactor bed temperature. The two CHERs approached and even exceeded the productivity of the CSTR for liquid hydrocarbons per gram of catalyst. At higher flow rates the methane selectivity increases at a lower rate for either of the CHERs than for the CSTR.

(114a) Advances in FCC Propylene Production and Recovery

Andrew Mezera, UOP, Des Plaines, IL

World propylene demand is projected to increase dramatically in the coming years. Conventional steam cracking and emerging technologies dedicated to propylene production will only be able to provide a portion of the demand. FCC Units are expected to contribute a significant portion of the shortfall. This talk will focus on two main aspects of propylene production from FCC Units: 1) increasing FCC propylene yields through operations, catalyst and equipment changes, and 2) improving propylene recovery in the Gas Concentration Unit.

(114c) An Experimental Study of Sulfur Release in Gas Products of Coal Steam Hydrogasification

Qian Luo, CEE, UCR, Riverside, CA

An Experimental Study of Sulfur Release in Gas Products of Coal Steam Hydrogasification

Qian Luo, Chan S. Park, Joseph M. Norbeck

Bourns College of Engineering

Center for Environmental Research and Technology (CE-CERT)

University of California, Riverside, California 92521-0425.

Steam hydrogasification reaction (SHR) is the thermochemical conversion of carbonaceous materials into synthetic gas in steam and hydrogen environment. It has been developed by the Bourns College of Engineering - CE-CERT at the University of California, Riverside (UCR). Like other thermochemical process, the formation of gaseous sulfur species from the solid sulfur in the feedstock becomes inevitable which necessitates the cleaning of gas to make it end-use possible. Therefore, a better understanding of the sulfur release in the gas phase in SHR could help to find the suitable method to remove sulfur containing species for the protection of downstream catalysts.

The temperature, steam and H_2amount have an effect on the distribution of the various gaseous sulfur species. Experiment results showed that sulfur in the feedstock is mainly converted to H_2S in the SHR process but COS and $CS_2$are undetectable. The $H_2$and steam environment in the SHR process are favorable for the formation of H_2S and suppress the COS and CS_2formation. Furthermore, increase of H_2S concentration was observed with the rise in temperature from 700°C to 800°C. However, the increasing the partial pressure of H_2 decreased the H_2S concentration released in the gas phase.

(114d) An Innovative Mixed Zinc Oxide Based Sorbent Regenerable At Low Temperature Used for the Desulfurization of Syngas in IGCC or XTL Processes

Vincent Girard, Catalyse and Separation Division, IFP Energies Nouvelles, Solaize, France, Arnaud Baudot, Catalysis and Separation Division, IFP Lyon, 69360 Solaize, France, David Chiche, Catalyse and Separation division, IFP Energies Nouvelles, Solaize, France, Delphine Bazer-Bachi, GMD Division, IFP Energies Nouvelles, Solaize, France and Christophe Geantet, IRCELyon, Villeurbane, France

Integrated Gasification Combined Cycles (IGCC) and the second generation biofuels processes, and more generally Fischer-Tropsch based XTL processes, are attractive alternatives for future power production[1-3]. In order to prevent the corrosion of the industrial units, especially the combustion turbine blade used in IGCC processes or, in the others processes, to avoid the Fischer-Tropsch catalysts poisoning, deep desulfurization of syngas is achieved with metal oxides according to the following reaction :

$$MeO_{(s)} + H_2S_{(g)} \rightarrow MeS_{(s)} + H_2O_{(g)}$$

The large amounts of solid wastes produced are one of the major disadvantages of the in-situ desulfurization of syngas. The in-situ regeneration of the sulfided sorbent, back to the oxide, during the process could resolve this problem. From an industrial point of view, the oxidative regeneration is the simplest and most straightforward way to regenerate the sulfided sorbent, so it was mainly studied[4,5]. However, this reaction presents several drawbacks. The first one is its high exothermicity that can cause the sintering of the particles and the loss of reactivity due to the alteration of the textural properties[6,7]. Another drawback, which can be considered as the most important, is the formation of sulfates[5,8]. These refractory phases require high temperature to decompose, which causes important modification of textural properties and decrease of reactivity due to the sintering of particles. Moreover, the resulting temperature increase will impose the use of expensive alloys to make the reactors and will affect the thermal efficiency of the process. The formation of sulfate has to be avoided in order to maintain the sorbent durability and the temperature of the in-situ oxidative regeneration must be as low as possible, ideally the closest to the desulfurization step, to improve the global efficiency of this process.

The main goal of this work is to bring a new insight on the phenomena involved during the oxidative regeneration of the sulfided solids. The methodology developed is based on the confrontation between a global theoretical approach based on thermochemical data and an experimental study. This work is divided into two parts. The first one is the evaluation and the classification of the behaviors observed during both sulfidation and oxidative regeneration of simple oxides, in particular ZnO. The second one is the evaluation of the mixed oxides, selected thanks to the results of the first part and by the anticipation of potential synergetic effects. The synergetic effects are expected to avoid the formation of refractory sulfates and the resulting increase of temperature.

The first part led to a classification of the simple oxides based on the stability of their sulfide and sulfate phases. This classification led to a selection of different simple oxides whose mixture could bring synergetic effect. Finally, this methodology led to an innovative material which composition allows a fast regeneration, at a temperature 150°C lower than ZnO regeneration temperature and without sulfates formation.

Reference List

1. Daniel Vallentin, *Energy Policy*, 2008, **36**, 3198-3211.

2. Osamu Shinada, Akira Yamada, and Yoshinori Koyama,*Energy Conversion and Management*, 2006,**43**, 1221-1233.

3. Mark J. Prins, Krzysztof J. Ptasinski, and Frans J. J. G. Janssen,*Fuel Processing Technology*, 25-1-2005,**86**, 375-389.

4. R. V. Siriwardane and S. Woodruff,*Industrial & Engineering Chemistry Research*, 1995,**34**, 699-702.

5. W. F. Elseviers and H. Verelst,*Fuel*, 1999,**78**, 601-612.

6. T. Grindley and G. Steinfeld,*Proceedings of the Second Annuals Contractors Meeting on Contaminant Control in Hot Coal Derived Gas Stream.*, 1981.

7. J. Huang, J. Zhao, X. Wei, Y. Wang, and X. Bu,*Powder Technology*, 14-1-2008,**180**, 196-202.

8. R. V. Siriwardane and S. Woodruff,*Industrial & Engineering Chemistry Research*, 1-12-1997,**36**, 5277-5281.

(122a) Tuning Equation-of-State to Match Experimental Data for near Critical Fluid: A Case Study of NKJ

Adel Elsharkawy, Petroleum Engineering, Kuwait University, Kuwait, Kuwait and Bashayer . Al-Enize, Kuwait Oil Co, Kuwait, Kuwait

Tuning equation-of-State to Match Experimental Data for Near Critical Fluid: A Case Study of NKJ

Bashayer h. Al-Enize and Adel Elsharkawy

College of Engineering & Petroleum, Kuwait University

Po Box 5969 Safat 13060 Kuwait

Abstract

Equations-of-state (EOS) are widely used in compositional simulation and surface facility design. Volatile oils and gas condensates require special treatment due to their near-critical nature; the classical material balance cannot be applied to these near-critical fluids because their molar compositions change as the reservoir is depleted. Thus, compositional simulation is necessary to determine the best economical and technical exploitation. However, the performance of cubic equations of state is questionable in their predictive mode when they applied to petroleum mixtures. Tuning of the EOS is, generally, necessary for improving the predictions of compositional reservoir simulators, since the plus fraction of reservoir fluids has some uncertainty in its molecular weight and critical properties, equation-of-state, EOS, are generally not predictive without tuning its parameters to match experimental data.

We describe the optimum characterization and regression techniques that could provide tuned Equation-of-state for the near-critical volatile oil and gas condensate from the North Kuwait Jurassic reservoirs (KNJ). The PVT simulator PVTsim is used for the purpose. The results show that the implemented methods provide a reliable and efficient tool for representing PVT properties and phase behavior of the complex reservoir fluids. In this paper detailed of two samples is provided: one gas condensate and a volatile oil sample. We discuss the optimum sampling technique in this region and the quality control needed to perform the PVT analyses in order to ensure the validity of data to be used in EOS tuning.

Equations of state EOS parameters such as critical temperatures, critical pressures and acentric factors of the pseudo-components are matching parameters used to EOS tuning. In this study A characterization procedure is described to simulate the plus fractions of reservoir fluids and optimal selection of the regression parameters in determining the quality of the tuned fluid model.

(122b) Dynamic Simulation: Value During LNG Process Design and Beyond

Kenneth V. Allsford, Honeywell Process Solutions, Honeywell, Houston, TX, Vijay KumarPatel, Honeywell Process Solutions, Honeywell International, Houston, TX and Robin K Conwell, Honeywell Process Solutions, Honeywell International, Lakewood, CO

A contributor to complex/large project management difficulty is uncertainty gaps – gaps between the information required and that readily available within the project organization. Moreover, for large projects, information is generally spread among many professional personnel across multiple organizations.

In the process industries, LNG engineering teams are finding that – both in front-end engineering design (FEED) and during detailed design – dynamic simulation is a cost effective tool to reduce uncertainty gaps and facilitate information communication. This is all the more so as both inherent functionality and exploitation capability of commercially available dynamic process simulation software expands, enabling the scope of rigorous dynamic modeling to cover from wells to export jetty within a single software application.

But where should engineering effort be focused? What are the expectations from dynamic simulation during the design phase? And what happens after the design phase completes? To address these questions, the authors will reference the literature and discuss some of their generalized experience. Technical focus items will include:

- Representing wells, pipelines and more

- Assurance evaluation of high integrity pressure protection system and/or pressure relief valve scenarios

- Defining and evaluating control strategies associated with tight thermal integration, including for anticipated (even if rare) off-normal operations

- Checking the integrity of available operating procedures and/or providing logistic information that can feed into sequencing start-up, shutdown and other operating procedures for the fully integrated facility

- Example of exchanger tube rupture unanticipated potential consequences

Evaluation of Reverse Osmosis - Electrode-ionization (RO – EDI) Pilot Project in PT Badak NGL

Fany Arfianto, PT Badak Natural Gas Liquefaction, Kalimantan Timur, Indonesia

Background

PT Badak NGL utilizes groundwater for Plants and Community water consumption. To produce ultrapure water for Boiler Feed Water (BFW), PT Badak NGL services demineralization process. Figure 1 shows the simplified water treatment process diagram of PT Badak NGL. The Demineralization unit uses ion exchange process. For regeneration, it utilizes sulphuric acid and caustic, which has high safety risk. The sulphuric acid sometimes leaks out from the pipe and connections that can endanger the worker and plant facilities around the leakages due to its aggressiveness.

Figure 1. Simplified Process Diagram of Water Treatment Facilities

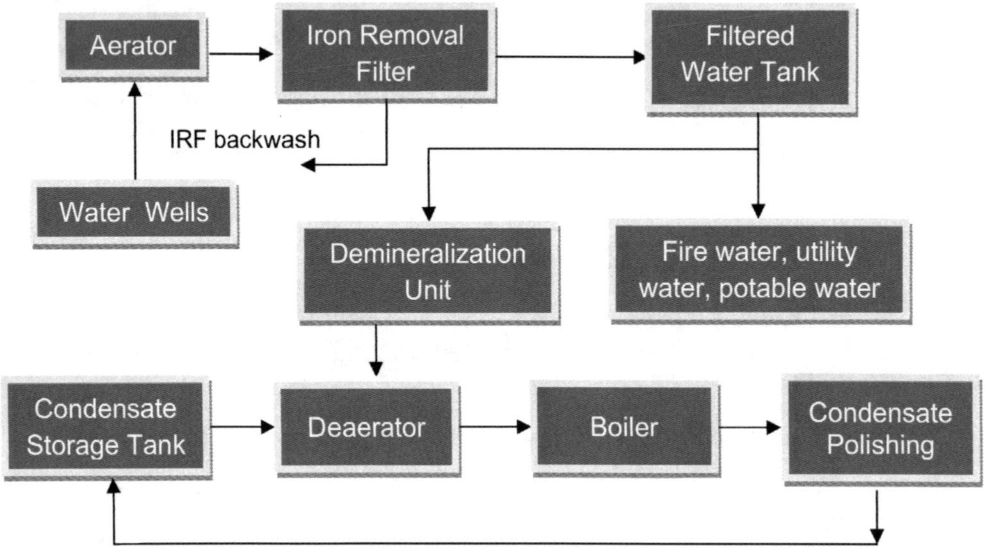

PT Badak NGL intends to reduce the risk of sulfuric acid usage by implementing a more safe process that use less chemicals such as RO-EDI process. The RO-EDI is a new water treatment technology that utilizes less hazardous chemicals.

General Process of RO – EDI

The implementation of new technology is high risk and high capital investment, therefore PT Badak NGL conducted a literature study and a pilot project of RO – EDI in order to ensure that the parameter of safety, product quality, operability and feasibility are acceptable. The previous RO – EDI study literature was conducted in 2009. This study was intended to develop the methodology of membrane technology utilization in PT Badak NGL waste water and demineralization unit. The study recommended implementing pilot project of RO – EDI process in Demineralization unit.

RO – EDI is a process to produce pure water by combining a membrane separation process and an Ion Exchanger. The membrane separations are used in a wide range of industrial processes to separate biological macromolecules, colloids, ions, solvents and gases. RO is a part of membrane processes which utilize porous polymeric or inorganic films. EDI process is removing ionized species from water using a combination of ion exchange membranes, ion exchange resins and a DC electric potential.

Figure 2. Simplified Process of RO – EDI

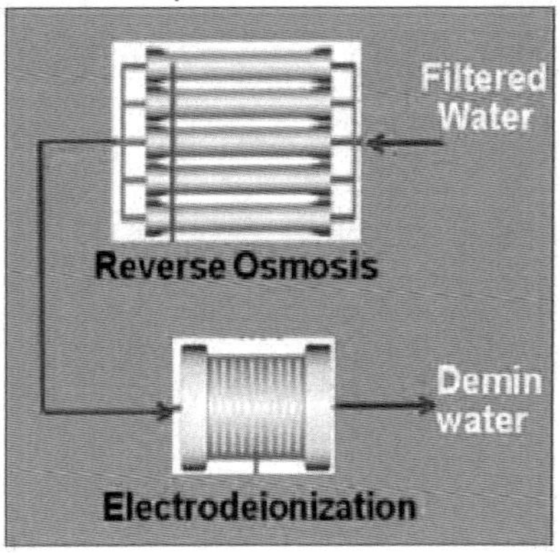

To fulfill the PT Badak demineralization water requirement, RO – EDI pilot plant PT Badak NGL Water Treatment Plant consists of RO module and EDI unit. Single-stage of RO module still need a support from polishing unit to ensure that the produced water meet the specification. The polishing unit is Electrodeionization (EDI). According to previous study, RO water product quality does not meet with PT Badak NGL demineralization water specification (see Table 1), specifically in conductivity and Silica. Therefore combination of RO and EDI is required.

Table 1. Demineralization Water Quality

Water Product	Conductivity (μmhos/cm)	pH	Silica (ppm)	Na (ppm)
PT Badak Spec.*	**< 5.0**	**6.5 – 8.5**	**< 0.12**	**< 0.3**
Ion Exchange**	**0.87**	**6.1 – 7.1**	**0.03**	**0.2**
RO	2.50	5.54	0.29	0.18
EDI	0.30	6.20	0.01	0.00

Note: *) PT Badak NGL demin water specification
**) actual Demineralization unit quality

Pilot Project of RO – EDI

The simplified process diagram of RO – EDI pilot plant can be seen in Figure 3 and the related mass balance is shown in Table 2.

Table 2. Mass Balance of Pilot Project

Stream	1	2	3	4	5	6	7	8	9	10
Flow, m3/h	3.7	2.8	1.1	0.7	1.1	1.0	1.0	0.1	1.0	1.0

Figure 3. Schematic Process Flow Diagram of Pilot Project

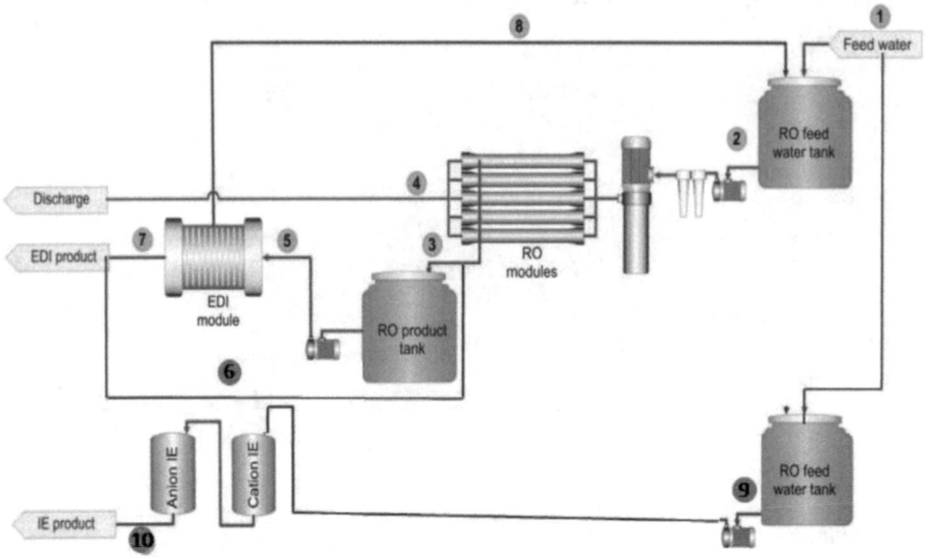

Key Performance Indicators

The water product quality was observed during pilot project field test observation, such as a product water recovery, pH, conductivity, silica and sodium contents, as well as product water recovery. The product water recovery is the ratio (percentage) of product to feed. The recovery in RO-EDI is proportional to the process efficiency. The target water product recovery of RO-EDI pilot project was between 80-90% and it was achieved 82.5% during the pilot project test.

The RO-EDI field test result can be seen in Table 3. The achievements of pH and conductivity were out of the target specification. However, the conductivity of RO – EDI product is still much lower than existing Ion Exchange unit product. The sodium target specification of the pilot project was 0.001 ppm, whereas the laboratory device detection limit is only 0.003 ppm. See detail field test result in Figures 6 – 9.

Table 3. Key Performance Indicators & Achievement of RO- EDI Field Test

Performance Items	PTB spec.	Pilot Project Results		
		Ion Exchange	RO-EDI	
			Target	Achievement
1. Product Recovery, %	N/A	94	80 – 90	82.5
2. Electrical Consumption, kwh/m^3	N/A	0.6	1.97	2.64
3. Specific Water Production Cost, $/m3	N/A	0.66	0.35	0.4
4. Demin water Quality				
- Conductivity, µmosh/cm	< 5.0	0.87	0.1	0.3 – 0.4
- pH	6.5 - 8.5	6.1 - 7.1	8 - 8.5	6.2 – 6.3
- SiO_2, ppm	< 0.12	0.03	0.005	0.00 – 0.01
- Na, ppm	< 0.3	0.2	0.001	<0.003*)

Note: *) the lowest analyzer detection level

The phenomenon of low pH of low conducitivity water are depicted in Figures 4 and 5, where conductivity level is correlated with pH of water. As shown in those figures, water with conductivity level of 0.3 – 0.4 , µmosh/cm will probably has pH of 6.2 or 8.2 (Figure 4); or even lower than 6 or higher than 8.5 (Figure 5).

Figure 5 shows the theoretical resistivity values of water following the addition of hydrochloric acid (HCl) or sodium hydroxide (NaOH). The resistivity of ultrapure water is 18.18 MΩ.cm and the pH is 6.998 (***Estelle Riché, et al. 2006***).

However, one cannot measure the pH of high-purity water simply by dipping the electrodes of a pH meter into a beaker full of water. This measurement is challenging due to the quasi-absence of ions that would enable electron transport between the measuring and reference sides of the pH electrode. It often results in erratic and meaningless pH readings. Pure water readily absorbs carbon dioxide (CO_2) when exposed to the atmosphere, which forms carbonic acid. ***(Estelle Riché, et al. 2006).***

Dissolved CO_2 in water (CO_2) is suspected substance which present in RO – EDI product water that influence pH. Ultrapure water such as EDI product readily absorbs CO_2 when exposed to the air in Aerator or during measurement which forms Carbonic acid. However, the lower pH can be solved by adjusting amine injection in water effluent of RO – EDI to meet the Boiler Feed Water pH specification of 6.5 – 8.5.

Figure 4. Relationship between pH and Conductivity

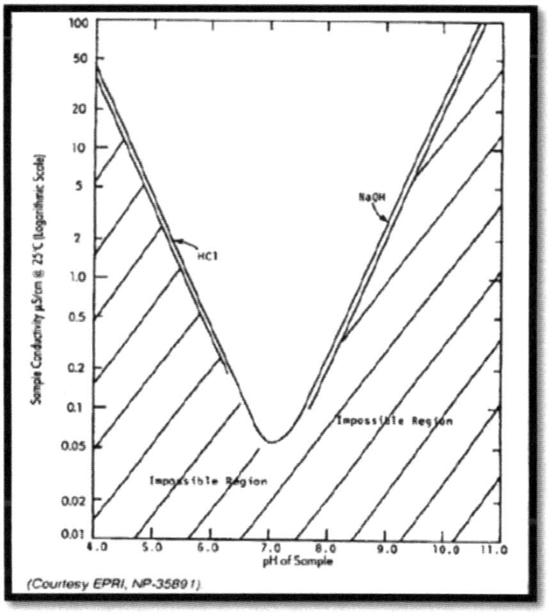

Figure 5. pH vs Resistivity of Ultrapure Water

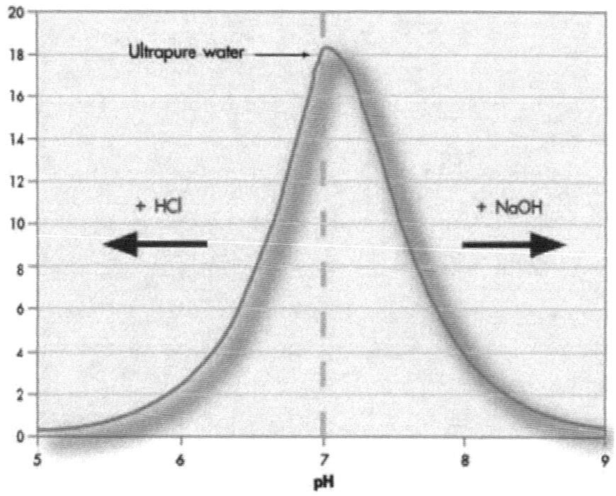

Meanwhile, conductivity, sodium and silica content of effluent RO – EDI is lower than water effluent Ion Exchanger. It means the product of RO – EDI is more pure. This condition is potential to improve boiler water quality. Based on the above comparison, RO – EDI produces water effluent quality better than Ion Exchanger.

The electric consumption was 2.64 kwh/m^3 which is 34% higher than the target, therefore it causes the specific water production increases up to 0.4 US\$/m^3 than the target of 0.35 US\$/m^3. However the RO-EDI specific water production cost of pilot project is remain lower than PT Badak specific water production cost (using Demineralization unit) of 0.66 US\$/m^3.

The water product recovery of Ion Exchanger is higher (94%) than RO – EDI (83%). This item indicates that Ion Exchanger is more efficient than RO - EDI for producing demineralization water with similar water flow rate and quality. Meanwhile, the product recovery of RO – EDI is still in range of target 80% – 90%.

EDI is continuous electro-chemical process of water purification. Therefore, electrical power contributes significant portion on RO – EDI operational. RO – EDI consumes electrical power of 2.64 kwh/m^3; higher than Ion Exchanger Unit (0.6 kwh/m^3). The higher electrical consumes power continuously for auto regenerating in cation and anion resin inside.

The demineralization water operating cost components are electricity, chemicals and maintenance. The higher operating cost of demineralization water produced by Ion Exchanger unit is contributed by chemical and maintenance. Meanwhile, significant operating cost of RO – EDI unit is contributed by electrical power. During field test, pilot project of RO – EDI has not consumed chemical for supporting its operational. However, for further operation, chemical cleaning and injection of antiscale agent should be performed. It shows that RO – EDI was not chemical free technology because it will require chemical consumption even though the chemical are less hazardous than Sulphuric acid.

Based on field test result, in overall the quality of RO – EDI effluent is good and from safety point of view, RO – EDI is potential to be implemented for replacing existing Ion Exchange unit regarding the utilization of hazardous chemical.

Sensitivity Analysis

The NPV at 10th year and IRR are simulated in various EDI capacities (see Figure 6). It concludes that EDI capacity significantly contributes to RO – EDI investment.

Figure 6. Sensitivity Analysis at Various EDI Capacities

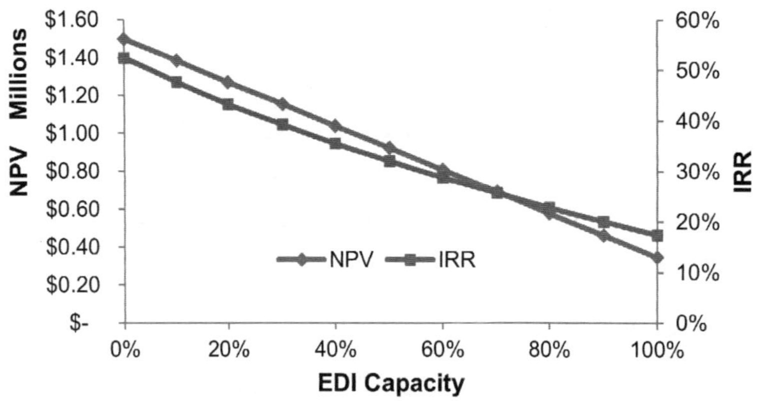

Variation of demin water quality and variation of EDI unit capacity for blending of RO product and EDI product also investigated. All of the variation resumed in Table 4. The minimum EDI unit capacity is 60 % to meet PT Badak NGL demineralization water specification.

Table 4. Variation of EDI Capacity and its Effect to Demin Water Quality

	Conductivity, mS/cm	**pH**	**Silica, ppm**	**Na, ppm**
RO	**2.5**	**5.54**	**0.29**	**0.18**
EDI	**0.3**	**6.2**	**0.01**	**0**
PTB Spec.	**<5**	**6.5 – 8.5**	**<0.12**	**<0.3**
EDI Capacity				
0	2.50	5.54	0.29	0.18
10	2.28	5.61	0.26	0.16
20	2.06	5.67	0.23	0.14
30	1.84	5.74	0.21	0.13
40	1.62	5.80	0.18	0.11
50	1.40	5.87	0.15	0.09
60	1.18	5.94	0.12	0.07
70	0.96	6.00	0.09	0.05
80	0.74	6.07	0.07	0.04
90	0.52	6.13	0.04	0.02
100	0.30	6.20	0.01	0.00

Cost & Benefit Analysis

Economical analysis was conducted to measure the feasibility of the project. In order to get the optimum capacity of RO – EDI, the analysis was performed for various RO – EDI design capacities.

RO – EDI application will reduce demineralization water specific operating cost from US$ 0.66 per m^3 to be US$ 0.40 per m^3 (the specific cost of RO – EDI included estimated chemical consumption cost and replacement of RO and EDI module cost). The margin of the specific water cost is gained from chemical cost and maintenance cost reduction. Meanwhile, total operating cost is calculated from sum of RO – EDI and Ion Exchange operating costs. Project feasibility is determined by the parameters of Payback Period (PBP), Net Present Value (NPV) and Interest Rate of Return (IRR), as shown in Table 5.

Table 5. Economic Analysis

RO – EDI Capacity, m^3/hr	**CAPEX, USD**	**PBP, years**	**NPV at 10th years, USD**	**IRR, %**
42	471,635.55	> 10	-2,465,953.84	-29.14%
77	667,407.31	> 10	-1,249,352.82	-10.92%
108	840,805.16	> 10	-398,558.96	5.79%
135	**991,829.08**	**5.82**	**294,096.08**	**15.89%**
165	1,159,633.45	5.93	278,047.18	15.43%

Based on the above analysis, RO – EDI with design capacity of 135 m^3/h is the optimum option compared to the others with higher Net Present Value (NPV) & Internal Rate of Return (IRR) and lower period of payback period.

The benefit analysis is developed to determine relative advantage and disadvantage of producing demineralization water, as determines in Table. 6. RO - EDI offers a variety of benefits compared with Ion Exchanger unit. The most obvious is the elimination of the regeneration process and its associated hazardous regeneration chemicals (acid and caustic). Since RO - EDI operates through a combination of ion-transfer across the resins and membranes, as well as electrochemical regeneration of a portion of the bed, the resins and membranes are always functional as long as the DC voltage is applied. The resin in an Ion Exchanger unit only purifies water when in its active (regenerated) form. It means RO – EDI does not require conducting the regeneration in separate process while Ion Exchanger does.

If RO – EDI is designed to supply make up demineralization water for supporting 4 Process Trains service with design capacity of 135 m^3/hr, it will replace the existing Ion Exchanger unit to supply make up demineralization water. It will leads to the elimination of the regeneration process and its associated hazardous regeneration chemicals (acid and caustic). So, the risk to personnel regarding the usage of hazardous chemicals can be reduced.

Meanwhile, RO – EDI implementation has a reliability concern due to short time of field test that cannot cover some maintenance program such as antiscale injection, maintenance activity (membrane and EDI unit replacement) and chemical cleaning activity. Beside reliability concern, limited information about operating experiences in big capacity of industrial applications should be considered. Detail comparison of Ion Exchanger and RO – EDI unit can be seen in Table 6.

Table 6. Benefit Analysis of Existing Ion Exchanger Unit and RO – EDI Application

Item	Ion Exchange	RO – EDI
Chemicals for regeneration, anti scaling and organic fouling	Hazardous chemicals (NaOH & H_2SO_4)	Less hazardous chemicals (Citric acid)
Safety and Environment Aspect	High risk of hazardous chemicals impact	Low risk of hazardous chemicals impact
Acid regeneration infrastructure	Should be maintained	Not required
Temperature	High temp. (60°C)	Ambient temp.
Pressure	Low pressure (3-5 kg/cm^2)	High pressure (12 kg/cm^2)
Feed quality	More stable against feed fluctuation	Sensitive against feed quality fluctuation
Unit capacity	Fixed	Expandable

Conclusions

1. The implementation of RO – EDI potentially improves boiler water quality.
2. Safety and environment issues can be eliminated since no hazardous chemical handling.
3. The optimum RO – EDI capacity to be installed at PT Badak NGL for supporting 4 Process Trains is 135 m^3/h.

(122d) Reducing Black Smoke and CO2 Emissions in PT Badak's Ground Flare

Ferry A. Perdana, Technical, PT Badak NGL, Bontang, Indonesia

Reducing Black Smoke and CO_2Emissions

in PT Badak's Ground Flare

Ferry Adhi Perdana

Process Engineer

PT Badak Natural Gas Liquefaction

Bontang, Kalimantan Timur, Indonesia

E-mail :ferry@badaklng.co.id

During plant upset (such as start up or shutdown), the waste liquid hydrocarbon is burned at ground flare and results in black smoke and CO_2emission. Approximately 35 kilotons / year of CO_2emission is generated only from sweep gas burned to maintain ground flares in service. To support PT Badak's commitment to the environment issues and government regulation, the study is required with the intention to evaluate the possibility eliminating the black smoke and reduce CO_2emission at ground flare.

PT Badak has five ground flares to burn the waste liquid hydrocarbon from all plant area such as process trains, LNG & LPG storages and loading dock. Once plant upset, firstly the liquid hydrocarbon is sent to Hydrocarbon Condensate Stabilizer unit for its recovery by delta pressure as a driving force. After the delta pressure is low, the remaining liquid hydrocarbon is sent to ground flare by gravity and burned. At ground flare, combustion reaction is occurred between hydrocarbon and air with carbon dioxide and water as product. When air supply for combustion is lower than requirement, incomplete combustion is occurred and it causes black smoke. There are three options to prevent black smoke namely Steam Educator Assisted, Liquid Atomizing Burner and Liquid Hydrocarbon Recovery. From the evaluation, Steam Educator Assisted and Liquid Atomizing Burner will still produce CO_2emissions and it will not offer an additional benefit. Meanwhile, for Liquid Hydrocarbon Recovery (LHCR), it will provide advantages in eliminating black smoke and CO_2emissions, saving sweep gas and also recovering liquid hydrocarbon. So, based on the evaluation, LHCR is the preferred option to reduce black smoke and CO_2emissions in PT Badak's ground flare. In LHCR, the waste liquid hydrocarbon from plant upset will be collected in the surge drum and processed in Condensate Stabilizer unit and Fuel Gas system. LHCR will reduce CO_2emissions around 35 kilotons/year and save sweep gas approximately 1,818 kNm^3/hr.

Keywords: waste liquid hydrocarbon, black smoke, CO_2emissions

Modeling of Ground Flare Pits

Nicolas Ponchaut, Ph.D., PE, CFEI [1,2]
Harri Kytömaa, Ph.D., PE, CFEI, CFI [2]
Christopher DesAutels [2]
Shoichi Kaganoi [3]
Olivier Becu [4]
Gareth Cardiff [3]

Abstract

Some new LNG liquefaction facilities have been configured to include large ground flares to allow for the safe combustion of large volumes of combustible gasses in the event of extremely rare, short duration, facility upset conditions. A recent example is the planned INPEX LNG development near Darwin, Australia. Ground flares can exceed the size of football fields, measuring greater than one hundred yards in length. A typical ground flare consists of several hundred burners, surrounded by porous (slotted) radiation protection fences. Questions have been raised as to whether the plumes of hot combustion gasses may impact the facility, the work force, or nearby roadways in the event that the wind happens to be blowing in the wrong direction at the wrong speed.

This study examines the use of Computational Fluid Dynamics (CFD) to model plumes from ground flares and the downwind temperature distribution. The CFD model is first validated by comparing its results with commonly used dispersion models such as PHAST (Process Hazard Analysis Software Tool) and CALPUFF. The flow features of a ground flare plume rising in a crosswind are then discussed. In particular the plume acts as what is commonly referred to as a "jet in cross flow." In this type of flow, counter rotating vortices are known to form. These vortices are not represented in commonly used Gaussian plume models.

The use of two adjacent flare pits is then investigated. The interaction of the two pairs of counteracting vortices is discussed, as it influences the downwind temperature distribution. The results show that the distance between adjacent flare pits is of importance when planning multi-flare pit operations.

Introduction

Some new LNG liquefaction facilities have been configured to include large ground flares to allow for the safe combustion of large volumes of combustible gasses in the extremely rare, short

[1] Corresponding author: nponchaut@exponent.com
[2] Exponent Inc. Natick, MA, USA
[3] INPEX Operations Australia Pty Ltd Yokohama, Japan
[4] TOTAL Paris, France

duration event of facility upset conditions. These ground flares, an example of which is shown in Figure 1, typically consist of several hundreds of burners surrounded by porous (slotted) radiation protection fences.

Figure 1: Typical ground flare pit. The flare pit is composed of many individual burners. It is surrounded by radiation fences that are not impermeable.[5]

The side length of these flare pits can reach several hundred feet. During upset conditions, these large pits are able to burn thousands of tons of gas every hour, generating tens of thousands of MW of heat.[6] As an example, the planned LNG development near Darwin, Australia is expected to include 4 large flare pits (see Figure 2), designed to be able to burn more than 2,500 tons of gasses per hour.

Given the very large amount of heat that can be released by these flare pits; concerns have been raised that these upset conditions may result in elevated downwind temperatures near the ground. In particular, the downwind temperatures are of a concern for human safety. This publication quantifies the downwind ground and elevated temperature for worst-case flare pit operating conditions.[7] The flow features of the plume resulting from the operation of flare pits are discussed, and in particular the effect of simultaneously operating two adjacent flare pits is investigated.

---

[5] Multipoint Ground Flares. Hamworthy Combustion. http://www.hamworthy-combustion.com

[6] Note that these rare upset conditions only last up to 15 minutes and this high combustion rate would remain short lived.

[7] Elevated temperatures refer to temperatures above 2 meter (7 feet) elevations.

The results were obtained using CFD.[8] The CFD model is first validated by comparing its results to the results obtained by the dispersion models PHAST and CALPUFF, as well as to an analytical model published in literature.

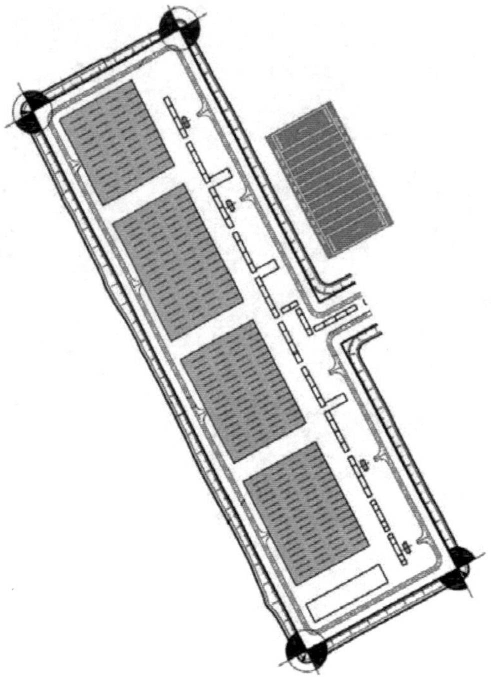

Figure 2: Size comparison between the four flare pits at the planned LNG Facility near Darwin, Australia (red), and a standard NFL football field (green).

CFD Model

In order to perform the CFD analysis, the flare pit geometry was built into a numerical model. The computational domain is centered on the flare pits and extends almost 890 meters radially from the pits. This wide domain ensures that upstream of the flare pits the wind profile is well developed, while a large distance downstream of the pits can be modeled accurately. The domain height was set to 1,250 meters, which is sufficiently high that the ceiling of the domain has no influence on the results (see Figure 3).

To simulate the boundary layer winds air enters on one side of the domain and leaves on the other side. Where air enters the domain the model specifies velocity, temperature and turbulence intensity distributions based on typical atmospheric conditions [1]. The mesh used in this analysis contain over one million cells with cell sizes varying from 3 centimeters in the boundary layers of the flare pits, to 60 meters far above the ground and away from the plume (see Figure 4).

---

[8] All CFD analyses presented in this publication were performed using the finite volume software STAR-CCM+ v.6.02.007 developed by CD-Adapco.

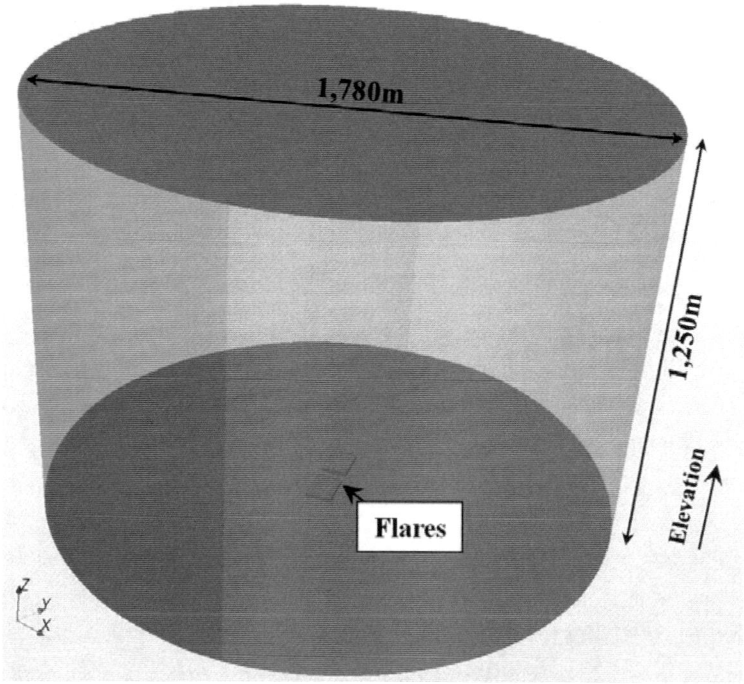

Figure 3: Typical CFD domain used in the plume modeling.

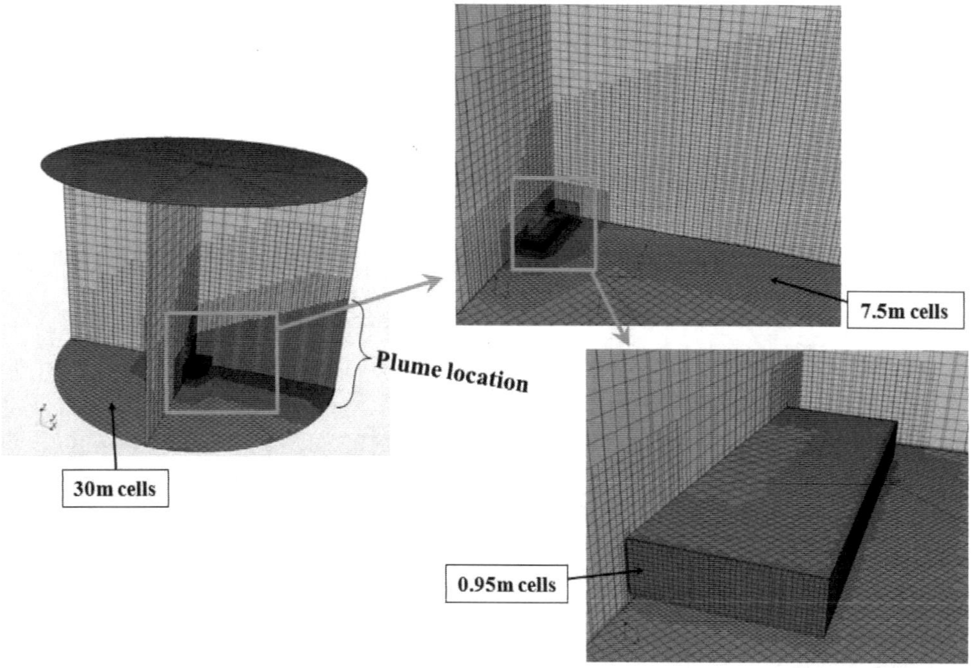

Figure 4: Typical CFD mesh.

Calculations are performed in an unsteady manner in order to capture any transient plume phenomena. The model is run for 300 seconds (5 minutes), but only the last 180 seconds are analyzed. The first two minutes (120 seconds) are therefore disregarded so that the results do not include any transient startup behaviors that are not representative of the actual plume behavior. 180 seconds is a sufficiently long duration to be able to capture all the periodic unsteady features of the plume, since these features have a characteristic period of about 10 to 20 seconds.

CFD Model Validation

In order to validate the CFD model the plume rise and temperature profile of a simple representative flare pit release was simulated using various models:

- Plume model by Chu and Goldberg [2]
- PHAST
- CALPUFF with the "Puff model" [3]
- CALPUFF with the "Slug model" [3]
- CFD (using Star-CCM+)

The model by Chu and Goldberg is a simple analytical model that simulates both the momentum driven and the buoyancy driven plume motion. The model assumes a wind speed that is invariant with elevation (in reality, the wind speed increases with elevation). To provide a relevant comparison, the wind speed at 350 meter elevation was used in the model since within the range of distances of interest; the plume motion is mostly influenced by wind speed between the 200 meter and 500 meter elevations.

PHAST is a risk analysis tool used to model the effects of hazardous releases. Since PHAST is not capable of modeling multiple rectangular flare pits, the validation case was run on a single circular flare pit. This simplification removes most of the three dimensional flow characteristics that only CFD is designed to model. The simulation parameters are summarized in Table 1.

CALPUFF is a Gaussian dispersion model commonly used in regulatory air quality modeling applications [3]. It has the capability to model an emitted plume either as a series of overlapping Gaussian "puffs" or as an elongated "slug." Both modes are tested in this validation. The flare pits have been modeled as two elevated rectangular area sources. CALPUFF contains only simple algorithms to account for the effects of obstacles. It cannot simulate the detailed flow between or around the flare pit structures.

Finally, Star-CCM+ was used to model the full three-dimensional plume. A k-ε model was used to simulate the turbulence in the wind, inside the plume and around the flare pits. In order to provide an appropriate comparison with PHAST and CALPUFF, a single circular flare pit was modeled. This simpler geometry removes some of the complex flow interactions that appear when running multiple flares pits simultaneously, and that cannot be modeled by integral models. The simulation parameters are summarized in Table 1.

Table 1: Atmospheric and flare pit running conditions used for validation purposes.

Atmospheric Stability Class	D (neutral)
10m elevation ambient temperature	27°C
10m elevation wind speed	14.9 m/s (54 km/hr)
350m elevation wind speed	24 m/s (86 km/hr)
Combusted gas flow rate	Around 200,000 tons/hrs
Combusted gas temperature	570°C

Given the fact that the plume models differ in their approach and assumptions, their results are somewhat different, especially in the vicinity of the flare pits where the details of the pit geometry are most significant. The Chu and Goldberg model does not consider any geometry details, while PHAST and CALPUFF do consider the plume source elevation. However, only CFD is able to accurately represent the details of the flow around the flare pits.

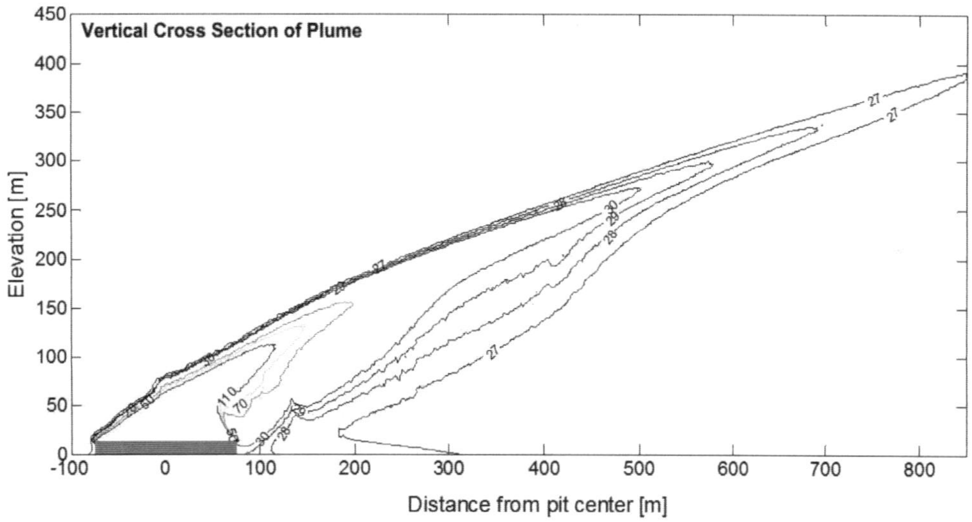

Figure 5: CFD results for the validation case conditions. Contours of equal temperatures are shown on a vertical plane aligned with the wind direction. The corresponding temperature of each contour is also shown.

An elevation view of the results of the CFD simulation is shown in Figure 5. The plume elevation and the plume temperature profiles were compared with the other models and the results of the comparison are presented in Table 2 and Figure 6. Given their differing approaches, the model results are in relatively good agreement in terms of plume rise.[9]

[9] Given all the uncertainties inherent in estimating atmospheric flows, we believe that a difference of only 20% in plume height prediction is very good agreement.

The plume temperature profile will depend on the turbulence, and on the detailed atmospheric wind speed and temperature profiles. The exact geometry of the pit also affects the temperature results since different geometries create vortices of different size and strength, which in turn result in different plume dilution rates. Given the very different assumptions included in the various models, temperature profile differences are not surprising and are expected. Note that as shown in Figure 6, the CALPUFF and the PHAST results are conservative and predict elevated ground temperatures for longer distances, when compared to the CFD results.

Table 2: Comparison of the validation test case results. In the case of CALPUFF, PHAST and CFD, at each cross-section, the plume height was determined to be the elevation of the warmest point in the plane of symmetry of the plume.

Model	Plume height, 400m downwind	Plume height, 800m downwind	2m elevation, 50°C downwind distance
CALPUFF puffs	275 m	430 m	220 m
CALPUFF slugs	275 m	430 m	170 m
PHAST	265 m		140 m
CFD	230 m	370 m	Nowhere
Chu and Goldberg	290 m	450 m	N/A

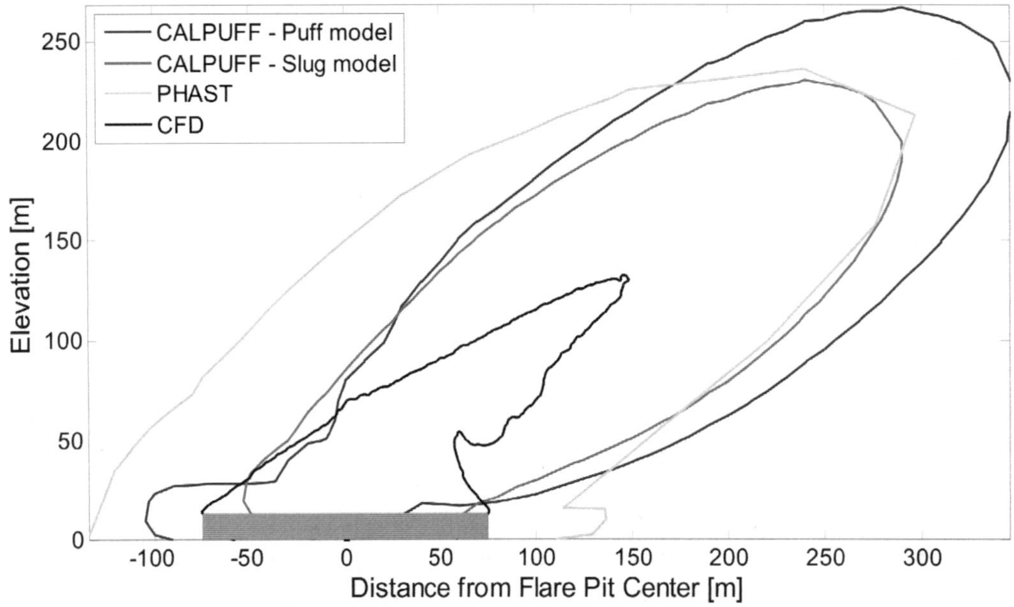

Figure 6: Vertical plume cross-section, as obtained using various models. The curves represent the location of the 70°C temperature. The Chu and Goldberg model is not included in the comparison since it is not meant to accurately represent plume temperatures.

Flow Features in a Flare Pit Plume

The CFD results obtained during the validation stage exhibit what is commonly referred to as a bifurcating plume. Because of the cross wind, the plume acts like a "jet in cross flow". The jet (*i.e.* buoyant plume of the flare pit products) enters into a cross flow (*i.e.* the wind). This phenomenon of counteracting vortices has long been recognized and is well known [4].

Bifurcating plumes have been studied in several publications [5, 6, 7]. This bifurcation and the existence of counteracting vortices has a major influence on the ground flare pit's downwind plume temperature distribution. Centrifugal forces in each of the counteracting vortices push the colder heavier gas outward, while keeping the warm lighter gas inside the vortex (see Figure 7). This results in high temperatures inside the vortices and colder gas outside of the vortices. The warmest downstream locations are therefore not on the plane of symmetry, but are symmetrically located on either side of the vertical symmetry plane.

These counteracting vortices cannot be captured by simpler integral models such as PHAST and CALPUFF and require that the full three-dimensional plume dynamics be modeled.

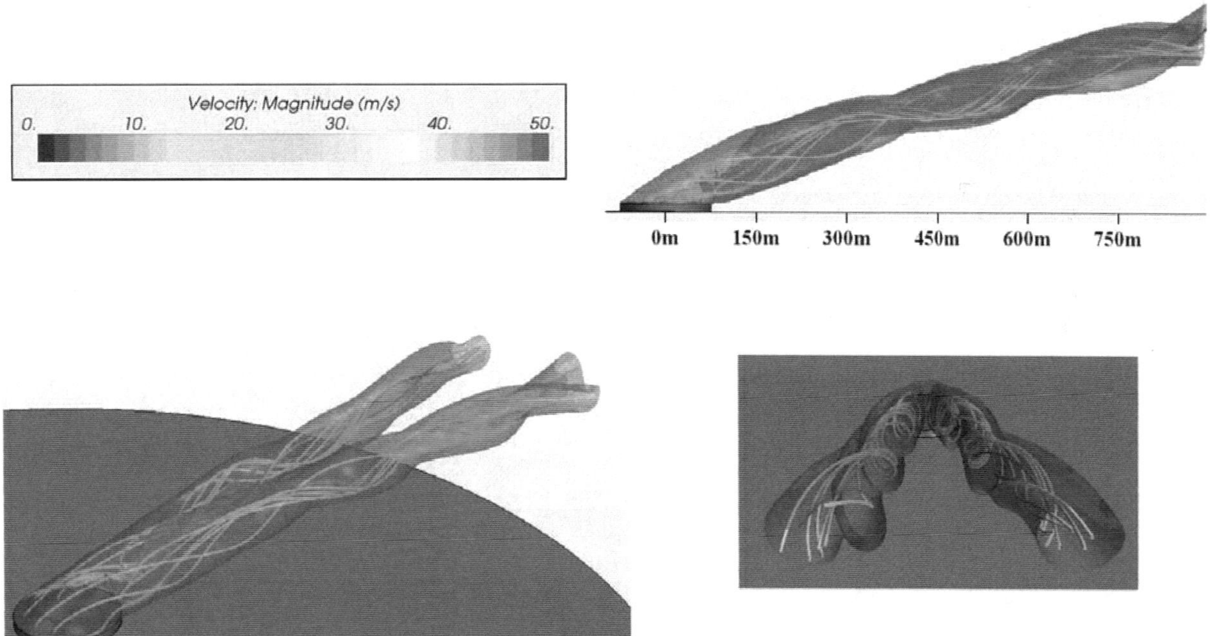

Figure 7: CFD results of the flow resulting from the operation of a circular flare pit. The flow clearly exhibits a pair of counteracting vortices. The blue surface represents the location of the 70°C isosurface. The curves represent streamlines which are color coded according to flow velocity.

Effect of Adjacent Flare Pit Operations

To investigate the effect of two flare pits, two adjacent flare pits separated by a 40 meter distance were modeled. The wind was blowing parallel to the vertical plane of symmetry separating the two pits. In this section, the CFD simulations are based on the conditions summarized in Table 3. A high wind speed and a relatively high product temperature were selected as these conditions correspond to worst-case scenarios. The results are illustrated in Figure 8 and show the existence of two pairs of counteracting vortices, one per pit.

Table 3: Atmospheric and flare pit running conditions used in the CFD model.

Atmospheric Stability Class	D (neutral)[10]
10m elevation temperature	27°C
10m elevation wind speed	14.9 m/s (54 km/hr)
Combustion gas flow rate	Around 85,000 tons/hrs
Combustion gas temperature	Around 1,200°C

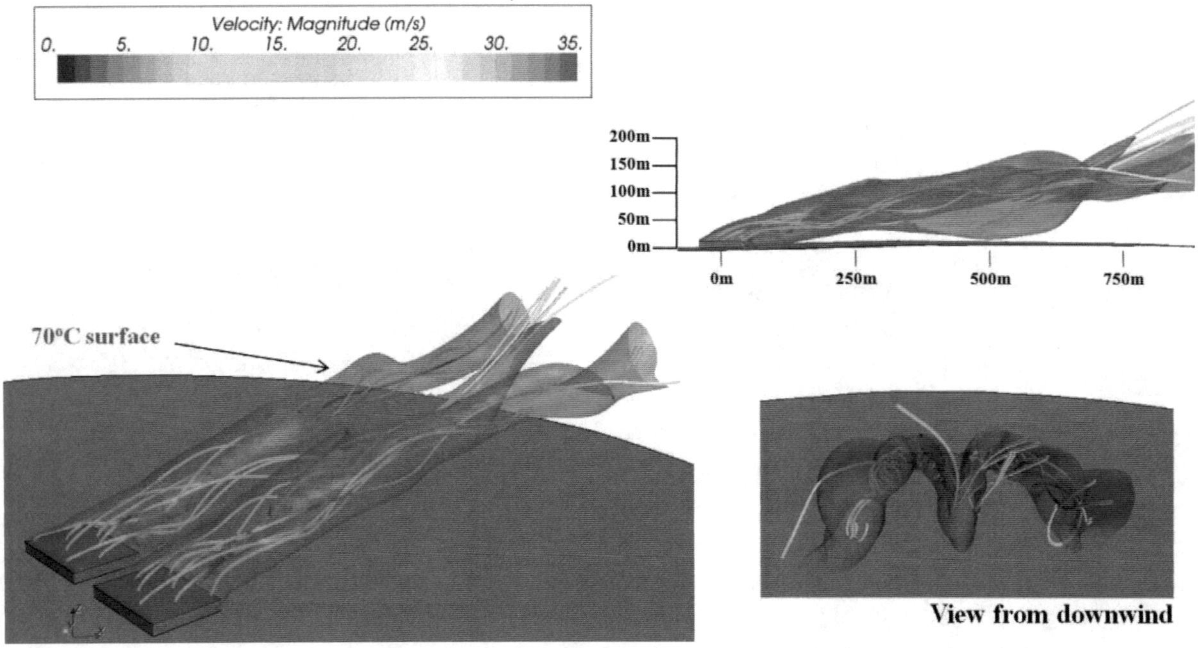

Figure 8: Instantaneous plume profile. The 70°C surface is represented in blue. Streamlines are presented with colors proportional to the local gas velocity.

[10] The selection of an atmospheric stability class D is in accordance with high wind weather conditions.

The existence of these counteracting vortices has significant implications on the ground temperatures. In the vertical plane of symmetry between the flare pits, the two middle vortices merge to create a significant downward flow that pushes the warm gases towards the ground, thereby increasing near-ground temperatures. The effect of this downward flow is most significant at about 500 meters downwind of the flare pits. A cross section of the plume, 500 meters downwind of the flare pits clearly shows the vortex structures (Figure 9). The temperatures on the vertical plane of symmetry and at 2 meter elevation are shown in Figure 10 and Figure 11. As seen in the results, due to the flow features around the walls of the flare pit, the temperature behind the pits is relatively high. In addition a peak 2 meter elevation temperature of 64°C can be observed at a 500 meter distance from the flare pits.

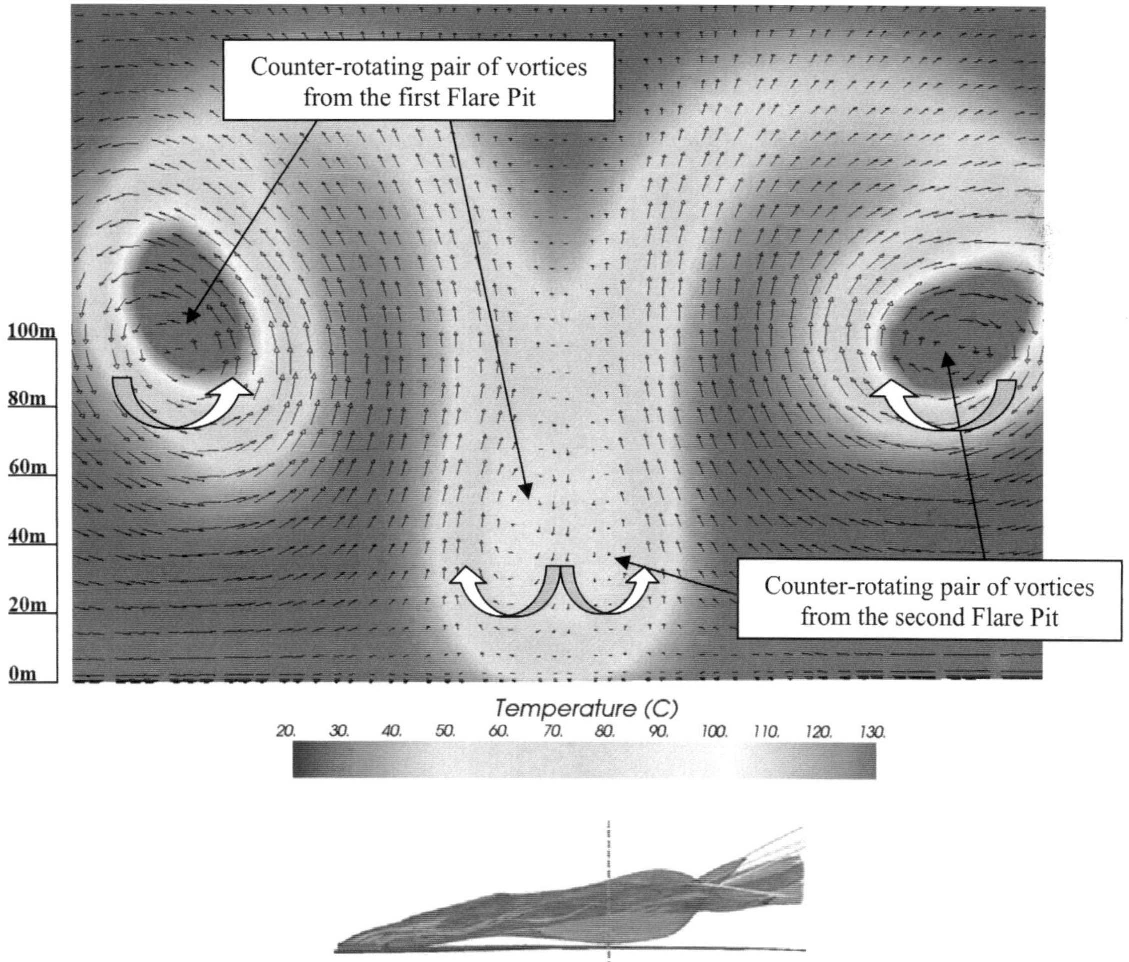

Figure 9: Plume cross-section, 500 meters downwind from the flare pits (see bottom sketch). The temperatures are represented by the colors, while the velocities are shown by the arrows. The longer the arrows, the higher the flow velocity.

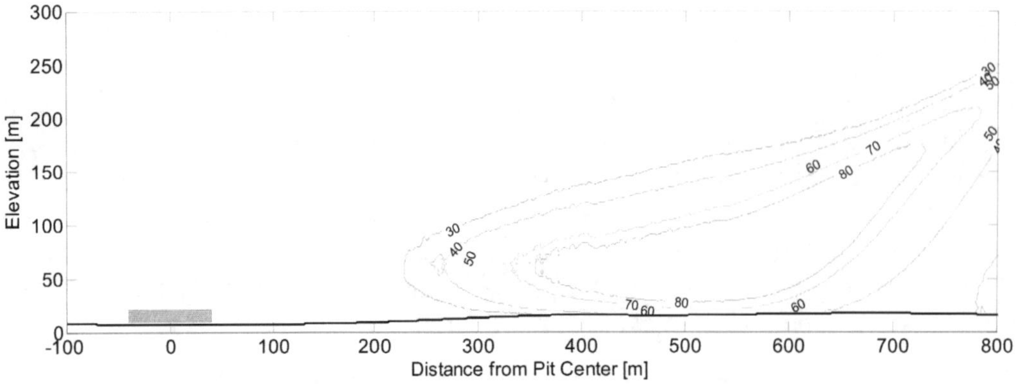

Figure 10: Average temperature distribution on the plane of symmetry (vertical plane) between the two flare pits. Temperatures are in degree Celsius.

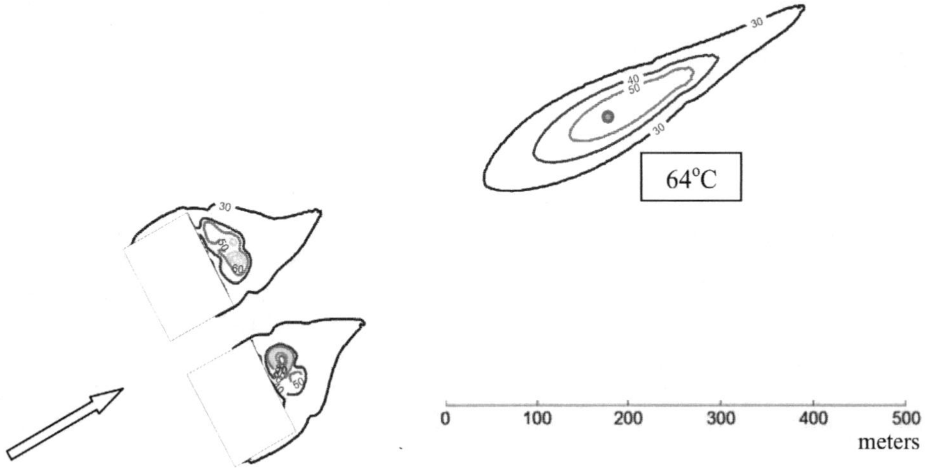

Figure 11: Average temperature distribution at 2 meter elevation. The two flare pits are represented by white rectangles. The wind direction is represented by the arrow, and the location of the maximum far-field temperature by a red circle.

Given the significant influence of the vortex interactions, the distance between flare pits is an important parameter to consider when designing or planning multi-flare pit operations. In particular, if two operating flare pits are far apart, their vortices do not interact and both flare pit plumes rise independently without merging. On the other hand, if the flare pits are touching, then the two flares act as a single larger one, and again, no vortex interaction occurs.

Between these two extremes, there is a distance between flare pits that will generate a maximum downward flow that pushes the combustion products toward the ground, thereby generating relatively high ground temperatures. In the present study, several distances were investigated and the results are presented in Table 4. These results highlight the importance of flare pit separation distance on the downwind ground temperatures.

Table 4: Maximum temperatures during steady operation of two flare pits, as obtained with CFD. These temperatures do not include the temperatures obtained in the immediate vicinity of the flare pits.

Elevation	20m pit separation	40m pit separation	150m pit separation	40m pit separation 45 deg winds
2 meters	71 °C	64 °C	<30 °C	32 °C
6 meters	84 °C	74 °C	<30 °C	32 °C
10 meters	84 °C	80 °C	<30 °C	39 °C
14 meters	90 °C	84 °C	<30 °C	40 °C
16 meters	90 °C	84 °C	<30 °C	40 °C
22 meters	100 °C	89 °C	33 °C	50 °C

The results in Table 4 represent a hypothetical worst case scenario where upset conditions would require flaring at the maximum flow rate, where the combustion gas temperature would be relatively high and where the wind conditions would be very exceptional (*i.e.* high winds blowing perpendicular to the flare pits). The likelihood that these conditions coexist is very remote, and the values in Table 4 constitute a very unlikely worst case scenario.

Note that these results use a wind that is blowing in a direction parallel to the axis of symmetry. This results in strong pairs of vortices that generate very strong downward currents when they interact. Other wind directions may generate greater temperatures in the immediate vicinity of the flare pits; however, the downwind temperatures remain significantly lower. As an example, the maximum temperature resulting from wind blowing at 45 degrees with respect to the symmetry plane is shown in the last column of Table 4.

Conclusions

CFD can play an important role in understanding the impact of large industrial ground flares and other similar releases. Simpler dispersion models will accurately calculate features such as plume height, but are not capable of resolving complex interactions between plumes, nor can they account for detailed flow around structures. CFD modeling demonstrates that there are complex interactions which can occur between adjacent flare pit plumes and which can have very significant localized effects on downwind temperatures and plume density.

The space between flare pits is shown to be a significant dimension which should be considered during plant design. With the increasing availability of CFD codes and computers capable of running them, CFD should serve as an important tool for plant design and industrial hazard planning.

References

[1] *K-ε Model for the Atmospheric Boundary Layer Under Various Thermal Stratifications.* C. Alinot, C. Masson. J. of Solar Energy Engineering. Vol 127. Nov. 2005

[2] *Fluid Mechanics for Industrial Safety and Environmental Protection.* T.K. Fannelöp. Elsevier Science. Pp 281-287. 1994

[3] A User's Guide for the CALPUFF Dispersion Model (Version 5). Earth Tech, Inc. Concord MA. 2000

[4] *Vortical Structures in the wake of a transverse jet.* T.F. Fric and A. Roshko. J. of Fluid Mechanics. Vol 279, 1-47. 1994

[5] *High-Resolution Numerical Models for Smoke Transport in Plumes from Widland Fires.* P. Cunningham and S.L. Goodrick. 2006

[6] *A Computational Model for the Rise and Dispersion of Wind-Bloan Buoyancy-Driven Plumes – I. Neutrally Stratified Atmosphere.* X. Zhang, A.F. Ghoniem. Atmospheric Environment. Vol 27A (15). 2295-2311,1993

[7] *Review of Vortices in Wildland Fire.* Journal of Combustion. Vol 2011. J.M. Forthofer, S.L. Goodrick. 2011

(122f) Anti-Surge Methods for the Base Load LNG Plant Propane Compressor

J. C. Kuo, Chevron Energy Technology Company, Houston, TX

The propane compressor represents a significant capital investment and operating expense of a LNG plant, and is an essential component in providing revenue streams of the plant. Therefore, operation and protection of the propane compressor are critical issues in the plant design and operation.

The propane compressor can operate under a wide range of conditions over the plant life cycle, especially during transient operations such as start-up, shutdown, restart and load variation. When the process condition changes, there is risk of surge where the compressor operates below its minimum stable flow, resulting in vibration and damage to the compressor and its components. Therefore, the propane compressor is required to have an anti-surge system for protection against surge.

The main difference between these methods is how the recycle stream is cooled. Different licensors/technology providers may provide different methods. This paper will summarize

- All the anti-surge methods which were used on some of the existing LNG plants. The LNG plants should be in operation or under the EPC stage, with the size varying from less than 3 to over 7 MMTPA. The driver should include GE's Frame 5, Frame 6, Frame 7, Frame 9 and the total motor driven case.
- Assessment and comparison of technical and economical benefits for different anti-surge methods,
- Indicating the advantages and disadvantages under the start-up, normal operation and trip scenarios.
- Assessment of any technical risk of using each method.

(131a) Getting the Most Out of Your Refinery Hydrogen Network

Alan Zagoria, CEOC, Des Plaines, IL; Optimization Services, UOP, Des Plaines, IL

The refinery balance is an important issue in almost all refineries. The operations of some refineries are actually constrained by the availability, the others have strong incentive to reduce the operating costs, energy consumption and CO_2 emissions associated with the refinery hydrogen network. We will discuss an overall, refinery-wide approach to improving the H_2 network configuration and operations. Plenty of time will be left for discussion and questions.

Heat Exchange Reforming for Increased Hydrogen Production

Niels R. Udengaard, Haldor Topsoe, Inc. Houston Texas & Jack H. Carstensen,
Haldor Topsøe A/S, Lyngby, Denmark

presented at

AIChE 2012 Spring Meeting in Houston TX on April 4, 2012

1 Introduction

Heat exchange steam reforming in hydrogen production is not a new development. Haldor Topsoe developed the first smaller scale steam reformers based on the heat exchange principle about 30 years ago for hydrogen units, where export steam was undesired. The capacity of the first hydrogen units ranged from 0.1 MM SCFD to 1 MM SCFD typically used for fuel cell applications. As the technology matured it became possible to increase the capacity and in 1997 the first industrial scale (5 MMSCFD) hydrogen units were successfully put into operation. The development of the heat exchange reformer technology has continued, and today Topsoe has licensed hydrogen plants including a heat exchange reformer with capacities of up to 185 MM SCFD. At the same time a number of different designs of heat exchange reforming technologies have been developed enabling Topsoe to tailor-make new hydrogen plants or revamp existing units to obtain maximum value for client. Today Topsoe has licensed more than 40 hydrogen units based on heat exchange reforming with more than 15 years industrial experience. Industrial feedback has confirmed that use of heat exchange reforming can save up to 20% on feed and fuel consumption (with corresponding savings in CO_2 emission) as compared to conventional steam reforming.

2 Heat exchange reforming - principles

Being an endothermic reaction steam reforming of hydrocarbons require a significant heat input to obtain the desired conversion into hydrogen. In the conventional steam reformer heat transfer takes place mainly by radiation which leads to a limited thermal efficiency as evidenced by a high flue gas temperature (typically more than 1800°F). The thermal efficiency of a conventional steam reformer is around 50% and the surplus heat is to a large extent used for steam production. Many clients have little or no use of the steam export generated in a hydrogen plant and it is therefore considered of low value.

Heat exchange reformers are characterized by being very compact and having a high thermal efficiency. In a heat exchange reformer the majority of the heat transfer takes place by convection with hot flue gas or hot process gas whereby the thermal efficiency can be increased by as much as 60-70% as compared to the radiant solution. This results in more of the waste heat being used to produce hydrogen instead of steam. Heat exchange reforming is well suited both as stand-alone units and as revamp option for capacity increase of existing plants.

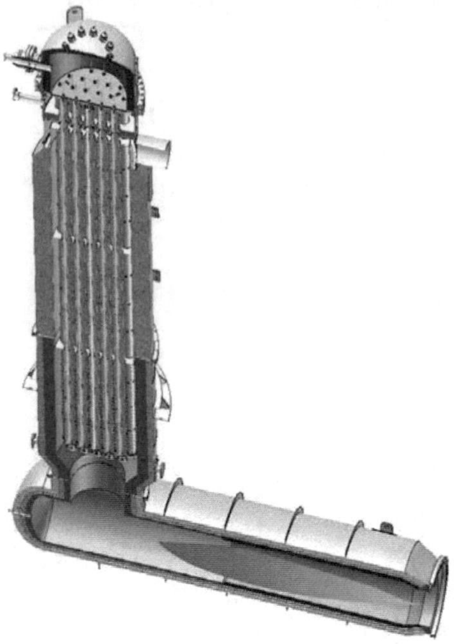

Figure 1 Heat Exchange Reformer

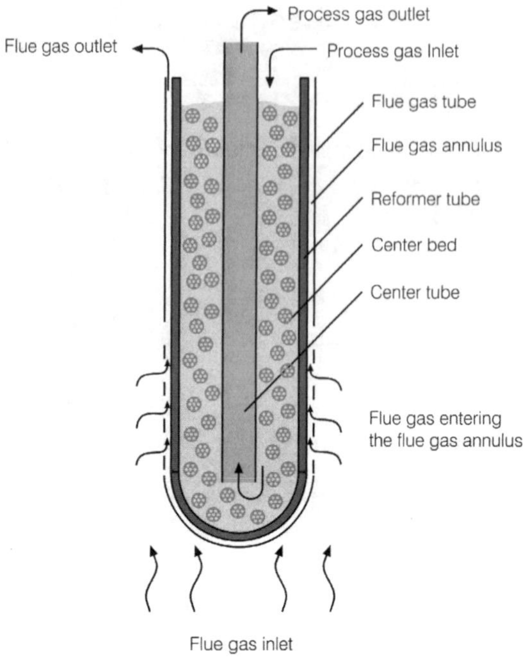

Figure 2 Heat transfer principle

3 Heat exchange reforming technologies

3.1 HTCR

The Haldor Topsoe Convection Reformer (HTCR) is a heat exchange reformer in which the process gas is heated mainly by hot flue gas. The HTCR is very compact and suited for new hydrogen units and for add-on revamps for capacity increase of existing plants.

The HTCR is shown in Figure 1 and the principle is shown in Figure 2. The reformer consists of a vertical, refractory-lined vessel, containing the tube bundle with bayonet tubes. The heat from the flue gas is transferred to the process gas inside the bayonet reformer tubes resulting in low fuel consumption and zero steam export. In the HTCR reformer the external heat input is provided by only one burner which ensures a very easy operation and a fast load response. The easy operation results in only a minimum of operator attendance is required and there are examples of unattended operation of HTCR plants.

The HTCR unit is to a high degree skid-mounted and the reformer is shop-lined minimizing installation time and cost on site..

Industrial experience with HTCR includes more than 30 plants and design capacity up to 27 MMSCFD.

Case story 1 - a grassroots 27 MM SCFD HTCR hydrogen plant

In connection with an extensive revamp of an existing refinery in the Russian Federation an analysis of the future hydrogen balance showed that additional 160 MM SCFD of hydrogen would be required. The major part of the hydrogen was needed for a new hydrocracking unit and the remaining part would be needed for hydrotreating purposes. The new hydrotreating unit was scheduled to come on-stream one year before the hydrocracking unit and the refinery decided to build two separate grassroots plants to cover the hydrogen requirement - one 27 MM SCFD unit and one 130 MM SCFD unit. For the smaller unit the HTCR technology was selected due to:

- Low feed and fuel consumption
- Fast implementation time
- No need for export steam as steam requirements would be covered by the larger hydrogen unit

Figure 3. HTCR tube bundle for a 27 MMSCFD hydrogen plant

Table 1 shows the consumption figures for a 27 MM SCFD HTCR hydrogen plant compared to a conventional SMR process typically used for this capacity range.

27 MM SCFD hydrogen plant	HTCR process	Conventional SMR process
Feed, BTU/SCF H_2	354	352
Fuel, BTU/SCF H_2	11	53
Feed + fuel, BTU/SCF H_2	365	405
Steam export, BTU/SCF H_2	0	60
Cost of feed and fuel, MM USD/year	18.5	20.5

Table 1. Consumption figures for a HTCR and a conventional 27 MMSCFD hydrogen plant

The above example clearly illustrates the advantage of the HTCR process in case of low or no value of steam export. Based on a cost of feed and fuel of 5.36 USD/MM BTU[1], the annual savings for a HTCR plant amount to USD 2 MM compared to a conventional steam reformer process. Furthermore the 11% lower consumption of feed and fuel results in a correspondingly lower emission of CO_2.

3.2 HTER

In the Haldor Topsoe Exchange Reformer (HTER) the reaction heat is provided by hot process gas. The HTER can be used in grassroots hydrogen plants in combination with a radiant wall steam reformer as well as an add-on unit for additional capacity in an existing plant.

The HTER utilizes a bayonet tube or a two-bed system (Figure 4) allowing for optimal utilization of the heat transfer area. The hot reformer effluent is used as heating medium and the high pressure results in a very compact design.

The inclusion of an HTER unit reduces fuel consumption and steam production and may increase capacity up to 30%. The HTER is well suited for both capacity revamps in existing plants as well as new units where factors such as low feed and fuel consumption and compactness are important.

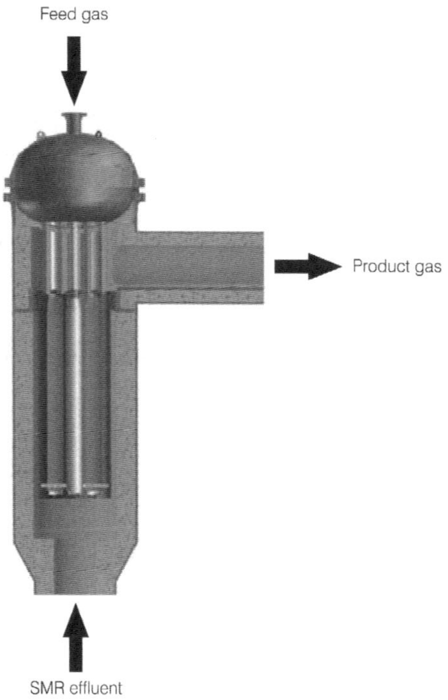

Figure 4. HTER principle

Figure 5. Plant with HTER

Since 2003 when the first HTER went into successful operation, the HTER technology has been adopted in several Topsøe licensed hydrogen plants.

Case story 2 – 3 x 185 MM SCFD grassroots hydrogen plants incorporating HTER

In connection with a large refinery project in the Middle East the refiner would need an additional 555 MM SCFD of hydrogen which was to be covered by three grassroots plants. The three plants should be designed to operate on a range of different feedstocks depending on cost and availability in the refinery.

Based on analysis of investment cost and consumption figures, the client chose a solution with three new grassroots plants of 185 MM SCFD, each plant with a state-of-the-art SMR process incorporating an HTER. Table 2 compares the key figures for the chosen HTER solution compared with an alternative without an HTER.

185 MMSCFD plant	State-of-the-art SMR process + HTER	State-of-the-art SMR process
Relative total installed cost (TIC)	100	100
Feed + fuel, BTU/SCF H_2	372	390
Steam export, BTU/SCF H_2	43	71
Feed + fuel minus steam, BTU/SCF H_2	329	319
Relative power consumption	86	100
Cost of feed and fuel, MM USD/year	129.9	136.4

Table 2. Consumption and TIC figures for a 185 MM SCFD hydrogen plant

For the same investment the solution based on an HTER results in a 5% saving in feed and fuel which, based on a price of $5.36/MM BTU[1], corresponds to annual savings of $6.5 MM, a correspondingly lower CO_2 emission and a lower power consumption.

3.3 TBR

The Haldor Topsoe Bayonet Reformer (TBR) is the newest addition to the Topsoe family of heat exchange reforming technologies combining the principle of convection heat transfer known from HTCR and radiant heat transfer known from the radiant wall steam reformer.

The TBR consists of a series of double tubes in a single row in a radiant furnace box similar to conventional steam reforming technology.

Feed gas flows downward through the catalyst bed located in the annulus between the two tubes as illustrated in Figure 5. At the bottom of the catalyst bed the gas turns and continues upward through the inner, empty bayonet tube. The bayonet tube arrangement is the same as that known from the HTCR technology. The gas exits at the top of the bayonet tubes and is collected in a common header.

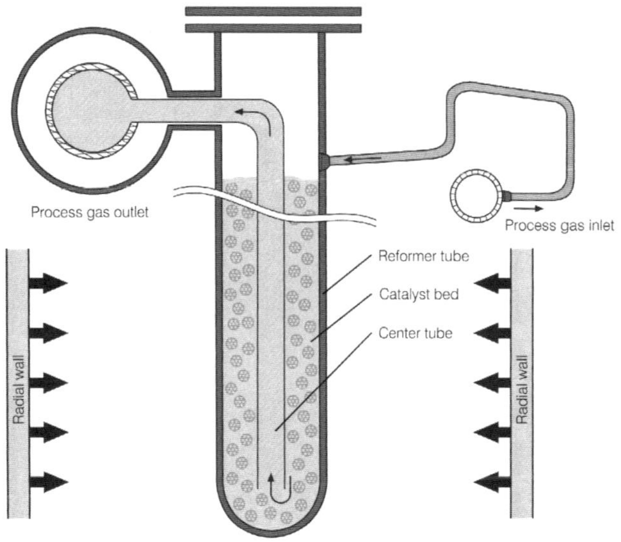

Figure 6. Simplified sketch of the TBR principle

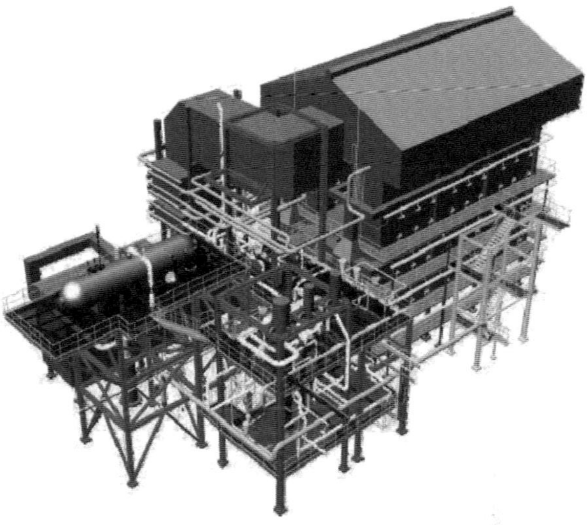

Figure 7. Hydrogen plant with TBR reformer

Improved heat utilization in combination with a high average heat flux in the TBR tubes significantly reduces the size and capital cost of the hydrogen plant and provides hydrogen production with low hydrocarbon consumption and little or no steam export.

Table 3 illustrates the main features for a 43 MM SCFD TBR hydrogen plant compared to a state-of-the-art SMR process. For the TBR process we have considered two cases: With and without steam export.

43 MM SCFD plant	State-of-the-art SMR process	TBR with steam export	TBR no steam export
Relative TIC	100	95	95
Feed + fuel, BTU/SCF H_2	390	361	350
Steam export, BTU/SCF H_2	71	30	0
Feed + fuel minus steam, BTU/SCF H_2	319	331	350
Relative power consumption	100	90	90

Table 3. Consumption and TIC figures for a 43 MM SCFD hydrogen plant based on TBR

The table illustrates that with little or no requirement for steam export, the TBR technology offers an ideal solution for the refinery with a 5% lower investment cost, and at the same time, up to 10% lower consumption of feed and fuel compared to a state-of-the art SMR process.

4 Revamping of an existing hydrogen unit

Revamping an existing hydrogen unit with an HTCR or an HTER is an economically attractive solution to increase capacity and reduce specific feed consumption.

When revamping with an add-on heat exchange reformer the capacity can typically be increased with up to 30% at a much lower investment as compared to a new plant. At the same time the compactness of a heat exchange reformer ensures that additional required plot area is kept at a minimum.

In order to determine the feasibility of a revamp Haldor Topsoe will perform a revamp feasibility study to provide the client with the optimal solution. The study will take into account the status of the existing unit, the various requirements of the client as well as prevailing conditions such as availability and price of feedstock, plot plan, available downtime for revamp etc.

4.1 Revamping with an HTCR

As a revamp option, the HTCR is a very flexible solution where the HTCR is operated in parallel with the existing reformer as illustrated in Figure 8. This lay-out will result in a revamp package consisting of few pieces of equipment, namely an HTCR with waste heat section, a process boiler and a combustion air blower that allows for independent operation of the two reformer sections.

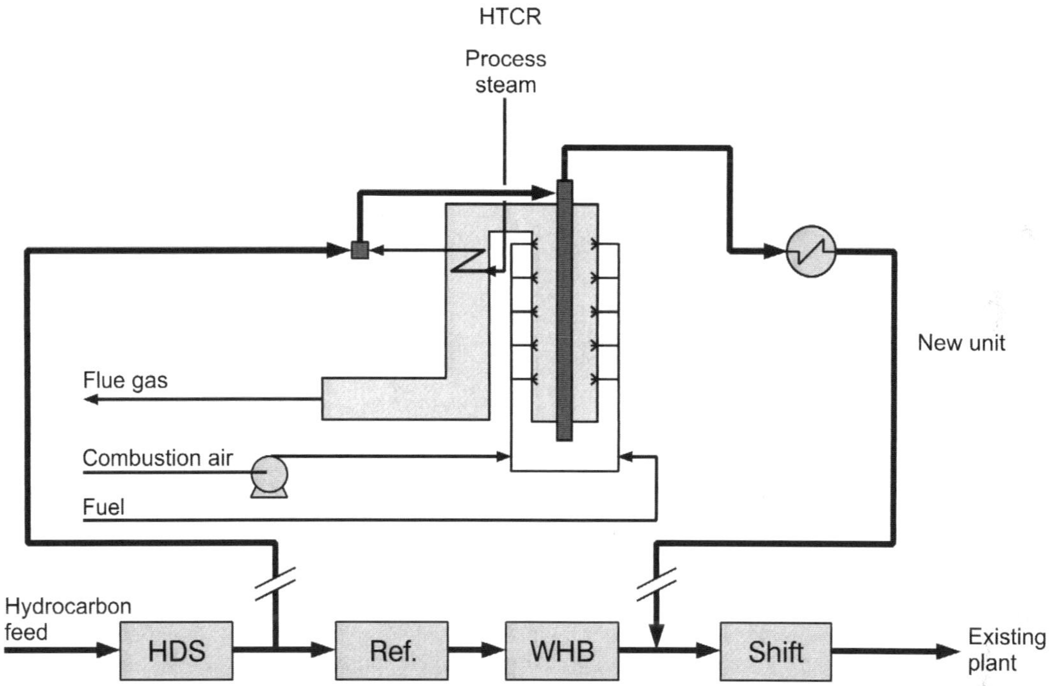

Figure 8. Schematic lay-out of an HTCR revamp solution

The on-site implementation of an HTCR revamp requires a minimum downtime as basically only three tie-ins need to be performed during a turnaround. The remaining construction can be performed during operation of the existing unit. An HTCR revamp can typically increase production capacity with up to 30%. Following a revamp the specific feed consumption will decrease as a result of the higher thermal efficiency of the HTCR.

Revamping with an HTER

Compared to the HTCR revamp an HTER revamp is completely integrated into the existing unit as illustrated in Figure 9, resulting in an economically attractive and less plot area requiring solution. The benefits of an HTER revamp are best utilized when the process gas exit temperature from the existing steam reformer is more than 1630°F. For an HTER revamp a longer downtime compared to the HTCR must be foreseen as the HTER is connected to the existing reformer.

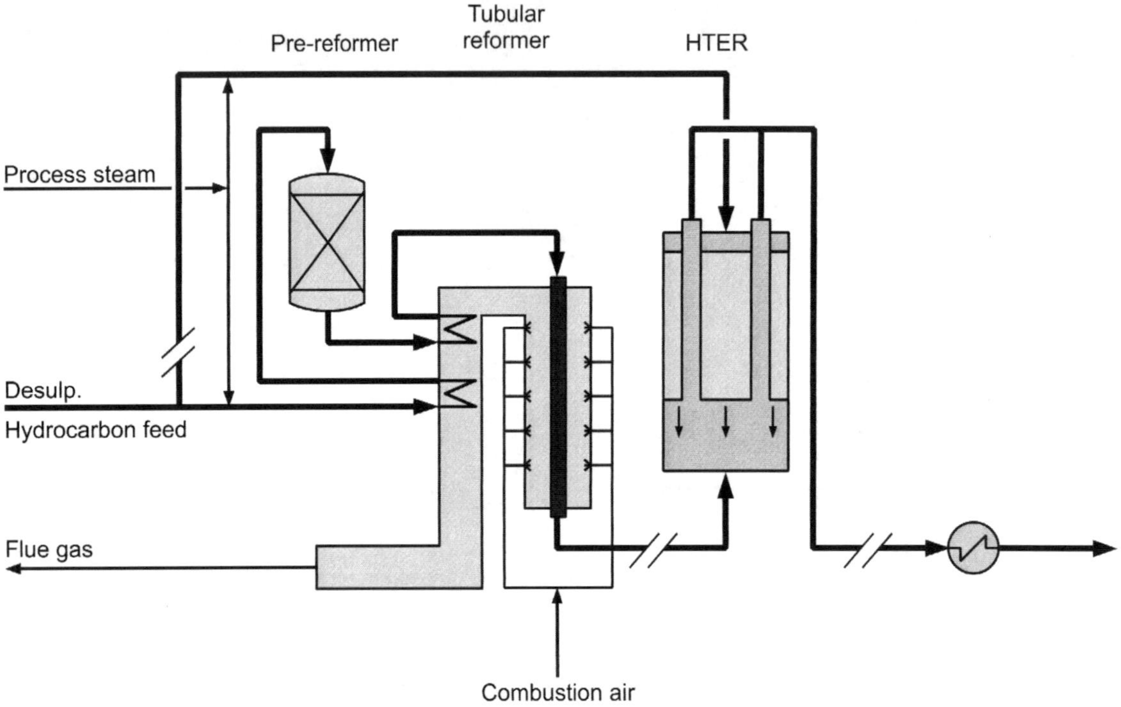

Figure 9. Schematic layout of an HTER revamp solution

An HTER revamp can typically increase production capacity by up to 30%. Integration of the highly efficient HTER results in a hydrogen plant with lower specific feed and fuel consumption and lower steam export as illustrated in the example below.

Case story 3 – increasing the capacity 25% by adding an HTER

A refinery in Asia operates a 65 MM SCFD, an 80 MM SCFD and a 110 MM SCFD hydrogen plant all designed by Haldor Topsoe. In 2009 the client wanted to increase hydrogen consumption by 27 MM SCFD. It was important that the solution could be implemented during a normal 4 weeks turnaround and that the additional plot area was minimized. The following revamp alternatives were considered:

- Adding an HTCR reformer to each of the smaller (and older) units
- Adding an HTER reformer to the 110 MM SCFD plant
- A new 27 MM SCFD grassroots plant

A feasibility study concluded that adding an HTER reformer could yield the 27 MM SCFD capacity increase at the minimum investment. The feasibility study also identified the required modifications/replacement of existing equipment necessary to achieve the desired capacity increase. Total installed cost (TIC) for the complete revamp including a new HTER and the required modifications/replacement of existing equipment amounted to approx. 60% of the TIC for a new separate hydrogen plant having the same capacity. The results in terms of energy consumption are show in Table 4 below:

110 MMSCFD plant revamp	Before revamp	After revamp
Capacity, MM SCFD	110	137 (+25%)
Feed and fuel consumption, BTU/SCF H_2	425	384 (-10%)
Steam export, BTU/SCF H_2	96	52 (-46%)
Net energy consumption, BTU/SCF H_2	329	332 (+1%)

Table 4. Consumption figures before and after an HTER revamp of a 110 MMSCFD hydrogen plant

The results show that the capacity could be increased by 25% with a 10% lower specific consumption of feed and fuel.

Figure 10. Installation of HTER during revamp

5 Conclusion

The innovative heat exchange reformer solutions developed and industrially demonstrated by Haldor Topsoe are optimal solutions for adding additional hydrogen capacity, either in the form of a stand-alone plant or as a revamp of an existing hydrogen unit. By utilizing the heat exchanger principle in hydrogen manufacture, it is possible to significantly reduce the consumption of hydrocarbon feedstock as compared to traditional radiation-based technology and reduce CO_2 emissions correspondingly. Furthermore heat exchange reforming is compact, characterized by easy operation and low investment cost. With the range of technologies presented in this paper, it is possible to provide tailor-made compact and energy efficient solutions for additional hydrogen production in all refineries.

References

1. Natural Gas Futures NYMEX price (Henry Hub) average for December 2009.
2. Dybkjaer, I., Winter Madsen, S., 'Compact Hydrogen Plants', Hydrocarbon Engineering, November 2004.
3. Gyde Thomsen, S., Han, P.A., Lock S., Werner, E., 'The First Industrial Experience with the Haldor Topsøe Exchange Reformer', AICHE, 2006.
4. Hedegaard Andersen, K., 'Hydrogen Agenda', Hydrocarbon Engineering, November 2006.
5. Winter Madsen, S., Olsson, H., 'Steam Reforming Solutions', Hydrocarbon Engineering, July 2007.

(131c) Contaminant Resistant High-Temperature Hydrogen Purification Membrane

Siu-Yue Tam[1], Andy Tsai[1,2] and Ying-Bing Jiang[2], (1)T3 Scientific LLC, Blaine, MN, (2)University of New Mexico, Albuquerque, NM

Hydrogen is a key building block of many chemical processes. Hydrogen can be produced in several ways, but economically the most important processes involve producing hydrogen from hydrocarbons. Fossil fuels including coal, oil and natural gas currently are the main sources of hydrocarbons for hydrogen production and biomass is an emerging one. Gas streams generated by hydrogen production from all of these hydrocarbon sources contain corrosive gases such as hydrogen sulfide (H_2S) and halides that attack metals on contact. Competitive hydrogen purification technologies such as pressure swing adsorption require high capital and operating cost. Membrane technology is the method-of-choice because of its low energy requirement, simplicity and a well established process. The most promising membrane materials for hydrogen purification are palladium and its alloys because of their high selectivity and permeability and high temperature tolerance. However, the critical downside is low tolerance for impurities, hydrogen sulfide, in particular. Conventional approaches to sulfur resistance improvement such as binary (e.g. Pd-Cu) or tertiary (e.g. Pd-Cu-Au) alloy often result in moderate gain at the expense of loss of permeance. The reason why these metals are active enough to transport hydrogen is also the reason why they react with reactive gases in the gas stream.

MembraGuard™technology is a different approach. MembraGuard™technology is a protective coating for hydrogen permeable metal membranes. MembraGuard™coating provides protection not only for sulfur componds, but for other reactive gases, such as halides and carbon monoxide. MembraGuard™coated membrane has high hydrogen selectivity and permeance under elevated temperature and yet is highly resistant to hydrogen sulfide. MembraGuard coated membrane was tested under elevated temperatures and pressures. Thermal resistance was demonstrated under repeated thermal cycles. Palladium membrane without any protective coating lost half of its permeability within 1 hour of contact with hydrogen sulfide. MembraGuard coated palladium membrane was demonstrated to resist hydrogen sulfide at high level with no degradation of performance for extended period. Mixed gas condition including H_2, CO, CO_2, H_2O and H_2S was also tested and MembraGuard coating was proved to be effective under the mixed gas condition.

(131d) Composite Catalytic-Permselective Membranes for H2 Purification

Daejin Kim[1], Elva Romero[1] and **Benjamin Wilhite**[2], (1)Artie McFerrin Department of Chemical Engineering, Texas A& M University, College Station, TX, (2)Artie McFerrin Dept. of Chemical Engineering, Texas A& M University, College Station, TX

Our research group is developing a novel composite catalytic-permselective membrane design approach, in which a permselective film (e.g., polymeric, metallic) is integrated with a porous catalytic film such that reaction and diffusion within the catalytic layer directly modifies the gas composition at the surface of the underlying permselective barrier [1]. In this talk, we will review general design rules for implementing this strategy for hydrogen purification from diesel reformate mixtures using water-gas-shift as the enhancing catalytic reaction. Detailed two-dimensional simulations and preliminary experimental results will be presented to demonstrate this strategy for (i) enhancing the permeance of palladium membranes, and (ii) enhancing the permselectivity of polymeric membranes.

[1] B.A. Wilhite. Composite Catalytic-Permselective Membranes: A Novel Strategy for Enhancing Selectivity and Permeation Rates via Reaction and Diffusion. *Ind. Eng. Chem. Res.,* **2011**, 50(17), 10185-10193.

(137a) Catalytic Strategies for Upgrading Pyrolysis Vapors

Shaolong Wan[1], Xiaohan Zhang[1], Daniel Resasco[2], Lance Lobban[1] and R. G. Mallinson[1], (1)Chemical, Biological, and Materials Engineering, University of Oklahoma, Norman, OK, (2)School of Chemical, Biological and Materials Engineering and Carbon Nanotube Technology Center, University of Oklahoma, Norman, OK

Pyrolysis vapors are catalytically treated directly downstream in a coupled reactor. Dual function metal/metaloxide catalysts are used to convert small oxygenates to larger carbon number molecules with less oxygen and, simultaneously, deoxygenate phenolics and transfer methoxy methyl to the ring. The ultimate goal is to maximize liquid product (and carbon) yield while minimizing hydrogen requirements.

(137c) Fischer-Tropsch Synthesis Using a Pilot Scale Fixed Bed Compact Heat Exchange Reactor

Steven J. Vallee, Zhijun Jia and Hani Gadalla, Chart Energy & Chemicals, La Crosse, WI

The Fischer-Tropsch reaction is a highly exothermic reaction (157 kJ/mol CO converted) used to convert syngas (a mixture of CO and H_2) to liquid hydrocarbons. Conventional fixed bed reactor technology suffers from heat transfer limitations. Operating conditions must be limited to prevent reactor runaway, resulting in lower productivity per unit reactor volume. Non-isothermal operating conditions make the process more difficult to start up and control and can lead to increased catalyst deactivation and unwanted methane side-product formation.

A Compact Heat Exchange Reactor (CHER) combines a reactor and compact heat exchanger into a single device, and provides a solution to the heat transfer limitations commonly observed for conventional fixed bed reactors. A pilot scale CHER was designed and constructed by Chart Energy & Chemicals in La Crosse WI, for the purpose of synthesis of diesel fuel via the Fischer-Tropsch (FT) reaction from syngas. Results are presented for a pilot scale CHER reactor for Fischer-Tropsch synthesis for >1 bpd liquid fuel production that is larger than our previously presented CHER results.

(137d) Conceptual Design and Integration of XTL Plants

Mohamed Noureldin[1], Buping Bao[1], Nimir O. Elbashir[2]and M.M. El-Halwagi[1], (1)Artie McFerrin Department of Chemical Engineering, Texas A& M University, College Station, TX, (2)Texas A& M University - Qatar

ABSTRACT

Liquid transportation fuels may be produced through gas-to-liquid (GTL), coal-to-liquid (CTL), and biomass-to-liquid "BTL" technologies which are collectively referred to as XTL. This work looks at the conceptual design and integration of XTL with focus on thermal conversion technologies. The aim of this work is to establish benchmarks for the process and innovative ways in which the various technologies can be integrated. The performance benchmarks (targets) are identified ahead of detailed design. They also allow current process performance to be gauged against the established targets and determine how far from the targets the process is operating. A base case is used to develop insights into factors which reduce the production rate of liquid fuels and overall product yield of XTL processes. Overall this work helps to identify areas for improvement and innovation including improved mass integration (e.g., feedstock utilization, reduced CO_2emissions, reduced hydrogen input) and better energy integration. A case study is used to illustrate the developed concepts.

(144a) Evaluation of Dispersion Models for LNG Siting Applications

Andrew Kohout, Office of Energy Projects, Federal Energy Regulatory Commission (FERC), Washington, DC

The federal safety standards for LNG facilities under 49 CFR Part 193 approves the use of specific dispersion models for performing vapor-gas dispersion exclusion zones. In addition, provisions allow for the use of alternative models if approved by PHMSA.

In 2010, PHMSA, in consultation with FERC, issued Advisory Bulletin ADB-10-07 to provide guidance on obtaining approval of alternative models under Part 193. The guidance is based on the Model Evaluation Protocol (MEP) described in the Fire Protection Research Foundation Reports, Evaluating Vapor Dispersion Models for Safety Analysis of LNG Facilities (2007) and Guide to the LNG Model Validation Database (2009, 2010). This method has been applied to several dispersion models to establish limitations and safety margins, which would allow it to be used in Part 193 exclusion zone calculations.

This presentation examines the results of the reviews conducted by PHMSA and FERC, including the basis for the limitations and safety margins.

AIChE 2012 Spring Conference

Topical 6: 12th Topical Conference on Gas Utilization – LNG Plant Safety and Protection I

Modeling Dynamic Heat Transfer in Piping Systems using an Enhanced ASTM C-680 Algorithm

Debby Sielegar, P.E.
Process Engineering Specialist
Bechtel Oil, Gas & Chemicals, Inc.

David Messersmith, P.E.
LNG Technology Manager
Bechtel Oil, Gas & Chemicals, Inc.

Abstract

ASTM C-680 methodology is routinely used to model heat transfer in steady state systems. But not all applications can be treated as steady state. Some applications require modeling of systems with mixed phase streams in condensing or vaporizing services that are dynamically interacting with each other.

One of the common dynamic heat transfer problems is the relief of large quantities of low temperature fluid to the dry gas flare system followed by the immediate relief of a much warmer vapor close to its saturation temperature. The warm vapor comes in contact with the cold pipe surfaces and loses heat to the piping causing condensation on the inside surfaces of the pipe. The fluid thermally contracts which may result in air ingress into the flare header that can potentially create a flammable atmosphere. It has been typical to purge the flare header and raise the temperature of the piping following the relief of low temperature fluid to prevent condensation along the walls of the header. ASTM C-680 methodology is typically used to determine the required quantity of purge gas, the duration of purging, and the metal piping temperature for each pipe segment at every time step. Other heat transfer applications include the determination of insulation requirement, sizing of condensed fluid containment, natural cooling as a cost saving approach, etc.

ASTM C-680 algorithm allows the user to iteratively calculate the temperature of all the piping layers (including pipe wall, insulation etc.), which in turn is used to calculate the inside and outside heat transfer coefficients and thermal conductivity. In this paper, a few case studies are presented using an enhanced ASTM C-680 algorithm where the piping system is divided into multiple segments for accuracy, heat transfer is time-dependent, and the fluid properties are continually updated to reflect the impact of pressure drop.

Terminology

Nu	Nusselt number, dimensionless
Pr	Prandtl number, dimensionless
Re	Reynolds number, dimensionless
A	heat transfer area, ft^2
D_i	inside pipe diameter, ft
D_o	outside pipe diameter (including insulation thickness, if present), ft
C_p	fluid heat capacity, Btu/lb.°F
g	gravitational acceleration, 32.2 ft/s^2
h_{conv}	convective heat transfer coefficient, Btu/hr.ft^2.°F
h_i	inside heat transfer coefficient, Btu/hr.ft^2.°F
h_o	outside heat transfer coefficient, Btu/hr.ft^2.°F
h_r	radiation heat transfer coefficient, Btu/hr.ft^2.°F
k	fluid thermal conductivity, Btu/hr.ft.°F
k_m	metal thermal conductivity, Btu-in/hr.ft^2.°F
k_{ins}	insulation thermal conductivity, Btu-in/hr.ft^2.°F
L	pipe length, ft
LMTD	log mean temperature difference, °F
m	mass flowrate, lb/hr
MW	molecular weight
N	number of pipe segments
P_{in}	inlet fluid pressure, psig
P_{out}	outlet fluid pressure, psig
ΔP	pressure drop, psi
Q	heat gain or heat loss, Btu/hr
r	average radius, ft
r_i	inside radius, ft
r_o	outside radius (including insulation thickness, if present), ft
r_s	outside metal pipe radius in the presence of insulation, ft
t	time, s or hr time step
t_m	metal thickness, in
t_{ins}	insulation thickness, in
T_{in}	inlet fluid temperature, °F
T_{out}	outlet fluid temperature, °F
T_b	bulk fluid temperature, °F average of inlet and outlet fluid temperatures $(T_{in} + T_{out})/2$
T_i	inside surface temperature, °F
T_o	outside surface temperature, °F
T_s	interim temperature in the presence of insulation layer, outside metal or inside insulation surface temperature, °F
T_m	metal temperature, °F average of inside and outside surface temperatures $(T_i + T_o)/2$, or average of inside and interim temperature $(T_i + T_s)/2$ in the presence of insulation layer
T_{ins}	insulation temperature, °F average of interim and outside temperatures $(T_s + T_o)/2$ in the presence of insulation layer
T_∞	ambient temperature, °F
T_{avg}	air film temperature, °F average of outside and ambient temperatures $(T_o + T_\infty)/2$
U	overall heat transfer coefficient, Btu/hr.ft^2.°F
V	volume, ft^3
Wind	wind velocity, mph
ρ	fluid density, lb/ft^3
ρ_m	metal density, lb/ft^3
μ	fluid viscosity, cP or lb/hr-ft
μ_i	fluid viscosity at inside surface temperature (T_i), cP or lb/hr-ft
λ	latent heat of vaporization, Btu/lb
ε	emissivity
σ	Stefan-Boltzmann constant, 0.1713×10^{-8} Btu/hr.ft^2.°R^4

1. Introduction

1.1 Purpose

Modifying the ASTM C-680 algorithm as demonstrated in this paper is considered a simple approach to solve various heat transfer applications. As heat transfer calculation is iterative, algorithm for computer program is essential. Although some software programs provide a piping heat transfer calculation feature, options for selecting heat transfer correlations are limited. There are many recently published literatures presenting modified and proven heat transfer correlation(s) for a particular fluid and/or application.

1.2 Method of Calculation

ASTM C-680 provides an algorithm and computational methodologies to predict heat loss or heat gain in a one dimensional steady state system. The methodologies are based on standard heat transfer calculations that are commonly found in many heat transfer literatures. ASTM C-680 algorithm flowchart is shown in Figure 2. Although the algorithm may be used for both heating (heat gain) and cooling (heat loss) applications, this paper focuses on heat loss in piping systems only. Fluid flow, heat flow and temperature profile in a pipe segment are illustrated in Figure 1.

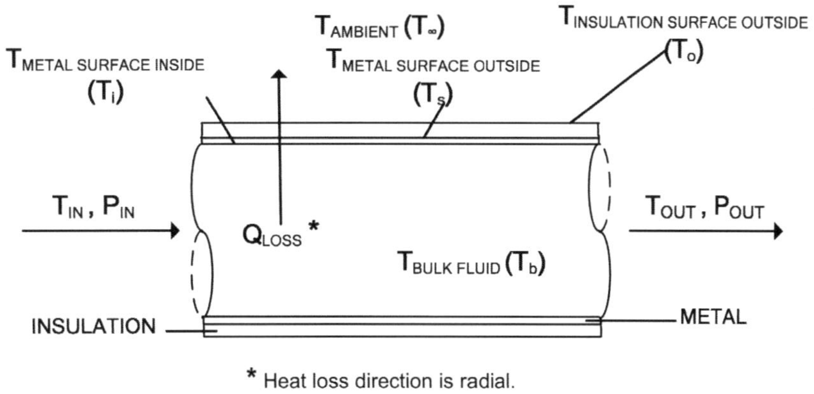

Figure 1. Heat Flow and Temperature Profile in a Piping System when $T_{AMBIENT} < T_{FLUID}$

Surface temperatures are important parameters for calculating the individual heat transfer coefficients. Individual heat transfer coefficients typically include inside heat transfer coefficient of the fluid (h_i), thermal conductivity of the metal (k_m), thermal conductivity of the insulation (k_{ins}), and outside heat transfer coefficient (h_o). Inside heat transfer coefficient (h_i) in a piping system is generally calculated based on forced convection heat transfer coefficient correlations shown in equations (1) through (4). As physical properties are directly related to temperature, the inside heat transfer coefficient is thus temperature dependent.

Seider-Tate [5]

$$Nu = 0.023\ Re^{0.8}\ Pr^{1/3} \left(\frac{\mu}{\mu_i}\right)^{0.14} \qquad (1)$$

$$0.7 < Pr < 16700,\ \ Re > 10^4,\ \ \frac{L}{D_i} > 60$$

Dittus-Boelter [5]

$$Nu = 0.023\ Re^{0.8}\ Pr^{n} \text{ where n = 0.3 for heat loss and n = 0.4 for heat gain} \qquad (2)$$

$$0.7 < Pr < 120,\ \ Re > 10^4,\ \ \frac{L}{D_i} > 60$$

Nusselt Film-type Condensation [7] (*)

Horizontal: $$Nu = 0.73 \left(\frac{D_i^3 \rho^2 g \lambda}{k \mu (T_b - T_i)}\right)^{1/4} \tag{3}$$

Vertical: $$Nu = 0.943 \left(\frac{L^3 \rho^2 g \lambda}{k \mu (T_b - T_i)}\right)^{1/4} \tag{4}$$

(*) Condensed liquid properties should be used. Reynolds number is based on the condensing film characteristics and is less than 2,100.

Metal and insulation thermal conductivities are also as a function of temperature.

[6] Perlite $k_{ins}(T_{ins}) = 0.4030 + 6.38E\text{-}4\ T_{ins} - 3.56E\text{-}7\ T_{ins}^2 + 3.53E\text{-}10\ T_{ins}^3$ (5)

[6] Mineral wool $k_{ins}(T_{ins}) = 0.228 + 3.72E\text{-}4\ T_{ins} - 6.0E\text{-}7\ T_{ins}^2$ (6)

The outside heat transfer coefficient is a function of the convective and radiation heat transfer coefficients as shown in equations (7) through (9).

Outside heat transfer coefficient [1] $$h_o = h_{conv} + h_r \tag{7}$$

Convective [1] * $$h_{conv} = 1.016 \left(\frac{1}{12D_o}\right)^{0.2} \left(\frac{1}{T_{avg}}\right)^{0.181} (T_o - T_\infty)^{0.266} \sqrt{1 + 1.227\, Wind} \tag{8}$$

Radiation [1] ** $$h_r = \frac{\sigma\, \varepsilon\, (T_o^4 - T_\infty^4)}{T_o - T_\infty} \tag{9}$$

* For piping diameter greater than 2 ft or 24 in, use D_o = 2.

** T_o and T_∞ for calculating radiation heat transfer coefficients are to be in °R unit.

Once individual heat transfer coefficients are determined, overall heat transfer coefficient and heat transferred are then determined using equations (10) and (11), respectively.

Overall heat transfer coefficient $$\frac{1}{Ur} = \frac{1}{r_i h_i} + \frac{ln\left(\frac{r_s}{r_i}\right)}{k_m} + \frac{ln\left(\frac{r_o}{r_s}\right)}{k_{ins}} + \frac{1}{r_o h_o} \tag{10}$$

Heat loss $$Q = U\,(2\pi r L)\, LMTD \quad \text{where} \quad LMTD = \frac{(T_{in} - T_\infty) - (T_{out} - T_\infty)}{ln\left(\frac{T_{in} - T_\infty}{T_{out} - T_\infty}\right)} \tag{11}$$

In piping systems, the fluid outlet temperature can be calculated using equation (12).

$$Q = m\, C_p\, (T_{in} - T_{out}) + m\, \lambda \tag{12}$$

The surface temperatures can be back-calculated using the individual heat transfer coefficients and heat loss rate.

Inside metal surface temperature $$T_i = T_b - \frac{Q}{A\, h_i} \tag{13}$$

Outside metal surface temperature $$T_s = T_i - \frac{Q}{A\left(\frac{k_m}{t_m}\right)} \quad \text{or} \quad T_s = T_\infty + \frac{Q}{A\left(\frac{k_{ins}}{t_{ins}}\right)} \tag{14}$$

Outside insulation surface temperature $$T_o = T_\infty + \frac{Q}{A\, h_o} \tag{15}$$

The calculations are iteratively repeated until the calculated surface temperatures from equations (13) through (15) match closely with the inputted surface temperatures.

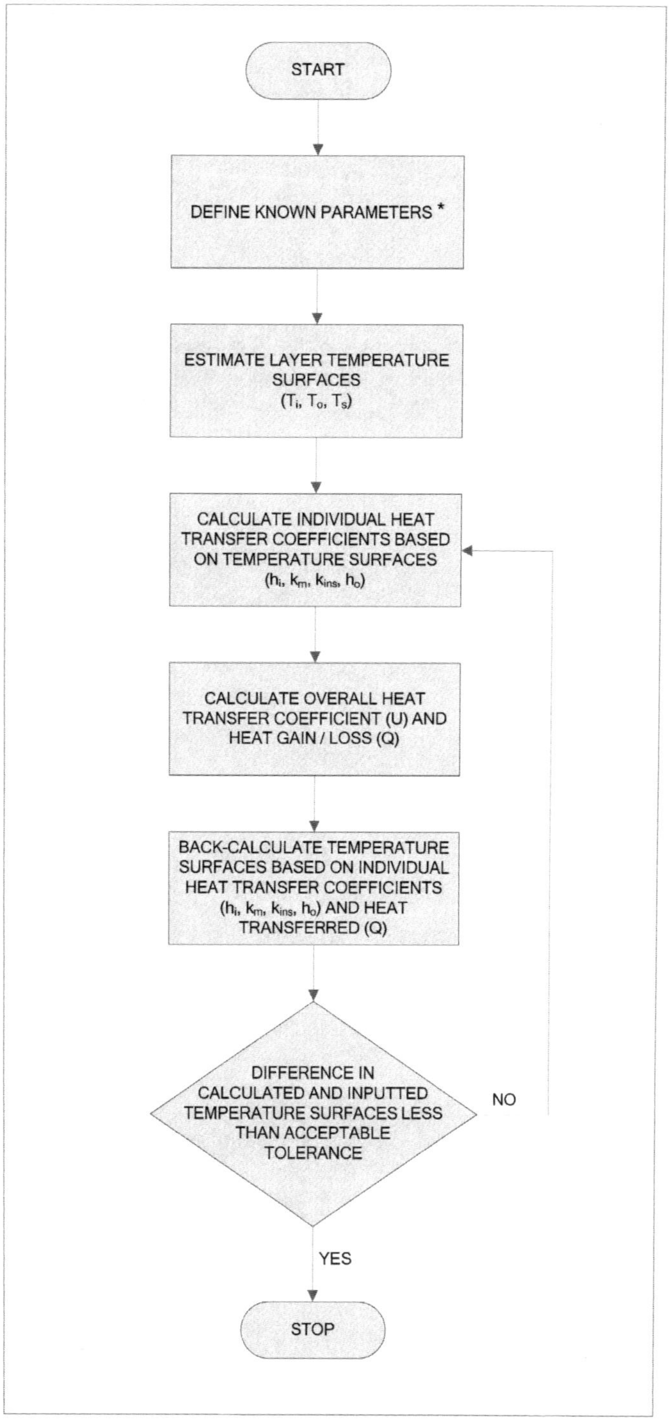

* Parameters may include initial process fluid conditions and properties, ambient temperature, system dimensions, etc.

Figure 2. ASTM C-680 Algorithm Flowchart [1]

2. Case Study Applications

In this paper, the following two case studies are presented:

1. Enhanced algorithm for a time-dependent application
2. Enhanced algorithm to quantify impact of pressure drop in a two-phase stream

2.1 Enhanced Algorithm for a Time-Dependent Application

Case study problem statement (1):

Assuming that a large quantity of low temperature methane liquid relief occurs for a considerably long period of time, the flare header metal piping temperature will drop and eventually reach the fluid temperature. If this relief event is followed by a much warmer vapor relief such as propane, the saturated warm vapor will come in contact with the cold pipe surfaces and lose heat to the piping causing condensation. This creates a safety concern since the warm vapor can thermally contract and create vacuum, resulting in air ingress into the flare header and potential for flammable atmosphere. Based on relief analysis for an LNG plant, the following parameters are adopted for this case study.

Flare header length	5,000 ft
Outside / inside pipe diameter	54 in / 53.25 in
Flare header material	Stainless Steel
Header metal density	488 lb/ft^3
Header metal heat capacity	0.11 Btu/lb.°F
Low temperature methane liquid relief or initial T_m	-259.6°F
Saturated propane vapor relief parameters:	
Relief rate	2.44 x 10^6 lb/hr or 503 MMSCFD
Latent heat of vaporization	183.3 Btu/lb
Inside heat transfer coefficient correlation	Nusselt film-type condensation
Defrost gas parameters:	
Inlet temperature	125°F
Molecular Weight	16.80 lb/lbmole
Inside heat transfer coefficient correlation	Dittus-Boelter

2.1.1 Displacement Gas

One of the methods to protect the flare header from vacuum conditions is to provide a displacement gas, such as fuel gas, immediately following propane vapor relief. To quantify the required displacement rate, the following assumptions are established:

- Ambient air typically contributes to heating of the flare header. For simplicity as well as conservative design purposes, the ambient factor is not considered. Heat transfer occurs only between the metal film and the bulk fluid.
- As time lag between relief of saturated propane vapor and release of displacement gas can vary, the displacement gas rate required is determined based on the maximum heat loss rate along the pipe. The maximum heat loss rate occurs when the first segmental entry of propane vapor relief either is completely condensed or reaches the end of last pipe segment. Note that heat loss and the corresponding condensation rates decrease as a function of time, therefore using the first segmental entry of propane vapor relief is a conservative approach.

Refer to Table 1 for the estimated heat loss rates and the corresponding accumulated condensed propane rates. In general, accuracy of the results can be improved with larger number of segments or in other words, shorter segment length. For complete condensation application, it is essential that the cumulative quantity of condensed fluid at the last condensing segment is equal to the total mass of fluid entering the flare header. In this case study, the optimum number of segments is determined to be 210 segments. As the quantity of condensed fluid increases, the volumetric rate and the velocity decrease, resulting in increased time for the fluid to travel to the next segment.

Table 1. Heat Loss and Accumulated Condensation Rates as Functions of Time and Segment

Heat Loss (MMBtu/hr)								Accumulated Condensation Rate x 10^6 (lb/hr)					
Time (s)	Segment						Sum (MMBtu/hr)	Segment					
	1	10	...	75	85	210		1	10	...	75	85	210
0.1	5.25						5.25	0.03					
1.0	5.21	5.25	...				57.6	0.03	0.29	...			
5.0	5.09	5.11	...				227	0.03	0.28	...			
...	...	...	...	...			...	...	...	...	...		
15	4.97	4.97	...	5.23	...		382	0.03	0.27	...	2.15	...	
20	4.95	4.95	...	5.09	...		405	0.03	0.27	...	2.13	...	
32	4.94	4.94	...	4.98	5.25	--	**422**	0.03	0.27	...	2.11	**2.44**	--

The contraction rate (dV/dt) of saturated propane vapor relief is estimated to be 15.3 million ft^3/hr, which is derived from the total maximum heat loss rate, propane latent heat of vaporization, and difference in propane liquid and vapor densities. The required quantity of displacement gas varies depending on the gas temperature and molecular weight. Based on fuel gas density of 0.055 lb/ft^3 and MW of 16.80 lb/lbmole, 455 MMSCFD of fuel gas is required for this example. Note that this requirement may be larger than the maximum operating fuel gas demand in the plant.

The rapid contraction rate of propane requires the fuel gas system to be able to supply large quantities of displacement gas, which may not lend to an economical design. In addition, the probability of occurrence of cold and warm consecutive reliefs is considerably low.

2.1.2 Purging Gas

Another method to prevent vacuum formation due to cold and warm fluid consecutive relief is to purge the flare header with high temperature gas following the relief of low temperature fluid. The purge rate and time are important parameters to be established while engineering the system.

In this case study, defrost gas at 125°F was assumed to provide purge flow until the end segment of the header approached -50°F. The target temperature of -50°F is based on providing a reasonable margin above the dew point of propane vapor. To solve this problem, the flare header was divided into multiple pipe segments, and all the header segments were assumed to be at the methane liquid relief temperature of -259.6°F. The effect of ambient heat transfer was not considered and heat transfer was assumed to occur only between the metal film and the bulk fluid.

Refer to Figure 3 for the algorithm applied to produce the temperature profiles and estimate the purging time at each defrost gas flowrate run. As an example, piping metal temperature and defrost gas outlet temperature profiles for defrost gas flow of 120,000 lb/hr are shown in Tables 2 and 3, respectively. It can be observed from Tables 2 and 3 that the defrost gas continues to lose temperature to its surroundings while the piping metal temperature is gradually raised until the last segment reaches the target temperature.

Time step is defined as the time required for the purge gas to flow from one pipe segment to another. As the gas temperature drops, the density of the gas increases, thus the velocity decreases and the time step increases. For simplification purpose, time step may be assumed to be constant based on initial, final or average of the expected purge gas temperature, considering that error introduced may only occur at the first pass of the purge gas. Once calculations for first pass are completed, determination of time step is no longer critical. In addition, the number of segments should be further optimized such that increasing the number of segments does not alter the results significantly.

Table 2. Flare Header Piping Metal Temperature Profile

Time Step	Pipe Segment									
	1	2	3	4	5	6	7	8	9	10
0	-259.6	-259.6	-259.6	-259.6	-259.6	-259.6	-259.6	-259.6	-259.6	-259.6
1	-251.6	-259.6	-259.6	-259.6	-259.6	-259.6	-259.6	-259.6	-259.6	-259.6
2	-243.8	-256.2	-259.6	-259.6	-259.6	-259.6	-259.6	-259.6	-259.6	-259.6
3	-236.1	-252.8	-256.1	-259.6	-259.6	-259.6	-259.6	-259.6	-259.6	-259.6
4	-228.5	-249.4	-252.7	-257.9	-259.6	-259.6	-259.6	-259.6	-259.6	-259.6
5	-221.2	-246.0	-249.2	-256.2	-258.7	-259.6	-259.6	-259.6	-259.6	-259.6
6	-213.9	-242.5	-245.6	-254.5	-257.8	-259.1	-259.6	-259.6	-259.6	-259.6
7	-206.8	-239.0	-242.1	-252.8	-256.9	-258.6	-259.3	-259.6	-259.6	-259.6
8	-199.8	-235.5	-238.5	-251.0	-255.9	-258.1	-259.0	-259.4	-259.6	-259.6
9	-193.0	-232.0	-235.0	-249.1	-254.9	-257.5	-258.7	-259.2	-259.5	-259.6
10	-186.3	-228.5	-231.4	-247.3	-253.9	-256.9	-258.3	-259.0	-259.4	-259.5
↓	...	...	...	...	...	...	...	...	...	...
	...	...	...	...	...	...	...	...	...	...
	...	...	...	...	...	...	...	...	...	...
261	123.9	120.1	120.1	111.8	97.6	77.0	50.1	18.4	-16.0	-50.9
262	123.9	120.2	120.2	112.0	98.1	77.6	51.0	19.5	-14.8	-49.7

Table 3. Defrost Gas Outlet Temperature Profile

Time Step	Pipe Segment									
	1	2	3	4	5	6	7	8	9	10
0										
1	-75.3									
2	-71.3	-164.2								
3	-67.3	-160.8	-162.3							
4	-63.4	-157.4	-158.9	-205.7						
5	-59.5	-154.0	-155.5	-203.2	-228.2					
6	-55.8	-150.6	-152.1	-200.8	-226.5	-240.8				
7	-52.1	-147.2	-148.7	-198.4	-224.8	-239.6	-248.1			
8	-48.4	-143.9	-145.3	-195.9	-223.1	-238.4	-247.3	-252.5		
9	-44.9	-140.5	-142.0	-193.5	-221.4	-237.3	-246.5	-251.9	-255.2	
10	-41.4	-137.2	-138.7	-191.0	-219.7	-236.1	-245.7	-251.4	-254.8	-256.8
↓	...	...	...	...	...	...	...	...	...	...
	...	...	...	...	...	...	...	...	...	...
	...	...	...	...	...	...	...	...	...	...
261	124.4	122.0	122.0	116.5	106.2	90.3	68.4	41.3	10.4	-22.4
262	124.4	122.1	122.1	116.6	106.5	90.8	69.1	42.2	11.4	-21.2

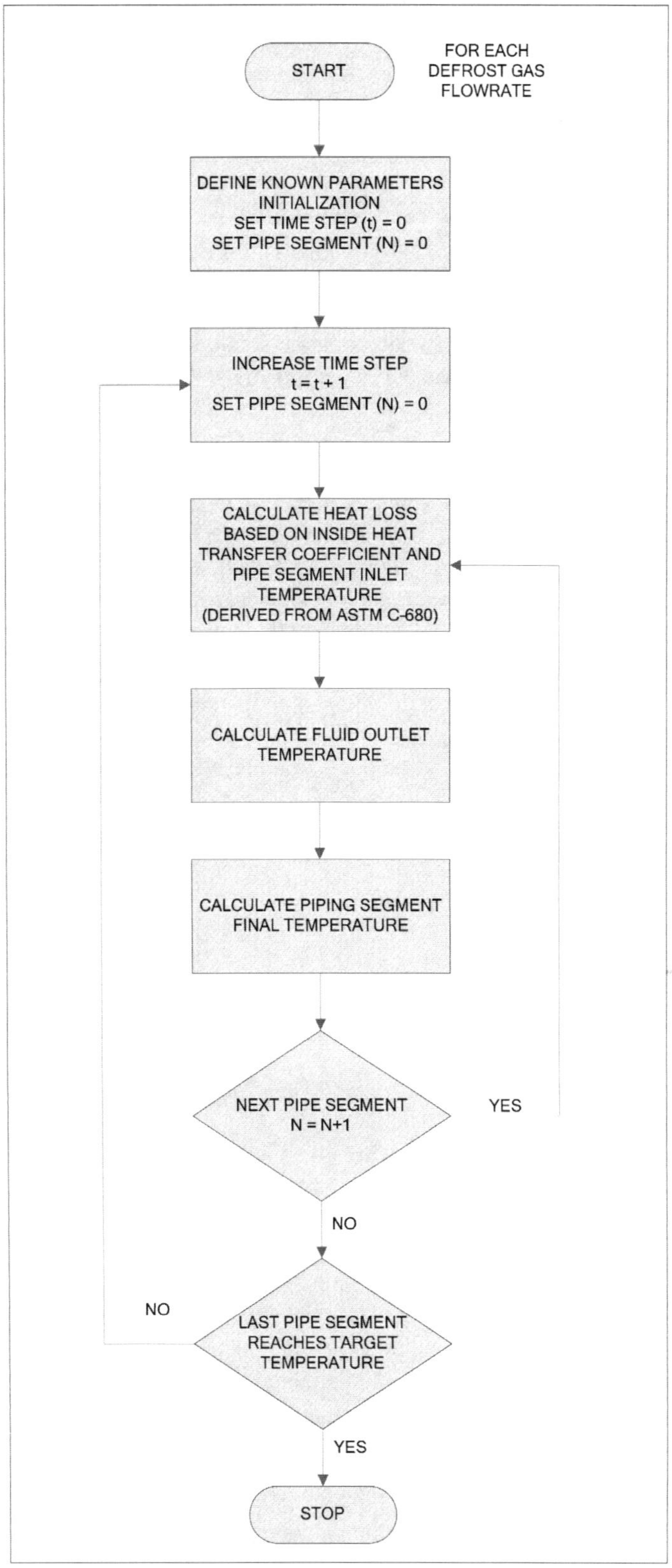

Figure 3. Flare Header Case Study with Modified Algorithm

Results from the algorithm and/or calculation methodology are summarized in Table 4.

Table 4. Purging Time and the Corresponding Purge Gas Flowrate

Gas flowrate, lb/hr	50,000	70,000	100,000	110,000	120,000	250,000	500,000
Gas flowrate, MMSCFD	27.1	37.9	54.2	59.6	65.1	135.5	271
Purging time, min	283.1	203.8	143.2	130.7	119.8	58.4	29.7

2.1.3 Discussions

Based on results of the two methods presented in the previous sections, the following conclusions can be drawn.

- Required displacement gas quantity is a function of the saturated propane vapor relief rate. At higher relief rates, propane vapor may completely condense further down the line. This increases the total heat loss rate and consequently the required displacement gas.
- Displacement gas may be introduced as early as possible to reduce the required quantity and to reduce vacuum formation potential.
- Required purging gas quantity is as a function of flare header length and is independent of the relief rates.
- Purging the flare header certainly lends to an economic design. However, a risk analysis must be performed if this method is selected. At higher purge rates, lesser time is required to warm up the flare header. The temperature profiles generated by the computer model would help with further analysis, especially for events when propane relief starts to occur even before the flare header is fully warmed to its target temperature.

2.2 Enhanced Algorithm to Quantify Impact of Pressure Drop in a Two-Phase Stream

Case study problem statement (2):

Acid gas exiting the amine absorber unit is saturated with water. Acid gas is sent to the H_2S scavenger prior to thermal oxidation. The reduction in pressure and temperature along the pipe causes water vapor condensation. The objective of this case study is to define the containment size required to capture the condensed fluids. The following parameters are adopted for this case study.

Flowrate	100,000 lb/hr
Composition	74.7 mol% CO2, 25.0 mol% H2O, 0.3 mol% hydrocarbon light ends with traces of H_2S and benzene
Inlet pressure and temperature	15 psig and 179.44^{0}F
Length of pipe	1,000 ft
Outside / inside pipe diameter	18 in / 17.25 in (uninsulated)
Pipe roughness	0.0002 ft
Wind speed	10 mph
Ambient temperature	40^{0}F
Inside heat transfer coefficient correlation	Nusselt film-type condensation

The modified ASTM C-680 algorithm for this case study is presented on Figure 4. For comparison purposes, results from a steady state model are compared with pseudo steady state results in Table 5.

Table 5. Comparison Methods for Cumulative Condensed Fluid at End of Pipe Segment

		End of Pipe Segment							
	Unit	1	2	3	...	8	9	10	% difference
Steady State	lb/hr	--	--	--	--	--	--	**2,486**	52.14
No ΔP correction	lb/hr	176.7	352.2	526.6	...	1,382	1,550	**1,717**	5.05 (Note 1)
ΔP correction	lb/hr	170.2	338.9	506.0	...	1,319	1,477	**1,634**	--
HTFS Hysys (Note 2)	lb/hr	171.4	341.3	509.9	...	1,332	1,493	**1,652**	1.08

Notes:

1. The percentage difference between pseudo steady state pressure drop and non-pressure drop corrected results varies with each case. Therefore, the percentage difference shown in Table 5 should not be used as an underlying basis for the accuracy of the two methods.
2. HTFS, the inside heat transfer coefficient correlation available in HYSYS for two-phase applications, was run for this case study.

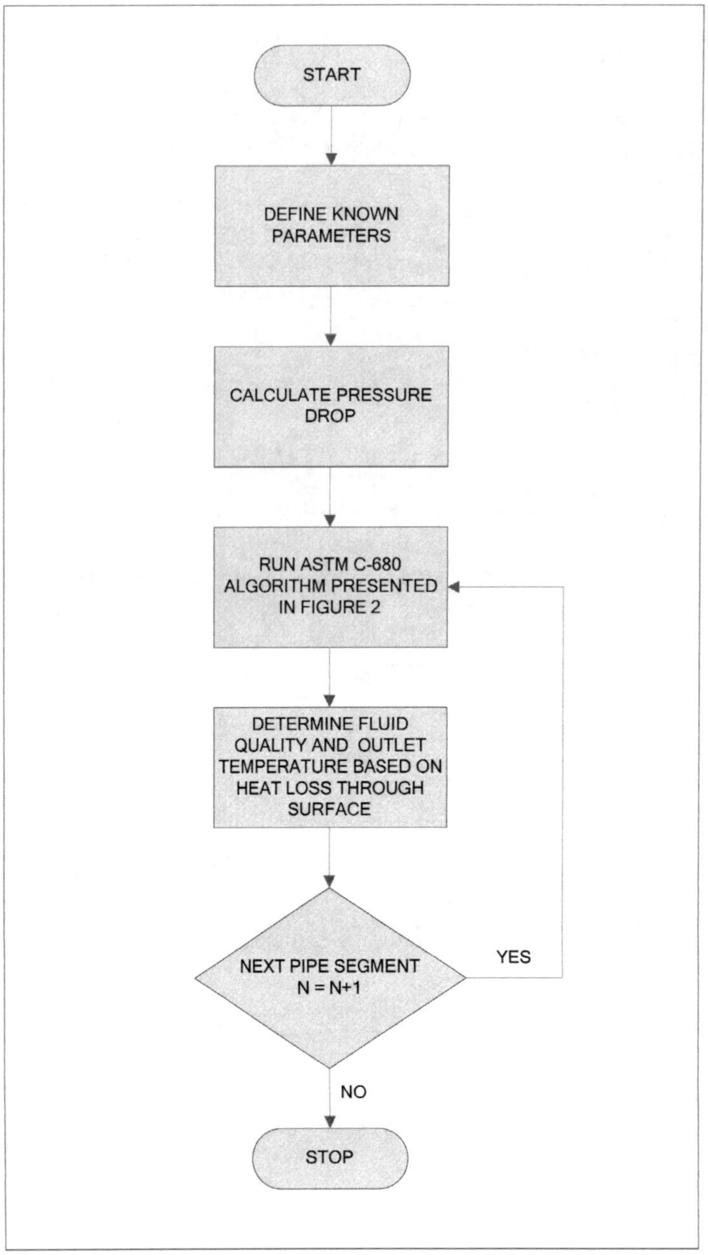

Figure 4. Modified Algorithm with Pressure Drop Factor

3. Conclusions

Based on the two case studies outlined above, it is demonstrated that ASTM C-680 algorithm can be modified to solve iterative heat transfer calculations in a piping system for various applications, such as for a time-dependent application and for a system requiring continuous update to its equilibrium state.

4. Acknowledgement

The authors would like to thank Vinod Rajkumar for his continuous support and constructive inputs. The authors would also like to thank Terry Timmerman for reviewing the paper.

5. References

1. ASTM C680, Standard Practice for Estimate of the Heat Gain or Loss and the Surface Temperatures of Insulated Flat, Cylindrical, and Spherical Systems by Use of Computer Programs, 1971 - 2010

2. Bornt, B., Spreadsheets for Heat Loss Rates and Temperatures, 1995

3. Goudar, C.T., Comparison of the Iterative Approximations of the Colebrook-White Equation, 2008

4. Konak, A.R., Calculate the Heat Loss from Pipes, 1993

5. Lindeburg, M. R., Chemical Engineering Reference Manual, 2004

6. Nayyar, M.L., Piping Handboook, 1999

7. Perry, R. H., Perry's Chemical Engineering Handbook, 2007

6. Biography

Debby Sielegar, P.E. is a licensed engineer in the state of Texas. She completed her bachelor's degree in Chemical Engineering from Purdue University in 2007 prior to joining Bechtel OG&C, Inc. While working for Bechtel as a process engineer, she pursued her Master's degree in Chemical Engineering from University of Houston and completed the degree in 2011. In Bechtel, she has satisfactorily completed assignments on various projects such as lube oil refinery, LNG liquefaction and offshore crude production. In addition, she is also a certified six sigma yellow belt.

David B Messersmith, P.E. is Manager of the LNG Technology Group – for Bechtel Corporation in Houston, Texas. He has served in various lead roles in the LNG Projects for the 15 of the last 21 years. This work included the Atlantic LNG Project conceptual design through start-up as well as many other LNG studies, FEEDs and projects. Experience includes 11 years with Kellogg and 21 years with Bechtel in various LNG and Ethylene assignments. His LNG experience includes everything from conceptual design through commissioning, start-up and testing. He holds a Bachelor's of Science degree in Chemical Engineering from Carnegie-Mellon University and is a registered Professional Engineer in the state of Texas.

Theoretical Study of Expansion Foam Application on LNG Pool Using Computational Fluid Dynamics

Bin Zhang, Xinrui Li, Ammar Alkhawaldeh and M. Sam Mannan, Mary Kay O'Connor Process Safety Center, Artie McFerrin Department of Chemical Engineering, Texas A&M Univerisity, College Station, TX

LNG facilities could harm the personnel on site and jeopardize local communities nearby, due to the flammability and dense gas behaviors of LNG. Nowadays, the growing concern with chemical process safety leads to the timely establishment of regulations and standards regarding to consequence mitigation measures, for instance, NFPA 59 A requires that LNG storage tanks should have a dike or impoundment around to constrain the LNG spill and form a LNG pool, and the expansion foam has been one of the proposed safety provisions to minimize the consequence of LNG spill on the ground. However, because of the high cost and potential risk to people of performing field tests to evaluate effectiveness of foam application on LNG pool, just a few of experiments have been conducted, whose results vary as a result of uncertainties in experimental conditions.

In this research, computational fluid dynamics modeling was employed to model LNG spill scenarios on the ground with a dike, and mainly focused on the effect of expansion foam on the LNG vapor dispersion. The effectiveness was evaluated by studying several key parameters in the LNG spill and dispersion scenario. The simulation results were compared with medium-scale LNG spill field test results conducted by Mary Kay O'Connor Process Safety Center (MKOPSC) for validation. Moreover, a sensitivity analysis was conducted to a number of essential parameters of foam application. The outcomes of the analysis were used to guide the design of foam mitigation system and application method, which hopefully could be integrated into relevant regulations and standards for the sake of safer practice in the industry.

Guidelines for Jetting and Flashing LNG Vapor Exclusion Zone Analysis

Delmar "Trey" Morrison, Ph.D., P.E.
Exponent, Inc.
4580 Weaver Parkway, Suite 100, Warrenville, Illinois 60555
tmorrison@exponent.com

Ryan J. Hart, Ph.D.
Exponent, Inc.
4580 Weaver Parkway, Suite 100, Warrenville, Illinois 60555
rhart@exponent.com

Harri Kytömaa, Ph.D., P.E.
Exponent, Inc.
9 Strathmore Road, Natick, Massachusetts 01760
hkytomaa@exponent.com

Prepared for Presentation at
2012 AIChE Spring Meeting
Houston, TX
April 1-4, 2012

UNPUBLISHED

Guidelines for Jetting and Flashing LNG Vapor Exclusion Zone Analysis

Delmar "Trey" Morrison, Ryan J. Hart, and Harri Kytomaa
Exponent, Inc.
E-mail: tmorrison@exponent.com

Keywords: LNG Jetting and Flashing; LNG Release

Abstract

Liquefied Natural Gas (LNG) has been a growing energy resource in the United States, spurring significant import and storage infrastructure development over the past decade. The infrastructure for liquefaction and export has also been developing as a result of recent shale gas finds in North America. In the U.S., the construction and operation of LNG facilities are regulated by the Pipeline and Hazardous Materials Safety Administration (PHMSA) of the U.S. Department of Transportation (DOT), and the regulations are administered by the Federal Energy Regulatory Commission (FERC). As part of the regulatory process (Title 49, Code of Federal Regulations, Part 193.2059), regasification and liquefaction facility owners are required to perform prescriptive calculations to estimate the potential consequences of major releases of LNG prior to permitting. One of the criteria is that the ½-LEL boundary of resulting flammable vapor clouds cannot cross the property boundaries (exclusion zone). The consequence analysis requirements have recently been broadened to include new spill scenarios in addition to the previously required LNG spills into impoundment or "sumps" and to allow the use of alternative computational methods to calculate the ½-LEL exclusion zones for these scenarios. Specifically, in 2010 FERC issued a written interpretation regarding the flammable vapor exclusion zone requirements requiring analysis of "jetting and flashing" phenomena accompanying a pressurized release of cryogenic LNG, as well as spills into containment trenches and troughs. This paper focuses on the new requirement for the analysis of jetting and flashing events.

Jetting and flashing events can be modeled with relative ease by using integral models; however, the selection of appropriate scenarios can add complexity to the analysis. This paper presents typical scenarios that are required for this analysis, including high pressure flange and valve packing failure scenarios. We then parametrically present the consequence studies of a broad range of LNG jetting and flashing release scenarios to provide facility owners with a basis for selecting appropriate scenarios and estimates of the consequences of such releases. Finally, we review the benefits and limitations of the simplifying assumptions associated with integral models. The data presented in the paper can be used by facility owners as a preliminary aid in the decision making process for facility upgrades and modifications.

1 Introduction

Liquefied Natural Gas (LNG) is a primary energy resource in the U.S. and internationally. LNG has a long history of international import/export via ship tanker operations. As of January, 2012, the Federal Energy Regulatory Commission (FERC) listed thirteen existing U.S. import terminals, see Figure 1 [1]. Three of these terminals are currently authorized to re-export delivered LNG (Freeport LNG, Sabine Pass LNG, and Cameron LNG). From reviewing the Federal Dockets and industry news, several terminals are considering or actively pursuing approval from FERC to construct facilities to liquefy and export natural gas. The relatively recent non-conventional gas supply growth in North America has dropped the price of natural gas in the U.S., thus making export to other regions of interest to many terminals.

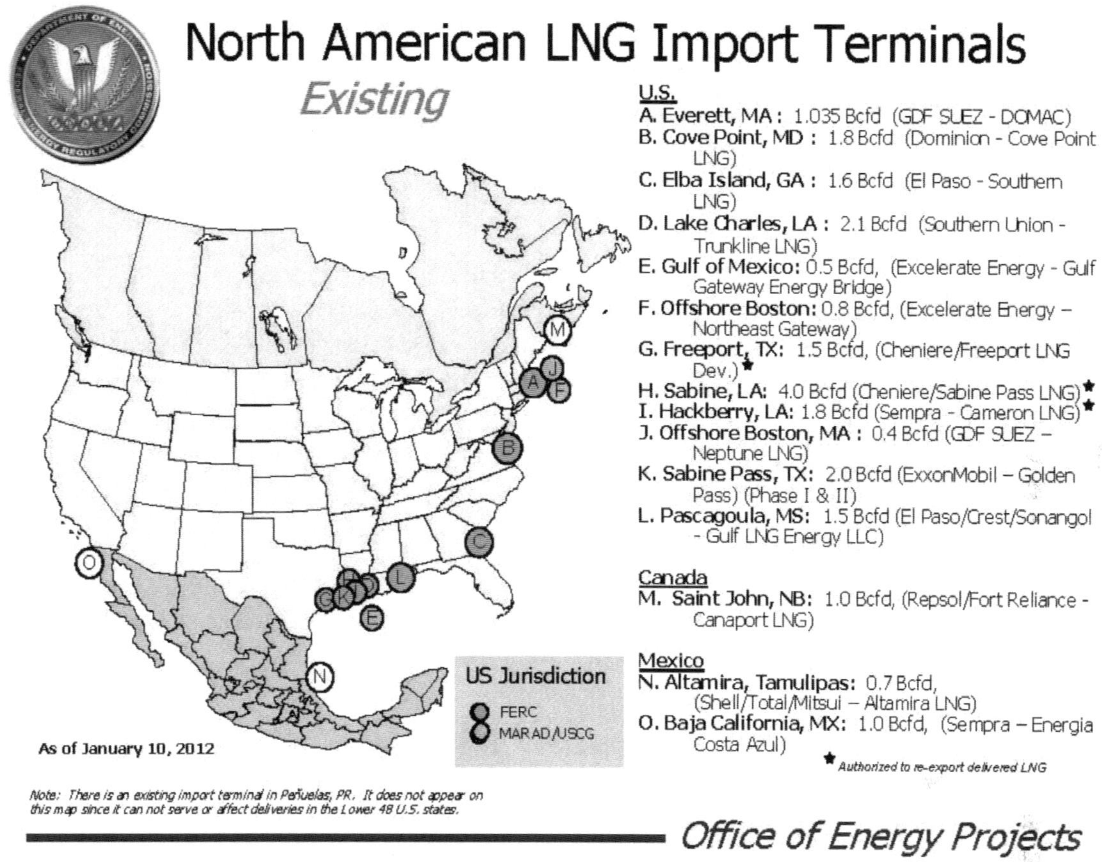

Figure 1. Existing North American LNG import terminals as of Jan. 2012.

In the U.S., the construction and operation of LNG facilities are regulated by the Pipeline and Hazardous Materials Safety Administration (PHMSA) of the U.S. Department of Transportation (DOT), and the regulations are administered by FERC. As part of the regulatory process (Title 49, Code of Federal Regulations, Part 193.2059), regasification and liquefaction facility owners are required to evaluate the potential consequences of major releases of LNG. This analysis may dictate equipment siting changes or design modifications within the proposed project. The regulations require a flammable vapor exclusion zone for flammable vapor clouds that falls within the property boundaries. The flammable range for methane is 5% by volume at the Lower Flammable Limit (LFL) and 16.5% by volume at the Upper Flammable Limit (UFL). The concentration of flammable vapors must remain below 50% of the LFL (i.e., 2.5% methane by volume) at the property boundary according to the regulations.

Modeling of the flammable vapor dispersion from a spill or release is used to demonstrate that the flammable vapor exclusion zone is maintained within the property boundaries that can be built upon (e.g., neglecting clouds over bodies of water). Only a few numerical models have been approved by PHMSA for this type of analysis. This paper applies one of those approved models [2], Process Hazard Analysis Software Tool (PHAST) Version 6.7, to a subset of release scenarios: high pressure jetting and flashing. Jetting and flashing events can be modeled with relative ease by using integral models; however, the selection of appropriate scenarios can add complexity to the analysis. In this paper, we present typical scenarios that are required for this analysis, including high pressure flange and valve packing failure scenarios. We then parametrically present the consequence studies of a broad range of LNG jetting and flashing release scenarios to provide facility owners with a basis for selecting appropriate scenarios and estimates of the consequences of such releases. Finally, we review the benefits and limitations of the simplifying assumptions associated with integral models. The data presented in the paper can be used by facility owners as a preliminary aid in the decision making process for facility upgrades and modifications.

2 Background

The regulatory requirements for consequence analysis have recently been broadened by FERC to include new spill scenarios in addition to the previously required LNG spills into impoundments. Specifically, in 2010 PHMSA issued a written interpretation to Downeast LNG regarding the flammable vapor-gas exclusion zone requirements calling for analysis of "jetting and flashing" phenomena accompanying a pressurized release of LNG [3]. Subsequently, in late 2011, FERC provided additional guidance on requirements for evaluating the potential for jetting and flashing of LNG releases in an Engineering Information Request to Sabine Pass LNG [4].

In an LNG terminal, LNG may be present in piping systems at high pressure in the vaporizer area where the pressure has been increased to meet pipeline pressure (e.g., on the order of 1000 psig) after vaporization and in the liquefaction area where natural gas has been compressed as part of the refrigeration and liquefaction process.

Saraf, et al, analyzed the jetting and flashing problem for LNG releases [5]. They analyzed unconfined jet releases from a 2-inch diameter orifice at 100 psig and 1000 psig source pressure with two-phase flashing flow of LNG. They found that immediate ignition of these releases could create jet fires of approximately 50-ft and 150-ft, respectively. Gavelli, et al, identified potential limitations of applying integral models to non-circular jet releases and in congested plant areas where obstacles may influence vapor cloud dispersion [6]. Computational fluid dynamics (CFD) codes are better-suited to model complex scenarios such as these, but with more refined models the level of conservativeness drops with modeling efforts.

2.1 Jetting and Flashing Basis by PHMSA/FERC

In the PHMSA July 16, 2010 letter of interpretation, the regulators identified jetting and flashing of LNG as a specific concern for modeling - in addition to liquid spills. Because LNG is a cryogenic liquid, low pressure/low momentum releases have historically been treated as liquid spills. PHMSA identified the concern that with a high pressure high momentum release, a jet of LNG may be propelled beyond the impoundment system (i.e., sprayed over trenches, dike walls, or other containment means). A jetting release of LNG can also result in mechanical fragmentation of the fluid which can then lead to aerosolization and conveyance of the fluid beyond the impoundment system. Concurrently, the flashing (rapid or instantaneous vaporization) of discharged LNG will contribute to fragmentation and aerosolization of the LNG; again, leading to potential conveyance of the fluid beyond the impoundment. The concern of certain release scenarios potentially leading to discharged material superseding impoundments has resulted in FERC and PHMSA to require exclusion zone analyses that consider possible jetting and flashing scenarios.

To guide in selecting appropriate potential jetting and flashing scenarios, FERC and PHMSA have identified low probability/high consequence releases as a conservative target for evaluation. The FERC staff has examined typical failure rates for piping and determined the following single accidental leakage sources are necessary to show compliance with 49 CFR Part 193:

a. For piping less than or equal to 6 inches in diameter, a full pipe rupture at any point along the line; and
b. For piping greater than 6 inches in diameter, the greater of:
    i. a full rupture of a connection less than or equal to 6 inches in diameter; or
    ii. a hole of 2 inches in diameter.

PHMSA has concurred with these criteria, and has requested that applicants provide a revised vapor dispersion analysis, including both flashing and jetting, based on single accidental leakage sources which meet the above criteria and yields the most significant vapor dispersion zone with respect to the plant property lines. Furthermore, the regulators have suggested that the analysis also consider release location, elevation, orientation, pressure, temperature, and flow rate.

2.2 Jet & Flash PHAST model

Physics of a flashing release – A pressurized flashing release of LNG from an orifice is a complex process that may involve at least the following sub-processes [7]:

- Droplet atomization
- Atmospheric expansion
- Two-phase dispersion
- Liquid rainout
- Pool formation
- Pool evaporation

PHAST Version 6.7 is commercial software developed and distributed by Det Norske Veritas (DNV). PHAST is widely used for the calculation of hazard distances from the release of several hazardous substances, including LNG. The PHAST code uses the Unified Dispersion Model (UDM), an integral model, to calculate the dispersion following a pressurized release. PHAST contains the following sub-models that are linked in the software to calculate complex release phenomena:

- Near-field jet dispersion;
- Non-equilibrium droplet evaporation and rainout, touchdown;
- Pool spread and vaporization;
- Heavy gas dispersion; and
- Far field passive dispersion.

The UDM allows for continuous, instantaneous, constant finite-duration and general time-varying releases. Additionally, the UDM allows for possible plume lift-off when a grounded plume becomes buoyant. The UDM has been validated extensively with experimental data and is the subject of several peer-reviewed scientific papers [8]. PHAST has received approval for use in evaluating jetting and flashing release scenarios necessary to comply with the FERC/PHMSA jetting and flashing requirements. However, the approval did specify limitations of using the PHAST-UDM for evaluating jetting and flashing dispersion [2]:

- the PHAST UDM is limited to circular LNG pools and impoundments with low-aspect ratios;
- the PHAST UDM is appropriate only for releases along the downwind direction;
- the PHAST UDM may not be used for scenarios with multiple coincident releases that may influence one another;
- and, in some cases, PHAST UDM may not be appropriate for the analysis of dispersion over varying or sloped terrain and instances where large obstructions are present.

Furthermore, it was decided that a recommended safety factor of 2 be applied to the LFL in exclusion zone analysis (i.e. the ½ LFL concentration will be used) in order to account for uncertainties that may arise from potential turbulence, source term specifications, dispersion over water, and the uncertainties associated with the experimental studies submitted for validation.

The regulators did determine that pure methane may be suitably used as a conservative substitute for LNG composition. Additionally, they determined that the guidance provided in 49 CFR 193.2059, concerning the ambient conditions to be used in the model, were appropriately conservative. The atmospheric modeling guidelines provided by 49 CFR 193 (and NFPA 59A) include [9, 10]:

- Pasquill-Gifford atmospheric stability: Category F
- Surface roughness factor of 0.03 m
- 50% humidity
- Average ambient air temperature for the region
- Ground surface temperature, assumed to equal ambient temperature
- Wind speeds that are exceeded less than 10% of the time, or 4.5 mph (2 m/s), whichever results in the longest predictable downwind dispersion distance

3 Parametric Analysis

In order to provide guidelines for rapidly screening possible LNG Jetting and Flashing scenarios, we performed parametric analyses on input parameters using PHAST Version 6.7. The parameters adjusted in this study include: source pressure, source temperature, leak (orifice) diameter, release elevation, release angle, and wind speed. To be conservative, the properties of pure methane were used in the model since it will have a slightly higher vaporization rate than LNG, which is a mixture of hydrocarbons. A summary of the range of parameters and values examined in this study is presented in Table 1.

Table 1. The parameters varied during the LNG Jetting and Flashing sensitivity analyses and the range of values examined in this study.

Source Pressure		Source Temperature		Leak Diameter		Release Elevation		Wind Speed	
(psig)	(bar)	(°F)	(°C)	(in)	(cm)	(ft)	(m)	(ft/s)	(m/s)
50	3.45	-265	-165	1	2.54	1.64	0.5	3.28	1
100	6.89	-256	-160	**2**	**5.08**	6.56	2	**6.56**	**2**
250	17.2	**-247**	**-155**	3	7.63	**16.4**	**5**	9.84	3
500	34.5	-229	-145	4	10.2	32.8	10	13.1	4
750	51.7	-211	-135	5	12.7	65.6	20	16.4	5
1000	68.9			6	15.2			23.0	7
1250	**86.2**							32.8	10
1500	103								

The values emphasized in Table 1 represent the "base case" model used in this study, for which all sensitivity parameters are deviated. The base case was chosen as being reasonably representative of a typical LNG Jetting and Flashing scenario. The remaining model boundary conditions are listed in Table 2.

Table 2. Baseline model boundary conditions.

Averaging Time	18.75
Orifice Discharge Coefficient	0.6
Pasquill Stability	F
Surface Roughness Length (mm)	30
Surface Roughness Parameter	0.068857
Atmospheric Temperature (°F)	68
Surface Temperature (°F)	68
Relative Humidity Fraction	0.7

The leak diameters were varied from 1 to 6 inches. The PHAST program was allowed to calculate the mass flow rate through each orifice size as a function of pressure. Figure 2 graphs the mass flow rate versus pressure for each orifice size. For higher pressures and larger orifices, the flows depicted may be unrealizable or unsustainable at a terminal if the pumping capacity is insufficient to maintain the source pressure. In these cases, the loss of backpressure on the pump system may lead to rapid depressurization of the system followed by low pressure high mass flow of LNG. This situation may require more detailed analysis than that presented in this paper. The results presented in this paper do not include these effects on jetting and flashing.

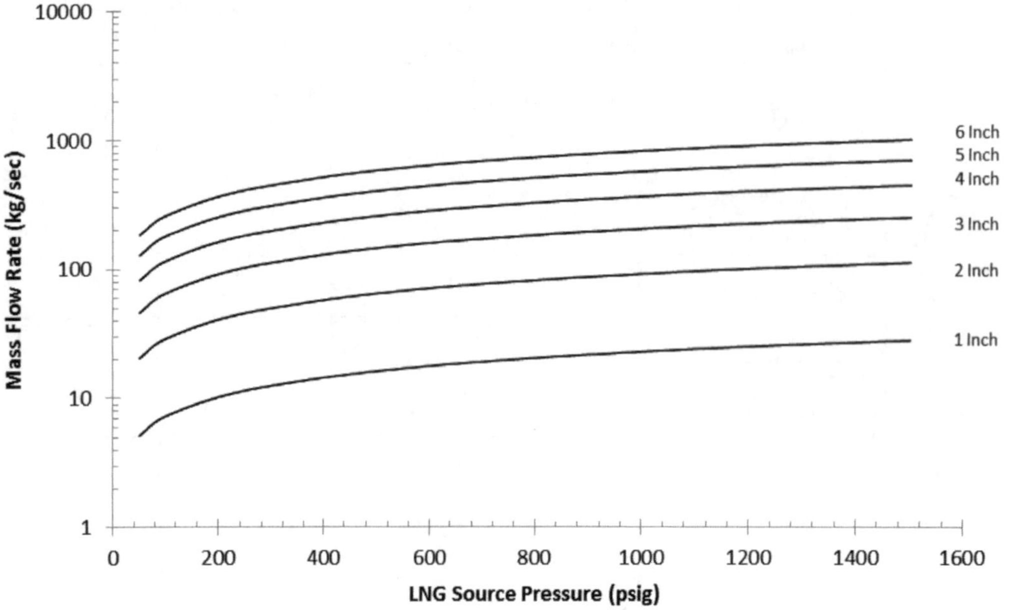

Figure 2. LNG mass flow rate through example orifice sizes at a source temperature of -247°F.

3.1 Droplet Formation

During a release of LNG from a pressurized source, the liquid jet will be fractured into a spray of droplets. PHAST uses built in correlations for calculating droplet diameter due to aerodynamic (mechanical) breakup and flashing breakup. For the leak model employed here, the program calculated the drop size according to both mechanisms and used the smaller of the two. From inspection of the data (see Figure 3), obvious trends in the droplet diameter were not apparent, which is a consequence of the program switching between mechanical and flashing breakup correlations. Droplet size was not affected by orifice size; it proved to only be a function of pressure and temperature in the models. Liquid rainout is affected by the rate of evaporation of droplets, which is related to release elevation, droplet size, and ambient temperature. The liquid rainout fraction is listed in the figure as x_L for the three cases where it occurred. For high values of rainout, collection and conveyance of liquid may be necessary; thus, simple jetting and flashing models such as that employed in PHAST may not be appropriate.

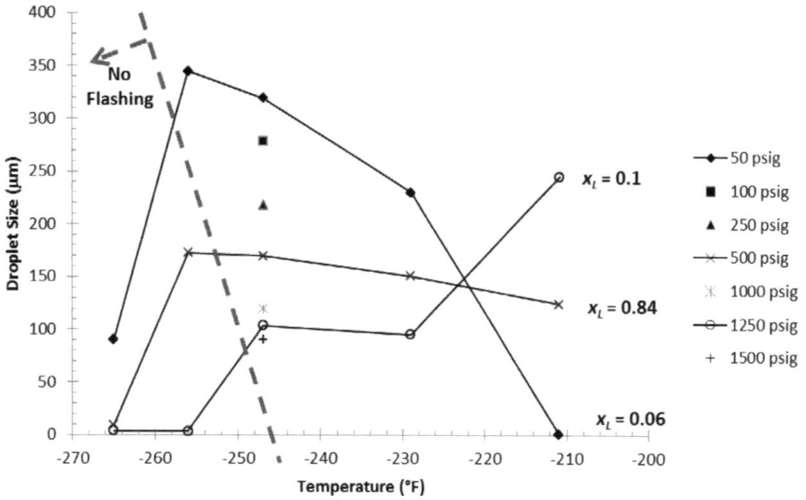

Figure 3. Methane droplet diameter versus temperature.

3.2 Variation of Orifice Size with Pressure

A parametric analysis was performed to vary leak orifice size and pressure within the ranges listed in Table 1. These releases were modeled at a height of 5 m and source temperature of -247°F. Figure 4 provides ½ LFL plume contours at 0.5 m elevation for a subset of the cases shown as the maximum steady dispersion plume. For the lower pressure, lower mass flow rate releases, the plumes demonstrate a wider aspect, thus potentially a larger flammable mass. In terms of the hazard posed by the vapor cloud, the measurable target is the boundary of the facility; thus, the flammable mass and overall size of the cloud is less important than the length. The representative centerline side view contours for the clouds shown in Figure 5 indicate that higher pressure releases create taller clouds due to the expanding jet. Figure 6 presents a summary of downwind dispersion distance for the ½ LFL boundaries for a range of orifice sizes at increasing pressure, fixed source temperature, and the baseline model/weather conditions listed in Table 2. These maximum distances are reported at an elevation of 0 m, indicating that the clouds remain close to the ground. An interesting feature of the result is the initial hump in dispersion distance at 50 psig that tails-off and appears to be independent of the source pressure. The longer distances for the 50 psig release pressure with leak diameters of 4, 5, and 6 inches is related to liquid rainout that occurs for these scenarios. The pool formation is followed by evaporation that results in cold dense vapors close to the ground.

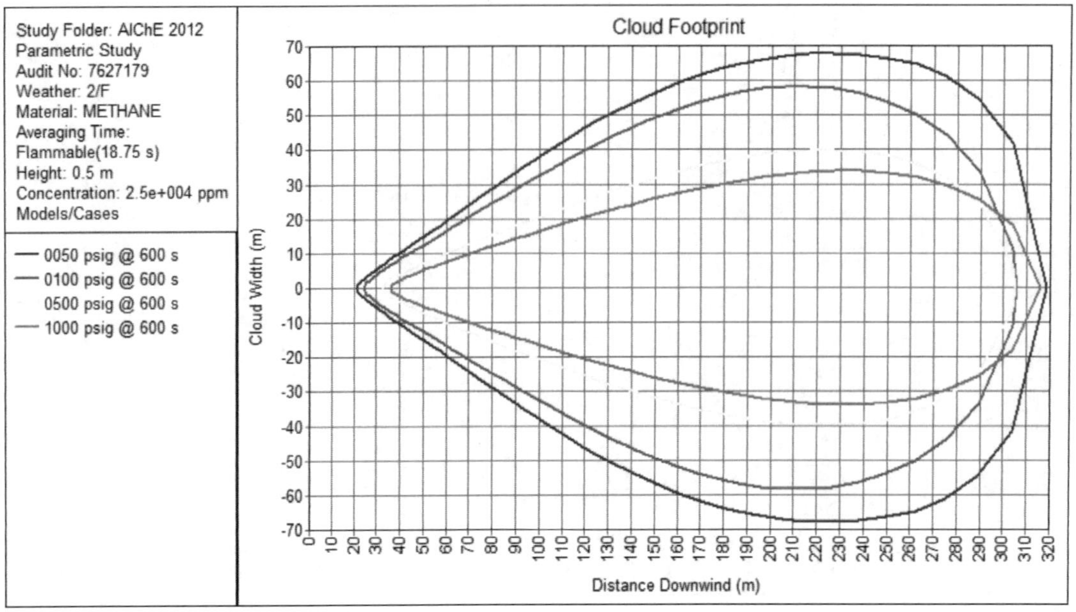

Figure 4. **Representative plume contours for 2-Inch orifice at 5 m elevation and -247°F.**

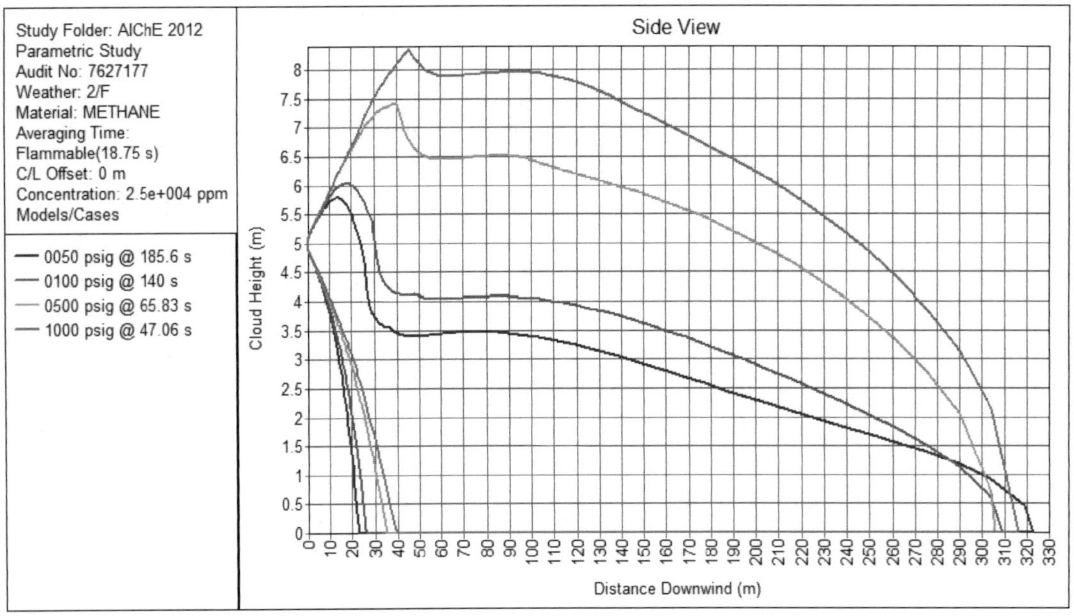

Figure 5. **Representative centerline side view contours for 2-Inch orifice at 5 m elevation and -247°F.**

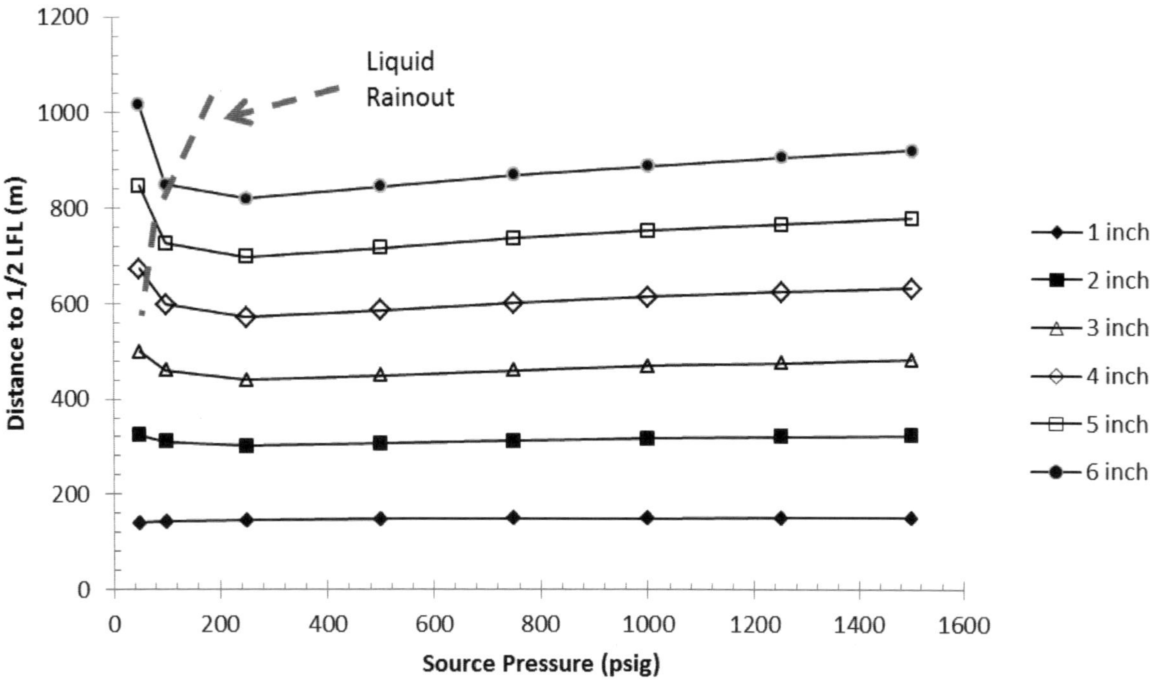

Figure 6. Dispersion distance to ½ LFL at -247°F and baseline weather conditions.

3.3 Effect of Release Elevation

The elevation above ground level for a release, in addition to the other release parameters, will greatly affect the size and length of the vapor cloud. Since LNG vapors are colder and denser than ambient air, low elevation releases may be inferred to create longer cloud lengths than high elevation releases. To test this inference, a series of elevations were evaluated for a range of source pressures at -247°F. The elevations spanned the range from 0.5 m (e.g., ground level piping), 5 m (e.g., low pipe racks), to 20 m (e.g., storage tank roof piping). The results of this parametric analysis were unusual. From Figure 7, for all pressures the downwind dispersion distance decreases with increasing elevation. However, there is slightly different behavior observed for the 100 psig case at the lower elevation release; the behavior then matches the trends observed with the other release pressures for elevations greater than 5 m.

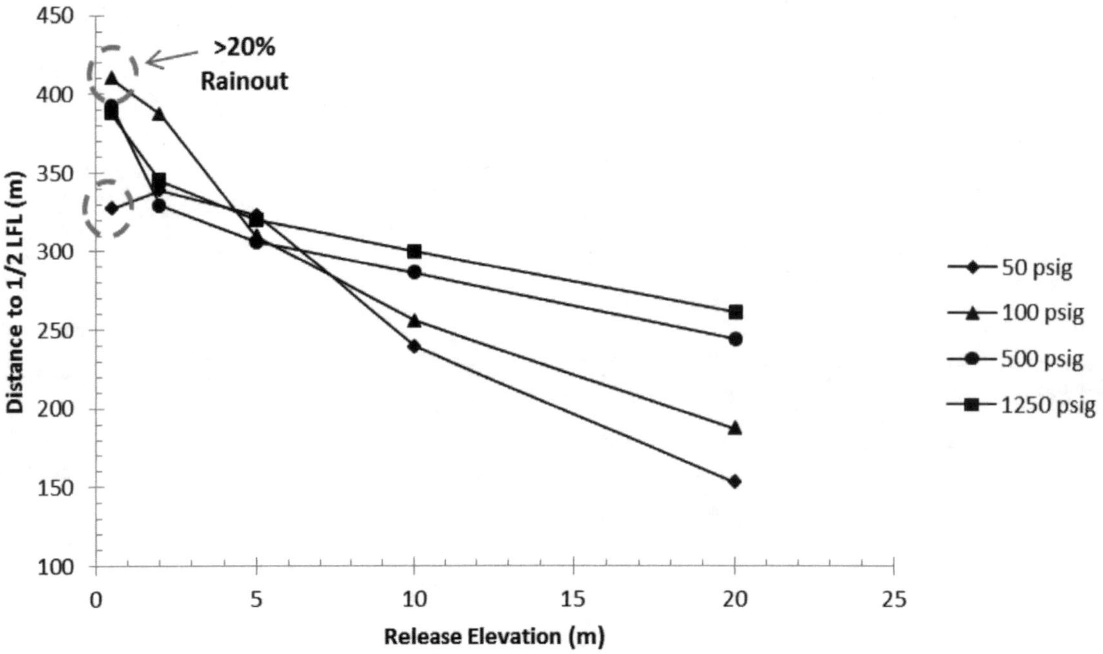

Figure 7. Trends in ½ LFL cloud downwind dispersion distance for a range of source pressures versus source elevation.

3.4 Effect of Release Angle

Three constant source pressure cases (50 psig, 500 psig, and 1000 psig) were considered to evaluate the effects of release angle on downwind dispersion. Four release angles were evaluated: down, horizontal, 45 degrees from horizontal, and vertical upwards. These are sketched in Figure 8 below. The example discussed here is for the following base model conditions: a source temperature of -247°F, at 5 m elevation, with 2-inch orifice, 2 m/sec wind speed, and F stability class. The relation between the maximum dispersion distance for the ½ LFL vapor cloud are consistent with experience and intuition for cold, dense vapor clouds (see Figure 9). For upward directed releases, complete evaporation occurs for all pressure cases with no liquid rainout predicted. Air entrainment and dilution subsequently prevents touchdown of the ½ LFL vapor cloud and diminishes the downwind dispersion distance (see Figure 10). The horizontal releases also don't predict any rainout at this temperature; however, the cold, dense cloud remains close to the ground and travels quite a distance. The downward oriented releases all produce significant rainout and liquid pooling. No bund wall was implemented in the model; consequently, the pool sizes grew to a size constrained by the vaporization rate. Due to the significant rainout, the downward facing jets do not meet the jetting and flashing criteria advanced by PHMSA/FERC. Instead, the effect of trenches, troughs, and impoundments must be considered on these liquid spills, vapor generation, and vapor dispersion. Depending upon the conveyance and impoundment design at the release location, PHAST may not be the best model for this analysis.

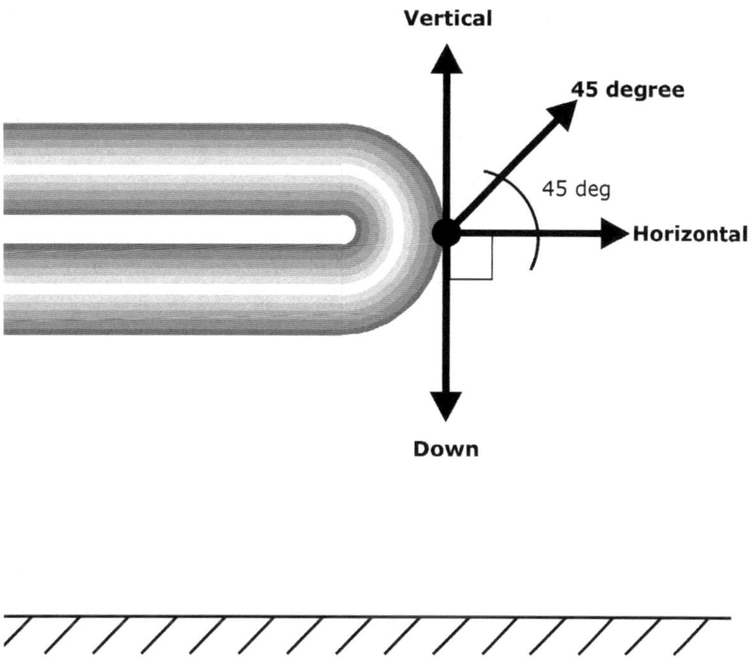

Figure 8. Release directions evaluated in the model.

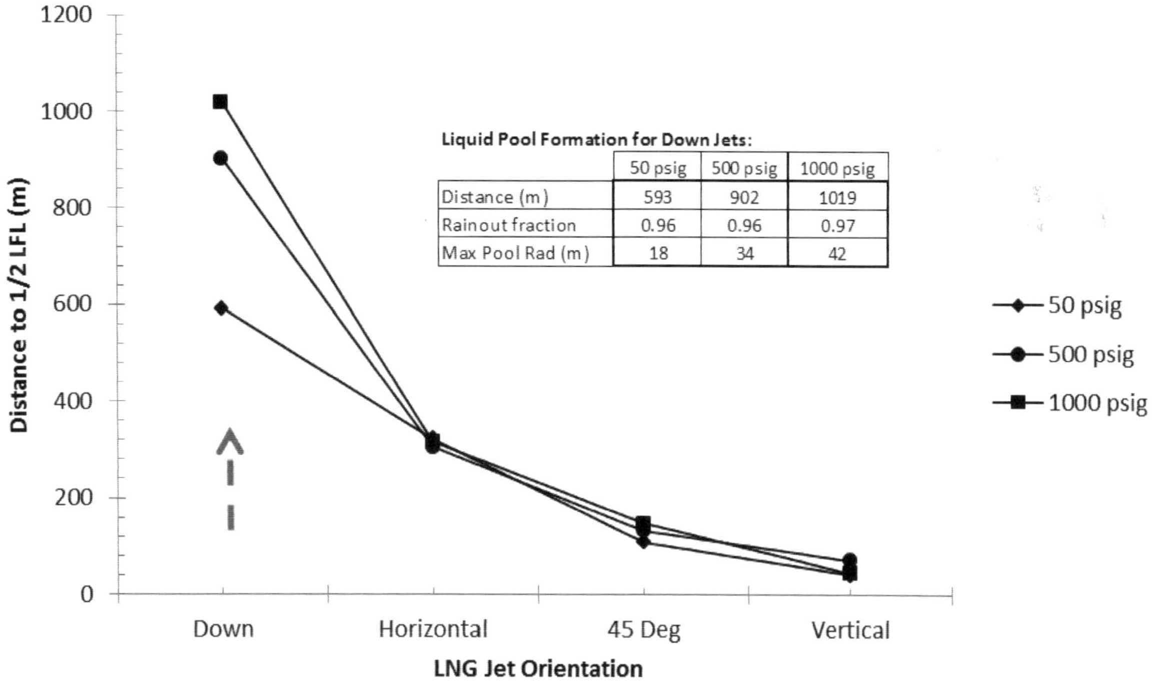

Figure 9. Maximum downwind dispersion distance to ½ LFL for varying release angle and source pressure.

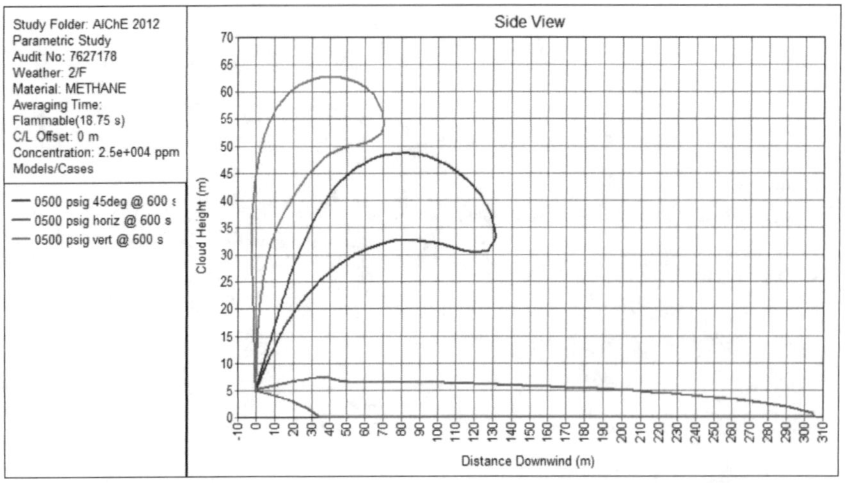

Figure 10. Comparison of side views for horizontal, 45 degree, and vertical releases at 500 psig.

3.5 Effect of Wind Speed

During a release, the vapor cloud dispersion will be influenced by the wind speed and stability class. The wind speed was evaluated at 2 m/sec and F stability class for most of the parametric studies. The regulations provide a choice between these conditions and those prevalent at the site; however, FERC has recently required a wind sensitivity analysis to demonstrate that dispersion results are reported with conservative wind conditions. Higher wind speeds can conceivably disperse a cloud faster than low wind speeds thus decrease the ½ LFL boundary distance. The wind conditions were varied from 1 m/sec to 4 m/sec for releases at -274°F to demonstrate the downwind dispersion distance versus source pressure. The results are depicted in Figure 11 for a 2-inch diameter orifice and in Figure 12 for a 4-inch diameter orifice. These results demonstrate an unusual trend. At a wind speed of 2 m/sec, the dispersion distance drops sharply to reach a minimum around 250 psig source pressure. At the 3 and 4 m/s wind speeds examined, the opposite behavior occurs with a maximum dispersion distance at this source pressure. A smoothly increasing function versus pressure at the low wind speed of 1 m/sec was also observed. All of the results appear to be converging upon a maximum dispersion distance versus pressure.

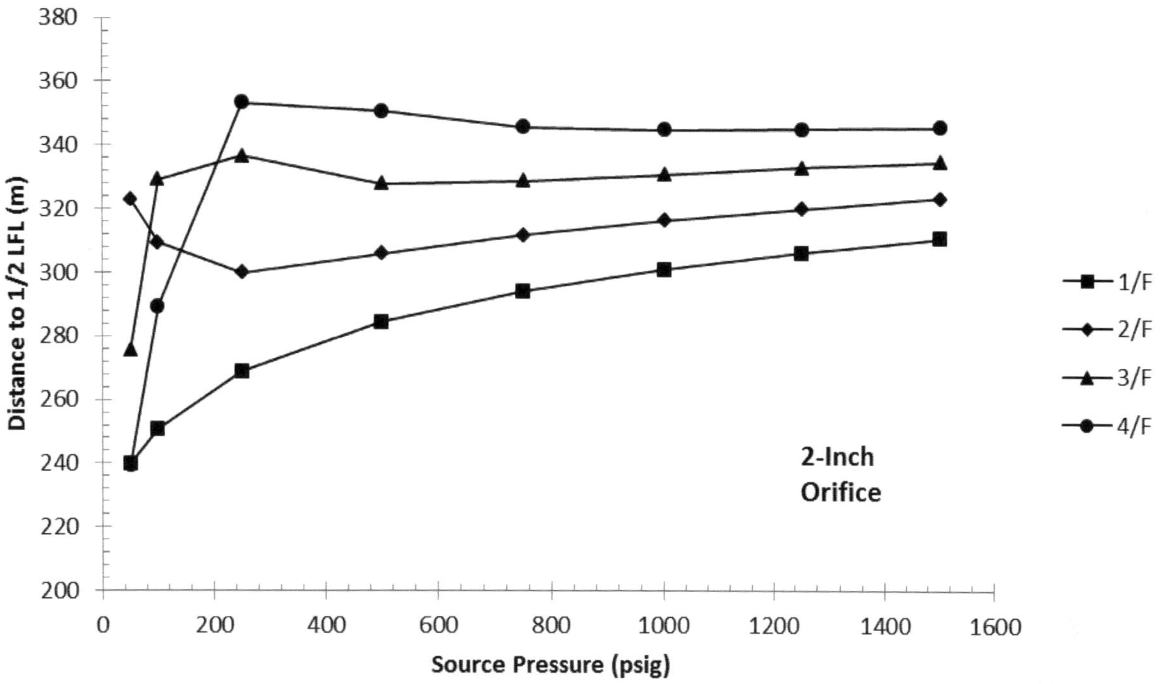

Figure 11. Variation of dispersion distance for a 2-inch orifice with baseline conditions as a function of wind speed.

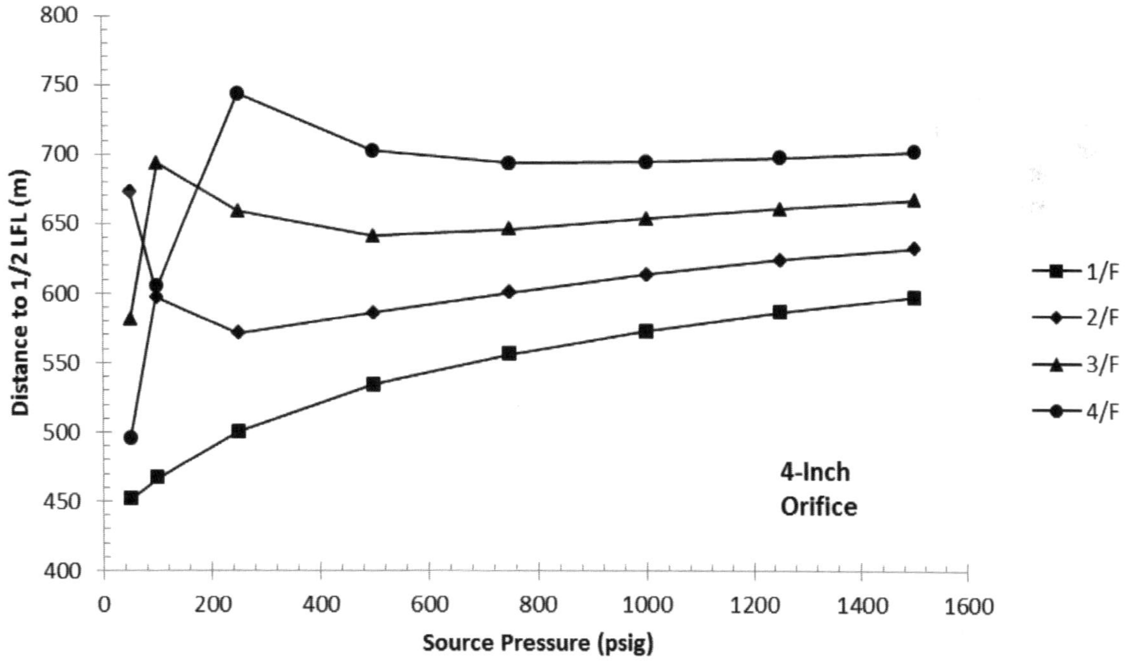

Figure 12. Variation of dispersion distance for a 4-inch orifice with baseline conditions as a function of wind speed.

3.6 Effect of Source LNG Temperature

To demonstrate the effect of LNG temperature on the downwind dispersion, a parametric study of source temperature versus source pressure was performed for a 2-inch diameter orifice at 5 m elevation above ground level. As the release temperature was raised for the higher pressure releases, the downwind dispersion distance dropped as would be anticipated due to lower cloud density (see Figure 13). Figure 14 provides the centerline side view contours for three temperature values in the 50 psig case. At high temperature, the jet completely disperses at a fraction of the distance for the colder cases.

These same source pressure and temperature cases were modeled at 1 m elevation to investigate what effect that elevation would have on the dispersion distance. The results are presented in Figure 15. For the low pressure case at 50 psig, the dispersion distance dropped with the lower elevation release likely due to warming of the cloud by the ground. Three cases were found to have significant liquid rainout of greater than 10% of the release mass; however, this did not appear to significantly affect the maximum dispersion distance. The dispersion distances for the higher pressure cases increased with a decrease in elevation, likely due to the reduction in air entrainment due to cloud location at ground level.

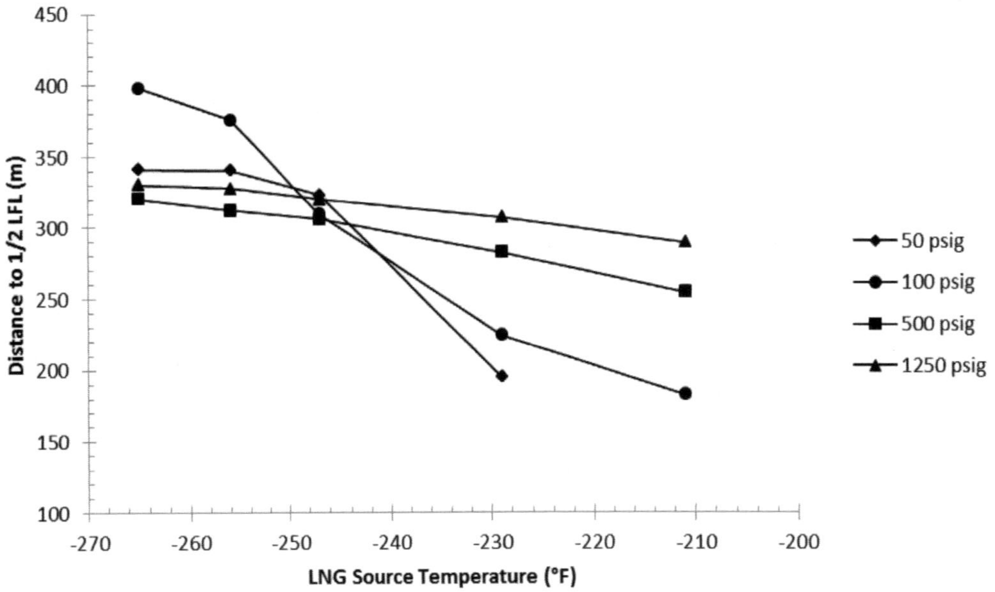

Figure 13. Comparison of distance to ½ LFL for a range of source temperature and source pressure values for a source at 5 m elevation. Note that cloud maximum distance for 50 psig/-211°F release does not reach ground level (see Figure 14).

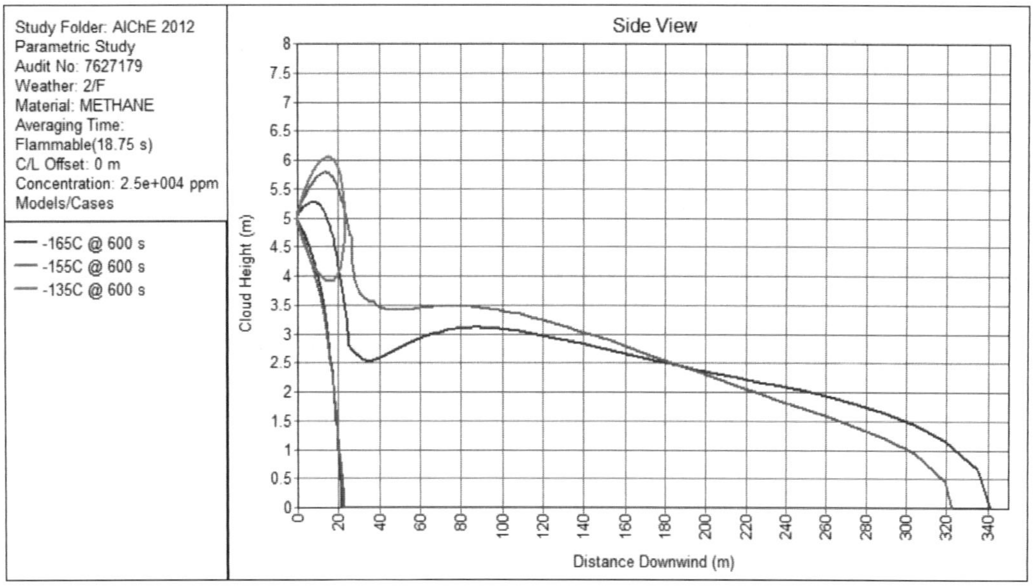

Figure 14. Representative side view contours for 2-Inch orifice at 5 m elevation and a series of source temperatures at 50 psig.

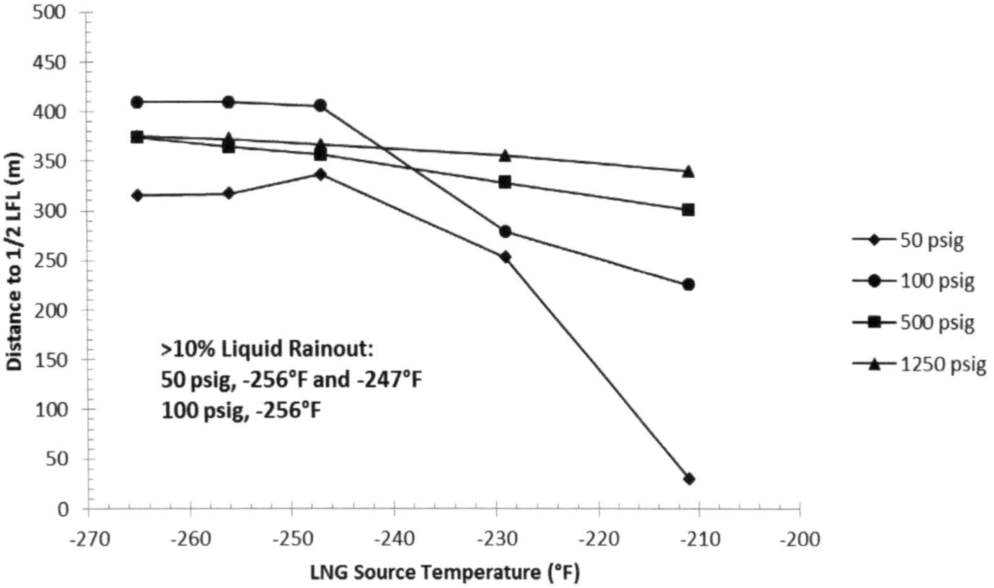

Figure 15. Comparison of distance to ½ LFL for a range of source temperature and source pressure values for a source at 1 m elevation.

4 Conclusions & Recommendations

This paper has presented the results of a series of parametric studies on pressurized LNG releases to aid facility owners in evaluating the maximum downwind dispersion distance of the vapor cloud. The PHAST program was used to perform the analyses. We found that the downwind dispersion distance was not dominated by any individual parameter, such as release temperature. In addition, we found that intuitive trends in release behavior may not be correct; thus, modeling for the specific conditions of the release is recommended. The results presented in this paper will give the analyst an indication of the release and dispersion behavior for conditions similar to those presented; however, the authors suggest that siting decisions be based upon modeling of the facility using appropriate parameters for the exiting/planned process operations.

5 References

[1] North American LNG Import Terminals, downloaded from http://www.ferc.gov/industries/gas/indus-act/lng.asp, January 10, 2012.

[2] PHMSA Docket No. 2011-0075, Final Decision on petition for approval for PHAST UDM Version 6.6 and 6.7, October 7, 2011.

[3] PHMSA Interpretation #PI-10-0005, Jul 16, 2010.

[4] FERC Engineering Information Request, Docket No. CP11-72-000, November 8, 2011.

[5] Saraf S,Medhekar, S.,Yang, E-H, "Have You Considered High-Pressure LNG Releases?" AIChE Spring Meeting, April 2009, 9th Topical Conference on Natural Gas Utilization, Tampa, Florida, April 26-30, 2009.

[6] Gavelli, F., Davis, S.G., and Hansen, O.R., "CFD Simulation of Vapor Dispersion from LNG Jetting and Flashing Releases," Paper presented at AIChE Spring meeting, Natural Gas Utilization Topical Session, San Antonio, USA, March 22-24, 2010.

[7] FLASHING LIQUID JETS AND TWO-PHASE DROPLET DISPERSION, Henk Witlox and Mike Harper, DNV Software, London, UK , www.dnvsoftware.com, Phil Bowen and Vincent Cleary, Cardiff University, UK, American Society of Safety Engineers Middle East Chapter, 7th Professional Development Conference & Exhibition, Kingdom of Bahrain, March 18-22, 2005.

[8] Witlox, H.W.M. and Holt, A., 1999, A unified model for jet, heavy and passive dispersion including droplet rainout and re-evaporation, International Conference and Workshop on Modeling the Consequences of accidental Releases of Hazardous Materials, CCPS, San Francisco, California, September 28 – October 1, pp. 315–344.

[9] 49 CFR 193.2059 "Flammable vapor-gas dispersion protection"

[10] NFPA 59A "Standard for the Production, Storage, and Handling of Liquefied Natural Gas (LNG)" (2009) Section 5.3.3.6

(144e) Effect of LNG Chemical Composition on Consequence Assessment

Xiaodan Gao[1], Yi Liu[1], Luc Vechot[2], Tomasz Olewski[2] and M. Sam Mannan[1], (1)Mary Kay O'Connor Process Safety Center, Department of Chemical Engineering,Texas A& M University, College Station, TX, (2)Chemical Engineering, Texas A& M University at Qatar, Doha, Qatar

Over the last ten years, LNG has become a popular fuel and its demand has been keeping the increasing trend. This leads to the construction of new LNG facilities around the world for import, storage and re-gasification. The formation of a vapor cloud following the LNG accidental spill presents potentially significant flammable hazards to both the facility and the surrounding communities.

Correctly predicted exclusion zones during the risk assessment process are critical to effectively mitigate the consequence of a LNG spill and protect the public from the flammable vapors. The accuracy of every step of consequence modeling, including LNG spilling, spreading, vaporization and vapor dispersion, influence the precision of exclusion zone estimation. Numerous publications have advanced mechanisms of the above mentioned steps to obtain a certain level of confidence, including comprehensive Computational Fluid Dynamics models to describe complex geometry and turbulence. Nevertheless, most of the available studies assume pure methane as the analog of LNG. This assumption may not be correct, since the vaporization process of mixture as LNG may be significantly different from the pure methane liquid. As a matter of fact, the methane fraction of LNG can vary from 83 to 99% mol, and neglecting this factor may lead to the inaccuracy of all the following steps. In 2007, the US Government Accountability Office had reported: the Public safety consequences of a liquefied natural gas spill need clarification (Report 07-633T, March 21st 2007). Under this report, the expert panel evaluated several research areas, one of which is evaluation of the effects of LNG composition on vaporization rate, thermal radiation from fire and explosive behavior.

This paper has investigated the influence of multi-component factor on the vaporization rate as well as vapor dispersion modeling. The studied mixtures involve five components: nitrogen, methane, ethane, propane and butane. Suitable equations of state and vapor-liquid equilibrium of LNG have been adopted to predict the vaporization rate, followed by the vapor dispersion modeling. The behavior of pure methane liquid and several different LNG compositions are compared. The results confirmed the significant impact of composition, which has emphasized the necessity of this factor to be involved in LNG consequence assessment process.

(144f) Source Term Modeling of Vapor Cloud Formation During a Cryogenic Liquid Spill Based on a Boiling Model by CFD Method

Yi Liu[1], Tomasz Olewski[2], Luc Vechot[2], Xiaodan Gao[1]and Sam Mannan[1], (1)Mary Kay O°Connor Process Safety Center, Texas A& M University, College Station, TX, (2)Chemical Engineering, Texas A& M University at Qatar, Doha, Qatar

The formation of a flammable vapor cloud following a spill of cryogenic liquids, such as ammonia and LNG (liquefied natural gas), on a ground or water includes significant toxic, fire or explosion hazards. The ability to predict the consequences of such an event is crucial to perform proper risk assessment. Numerous researchers have been focused on the modeling of the vapor cloud dispersion. However there is a lack of accurate source term model describing the vapor cloud source, which has obviously an effect on the predicted exclusion zones.

Webber et al. [Webber, DM; Gant, SE; Ivings, MJ; Jagger, SF.*LNG source term models for hazard analysis: A review of the state-of-the-art and an approach to model assessment.*2010] prepared a wide literature review of LNG source term models as well as the experimental data available to validate them. Eighteen models, most of which are integral and two of them are computational fluid dynamics (CFD) codes (FLACS by GexCon and Liquid Spill Modeling System by Cambridge Environmental Research Consultants) are briefly reviewed. All models differ primarily regarding assumptions, simplifications or/and correlations.

A boiling model, as one part of the source term modeling of cryogens, has been developed by using CFD code. Different boiling phenomena, including film boiling, transition boiling and nucleate boiling, are identified by the presented model. The vaporization rate can be calculated depending on the ground temperature. The boiling model by CFD method can give a more accurate result than integral models or empirical correlations, especially when a complex geometry of the spill and surrounding is considered.

Based on the boiling model, the pool formation and vaporization during cryogenic liquid spill on ground and/or water are studied in this paper. Liquid nitrogen is chosen as a safe analogue. The effects of the surface type (ground or water), hot surface conditions (temperature, slope, etc), weather conditions and two-phase boiling flow rate are discussed and compared against experiments or/and literature data.

LNG Terminal Vapor Dispersion Exclusion Zones to Protect the Public

Jerry Havens and Heather Walker
Chemical Hazards Research Center
Department of Chemical Engineering
University of Arkansas, Fayetteville, Arkansas 72701
Corresponding Author: jhavens@uark.edu

Abstract

The LNG import industry in the U.S. is at a crossroad. Speculation about shale gas supplies, with uncertain national/international energy economies, will likely dampen investment interest in LNG import facilities in the U.S. These uncertainties, arguably intensified by concerns for "worst case" accident consequences that might occur, were heightened by the Gulf Oil Spill in 2010 and by the Tsunami in 2011 that resulted in compromise of multiple nuclear reactors and nuclear-spent-fuel stores at Fukushima, Japan. It may take decades to complete a realistic assessment of the consequences, environmental and economic, of these events in the Gulf of Mexico and Japan. At the same time, concerns can only increase that terrorist threats be considered more carefully in siting potentially high-consequence, major hazard industries. LNG facility siting exemplifies the task faced – to require sufficient separation distance to ensure that the public is not in harm's way from unintended events that could occur at the facility. While this is not a new problem, it is to be expected that the stakes will grow higher for all major hazard facilities along with the concerns of the public. This paper addresses systematic determination of separation distances required to protect the public from the dangers of vapor cloud fires that could follow spills of LNG – that extent determined by the travel distance to which LNG vapor clouds with about 5% LNG vapor concentration in air could credibly result. A satisfactory and consistent way forward for LNG should include sound risk management practices applicable to siting any industrial facility that presents major hazard potential to the public.

Purpose

The purposes of this paper are:

1. To summarize the state of the art of vapor dispersion modeling to determine LNG vapor dispersion safety exclusion zones, and identify remaining areas of uncertainty.
2. To identify the input information required for dispersion models approved for use under 49 CFR 193.
3. To identify additional parameters required (not specified by the regulation) for the calculation of exclusion zones, and to consider the importance of and sensitivity to such parameters.
4. To identify important considerations for alternate model approvals.

State of the Art

The exclusion zone distance is defined as the maximum distance beyond which the near-ground-level concentration (in air) of LNG vapor (permitted by the regulation to be considered to be pure methane) downwind of a prescribed release of LNG will not support continued combustion (burning). This regulation-prescribed concentration, a time-averaged value intended to account for naturally occurring gas concentration fluctuations, is prescribed by the regulation to be 2.5% by volume, or one-half the accepted lower flammability concentration of methane in air.

At the beginning of 2011 there were two mathematical models approved by 49 CFR 193 for determining LNG vapor cloud dispersion exclusion zones; DEGADIS and FEM3A. DEGADIS, approved in 1990, was designed primarily for ground level, area source LNG vapor releases, with dispersion over flat unobstructed terrain (a special version of DEGADIS, applicable to dispersion of jetting releases, is not considered here). DEGADIS incorporated, for the first time in a regulatory model, methodology for determining the effects of (LNG vapor/air) density on dispersion. Prior to DEGADIS, only so-called "passive" dispersion theories had been employed to predict LNG vapor dispersion. However, application of passive dispersion models to LNG vapor dispersion were known to be incorrect because LNG vapor at ambient temperature is buoyant, causing the vapor to rise in the atmosphere, while LNG vapor at its boiling point (the temperature at which it would be evolved from an LNG spill) is about 1.5 times the density of air, causing the vapor to remain close to the ground and disperse less rapidly. Provision for the effect on dispersion of such higher gas density, as in DEGADIS, resulted in more realistic LNG vapor cloud exclusion zone predictions. FEM3A, approved in 2000, also accounts for the effects of gas density on dispersion, and considerable care was taken to ensure that the methods for describing the effects of density on dispersion were consistent in the two models (DEGADIS and FEM3A). FEM3A was also designed for ground level, area source vapor releases, but with added capability for predicting dispersion over/around mild terrain and/or obstacles such as vapor fences, impoundment dikes, buildings or storage tanks. The added capabilities of "complex" models such as FEM3A can result in either reduction or increase (compared with models such as DEGADIS) of dispersion model distances – reductions expected to result from increased mixing of LNG vapor with air caused by flow around obstacles, and increases from effects which cause the vapor cloud to be "channeled" between large obstacles or terrain features (causing flow confinement). For a variety of reasons, FEM3A calculations have not been submitted (to our knowledge) for consideration by FERC on any LNG terminal application to date. DEGADIS calculations submitted during this time period (1990 – present) typically utilized the SOURCE5 program (not part of, or integral to, DEGADIS) developed by the Gas Research Institute to determine the vapor rate input to DEGADIS. SOURCE5, designed to determine the LNG vapor overflow rate from a spill impoundment, assumed that the vaporization from the spilled LNG resulted directly from heat transfer between the impoundment floor and the spilled LNG, further assuming that the LNG vapor evolved filled the impoundment to overflow without mixing with air. This method typically resulted in a significant delay between liquid spill start and vapor overflow, during which the vaporization rate decreased significantly due to cooling of the impoundment floor. The result of this "vapor holdup" assumption was an unrealistic decrease in downwind LNG vapor rate that significantly shortened the vapor cloud exclusion zone distance requirement. Use of the SOURCE5 methodology is now prohibited. All LNG terminals built in the United States since about 1990 (import and peak-shaving) have been approved on the basis of DEGADIS vapor dispersion exclusion zone calculations, some with and some without the use of SOURCE5 to determine LNG vapor input rate.

LNG vapor cloud dispersion exclusion zone determination required by 49 CFR 193 is a three step process:

1. Determine the magnitude of the consequences that must be presumed credible – how big, and how rapid, might a spill of LNG be?
2. For the size and rate of the spill selected, determine the site specific conditions that would result in the maximum hazard extent; and
3. Predict, using an approved mathematical model, the separation distance from the spill center which would keep the public out of harm's way.

Regulation Specified Model Input Requirements

The primary data (inputs to the dispersion model) that control the vapor cloud exclusion zone predictions mandated by 49 CFR 193 are:

a) The rate at which LNG vapor enters the atmospheric (wind) flow. This rate can be either steady or time variable. The model input requirement is for specification of the vapor rate in kg/s as a function of time.
b) The environmental conditions of wind speed, air temperature and humidity, atmospheric stability (which importantly affects the rate of dispersion), and the "surface roughness" (a measure of the roughness of the ground surface in the area surrounding the spill).

<u>LNG Vapor Input Rate.</u>

49 CFR 193 prescribes guidance for "selection" (by the applicant) of so-called "design spills" to be considered. The design spill determines the vapor rate required as input to the dispersion model, and, along with specifications of the receiving impoundment thermal properties and temperature discussed in the next section, the vapor cloud exclusion zone distance. These <u>result-determining</u> design spills, assumed to be of 10 minute duration (or lesser time with prior approval by DOT based on the applicant's assurance of ability to shut the spill down before 10 minutes has elapsed) are for guillotine breakages of "selected" LNG transfer lines in the plant. Prior to 2000, when the NFPA 59A standard governing LNG safety practices was incorporated in 49 CFR 193, the LNG spill which was typically considered by applicants for determination of the vapor cloud exclusion zone was a guillotine rupture of the largest LNG-carrying line in the plant - the ship unloading line (SUL), typically carrying about 60,000 gal LNG per minute, for 10 minutes, or 600,000 gallons. This spill from the SUL determined (still determines) the minimum required spill impoundment volume. Consequently, most LNG import facilities approved prior to 2000 were required to show that the vapor cloud exclusion zone determined for the SUL spill did not extend beyond the plant boundary. As a result of incorporation of NFPA 59A in 49 CFR 193 the required "design spill" for the vapor cloud exclusion zone determination was reduced by allowing the applicant to specify (using "guidance" from NFPA Standard 59A) a smaller "accidental leakage break" spill. However, the requirement for the impoundment to be sized to contain the full-bore, ten-minute duration, SUL remained. As a result of the allowance of the smaller spill into the larger required sump volume, the liquid spilled occupied a smaller fraction of the sump volume provided. This change alone had important implications for the determination of vapor cloud exclusion zones, as numerous applications for LNG import terminals were submitted to FERC in which "vapor holdup" in the impoundment was claimed to delay the start of vapor flow down-wind. The delay period was determined as the time required

for the (assumed pure) LNG vapor to fill and overflow the impoundment volume. Because this delay period resulted in significant reduction of the evaporation rate at the time of overflow (due to rapid cooling of the impoundment by the LNG), the predicted exclusion zones were often substantially reduced. Indeed, in some of the earliest applications, even when the SUL spill was selected as the design spill, a larger impoundment was provided from which the applicant claimed the LNG vapor would never overflow because the total volume of (pure) vapor produced from the ten minute duration spill was less than the (additional) impoundment volume provided. In 2010, DOT disallowed the assumption of vapor holdup, and subsequent calculations are prohibited from assuming any vapor holdup in the impoundment. The disallowance of credit for vapor holdup in an impoundment resulted in an important increase in the vapor cloud exclusion distances (for a given spill amount and rate) obtained with either of the regulation prescribed dispersion models. However, by this time it had become applicant practice to "select" an "accidental leakage breakage" of a secondary (smaller) line attached to the main unloading line rather than a guillotine breakage of the SUL, and the practice continues today. The practice (accidental leakage break) has been criticized by Havens[1] as being inconsistent and unjustified, citing eleven terminal applications approved during the period 2004-2009 with design spills for the ship unloading line system ranging from (ten minute duration) 28,900 to 812,000 gallons, with essentially no quantitative risk assessment justification, even though all of the design spill selections were for terminal design specifications that were essentially identical.

Site Environmental Conditions.

The environmental conditions are required to account for local historical experience, but they are optionally allowed by 49 CFR 193 to be set at so-called worst-conditions-for-dispersion; designated by the regulation as 2 m/s wind at 10 meters elevation, F stability, relative humidity = 50%, atmospheric temperature = average in the region, and surface roughness 3 cm. This paper assumes this specification of environmental conditions for determining vapor dispersion exclusion zones.

Other Model Input Requirements Affecting Exclusion Zone Determination

Whether as a result of the disallowance of vapor holdup or of other factors, some applicants have considered additional means of reducing vapor cloud exclusion zone predictions. Although considerable research has been done to develop means for reduction using passive methods such as vapor fences and active methods such as water (or other chemical) curtains, to the authors' knowledge no applications for LNG import terminals[2] have been approved which include use of such methods for reducing the required vapor dispersion exclusion zones. To the authors' knowledge, the only method that has been used to decrease the dispersion distance for import terminals is to decrease the heat transfer rate from the impoundment floor to the spilled LNG by alteration of the thermal properties of the impoundment material, typically concrete. Modeling methods thus far approved by DOT for determining vaporization (boiling) rates of LNG (the so called "source[3] input" model, which is not part of either currently approved dispersion models) as a result of heat transfer between a concrete impoundment floor and the spilled LNG are entirely controlled by specification of the impoundment floor area[4], m^2, and the two impoundment thermal properties; k, thermal conductivity, W/m-K, and α, thermal diffusivity, m^2/s.

History of Model Approvals

49 CFR 193 has, since promulgation, provided for consideration of alternative models, requiring that such models be approved by the DOT Administrator. To date there have been three dispersion models approved by DOT for use under 49 CFR 193, the first appearing in the original promulgated regulation (~1980), and the second and third approved under the standing approval provision:

1) (1980 to 1992) The "MTB[5]" model appeared in 49 CFR 193 when the regulation was promulgated. The model utilized traditional atmospheric dispersion modeling theory developed for "passive" dispersion of atmospheric pollutants. Passive dispersion theory assumes that the pollutant/contaminant is of such low concentration in the atmosphere that it does not effectively alter the wind-flow direction, velocity, or mixing patterns. This assumption typically resulted in LNG vapor cloud exclusion distances that were severely conservative (over predicted) because the predicted LNG vapor cloud was very long and (laterally and vertically) thin (pencil shaped). The atmospheric dispersion of denser than air gases, especially following episodic releases in which the concentrations of the dense gas can be very high initially (for example, LNG vapor is emitted from a spilled pool at or near 100% concentration) resulted in vapor clouds with decreased vertical thickness (due to suppression of vertical mixing) but increased lateral dimension (resulting from gravity induced spreading). As a result, when lateral spreading was not prevented (by channeling), the typical result of a dense gas model was a shorter downwind (centerline) distance than was obtained with the MTB model. The MTB model description and computer coding was publicly available, at no cost.

2) (1992 to present) DEGADIS was developed at the University of Arkansas for the U.S. Coast Guard and the (then) Gas Research Institute. It was designed for general dense gas dispersion application with the specific requirement (by the Coast Guard) that it address the important factors involved in atmospheric dispersion of LNG vapor, including gas density and temperature (heat transfer) effects. DEGADIS was always recommended only for flat, obstacle-free terrain or water application. DEGADIS, which was subsequently used widely for general dense gas dispersion in addition to LNG application, is publicly available (user's description and computer coding) at no cost.

3) (2000 to present) The FEM3A model was developed by the Lawrence Livermore National Laboratory (LLNL). The model was further evaluated for 49 CFR 193 application at the University of Arkansas using comparisons with wind tunnel dense gas dispersion experiments, and a personal computer version was developed (the LLNL version at the time required super-computer technology not widely available). The work at the University of Arkansas, sponsored by the Gas Research Institute with added funding from an industry consortium, resulted in a GRI licensed personal computer based model that was approved by the DOT administrator in 2000. The model, to the author's knowledge, has not been used in either applications for LNG import terminal or peak-shaving plant siting approval. FEM3A is available to the public (user's description and computer coding) at no cost.

Potential for Variation in Predictions of Alternate Models

We analyzed the report Vapor Dispersion from Jetting & Flashing Scenarios and Spills into Trenches & Troughs Associated with the NGL Extraction Process Area, dated February 17, 2011, prepared by Exponent, Inc. and submitted to FERC for the Freeport LNG project. To our knowledge, this submission by Freeport was coincident with the first request (since 2000, when FEM3A was approved) for approval of an alternative vapor dispersion exclusion zone model (Flacs) under the regulation. The purpose here is to illustrate, by comparison of DEGADIS predictions for different design spill selections as well as comparison between DEGADIS and Flacs predictions of a specific vapor dispersion scenario, the model result variability that must be anticipated if applicants are allowed to choose from alternate models.

We considered the differences that occur in the determination of vapor cloud exclusion zones for three non-flashing liquid spill-into-impoundment scenarios. These specific spill scenarios were taken from the Weavers Cove and Freeport LNG dockets – all are guillotine spills from the line sizes given, and the environmental conditions are the "optional" one specified above (the same for all three):

- Weavers Cove - Tank Withdrawal Line Spill Scenario, 10 minute duration, 16" diameter line, 2,592,000 pounds LNG/hr.
    - LNG spill rate = 327 kg/s (Base)
    - Sump/Trench Floor Area = 261 m^2 (Base)
    - Sump/Trench Thermal Properties[6]: K = 1.44 W/m-K, alpha = 0.000000492 m^2/s (Base)
    - Dispersion Model – DEGADIS, using "Appendix T Method[7]" for vapor input
    - Vapor Cloud Dispersion Exclusion Distance = 955 feet

- Weavers Cove – Ship Unloading Line Spill Scenario, 10 minute duration, 30" diameter line, 13,185,743 pounds LNG/hr.
    - LNG spill rate = 1661 kg/s (Base)
    - Sump/Trench Floor Area = 1795 m^2 (Base)
    - Sump/Trench Thermal Properties: K = 1.44 W/m-K, alpha = 0.000000492 m^2/s (Base)
    - Dispersion Model – DEGADIS, using "Appendix T Method" for vapor input
    - Vapor Cloud Dispersion Distance = 2078 feet

- Freeport – Maximum NGL extraction project flowrate, 10 minute duration, 10" diameter line, 779,878 pounds LNG/hr.
    - LNG spill rate = 98.3 kg/s (Base)
    - Sump/Trench Floor Area = 945.3 m^2 (Base)
    - Sump/Trench Thermal Properties[8]: K = 1.1 W/m-K, alpha = 0.000001 m^2/s (Base)
    - Dispersion Model – Flacs, which predicts vapor input as well as (resulting) dispersion
    - Vapor Cloud Dispersion Exclusion Distance = 1000 feet

The following table shows the predicted vapor cloud exclusion zones for these three base spill scenarios. Alternate (alt) scenario calculations are shown for variations from the base input conditions specified for spill scenarios 1 and 2 as well as an additional calculation for spill scenario 3 which illustrates the difference in exclusion zone distance obtained using the Flacs and DEGADIS models with identical input data:

Scenario	Input	Exclusion Zone Distance, ft
1	Base, DEGADIS	955
2	Base, DEGADIS	2078
3	Base, Flacs	1000
1-alt	K = 1.1 W/m-k, $\alpha = 0.000001\ m^2/s$	665
2-alt	K = 1.1 W/m-k, $\alpha = 0.000001\ m^2/s$	1694
3-alt	DEGADIS/Appendix T	672

The two table entries Scenario 1: Weavers Cove Tank Withdrawal Line (TWL) Spill and Scenario 2: Weavers Cove Ship Unloading Line (SUL) Spill illustrates the overriding importance of the design spill selection. The SUL spill exclusion zone is more than twice (2.2 times) as long as the TWL spill.

The table entries Scenario 1 and Scenario 1-alt illustrate the importance of variation of the sump/trench thermal properties. The TWL spill exclusion zone distance is shortened using the insulating concrete properties (1-alt) by 30%. The table entries Scenario 2 and Scenario 2-alt indicate a reduction in exclusion zone distance for the SUL spill using the same insulating concrete properties of ~20%.

The table entries Scenario 3 and Scenario 3-alt show the difference in exclusion zone extent obtained for the Freeport NGL extraction project spill using Flacs (submitted by Freeport) and the DEGADIS/Appendix T. The DEGADIS predicted exclusion zone distance is shorter (than Flacs) by almost a third.

Conclusions

The primary determinants of vapor cloud dispersion distance resulting from modeling methods required by 49 CFR 193, assuming the optional worst case environmental conditions are used, are:

- Design spill selection
- Floor area and thermal properties of the spill impoundment
- Choice of model

Design Spill

The selection of design spill is typically the controlling factor, with wide ranging spill sizes adopted by applicants based on selection of guillotine failure of different size lines that are appurtenant to the spill unloading line. Lines with diameters from 1 inch to 6 inches have been selected, with the result that the ship unloading line design spill rates have ranged all the way from a 1 inch diameter pipe (2890 gal/min) to a full 36 inch SUL pipe (81,200 gal/min). In the authors' opinion, such variation, with minimal quantitative consideration of the probability of the releases occurring as a function of line size, is not justified. Indeed, such practices leave the appearance that the design spill selection process is being used to circumvent the intent of the regulation similarly to the now prohibited practice of taking unjustified credit for vapor holdup in determining the vapor input rate to the model. DEGADIS model comparisons (Scenario 1 and 2 above) of the exclusion zone required for a typical SUL spill (using design data specific to the Weavers Cove Terminal application) show that a full SUL spill into the regulation-required impoundment would be at least twice (2078 feet) as large as that obtained for the largest design spill selected in the application for the project (955 feet).

Impoundment Design

The impoundment floor area and thermal properties are determined by the specific terminal design. As shown above for Scenarios 1 and 2 above, significant reductions can be obtained by using insulating concrete. It is recommended that 49 CFR 193 specify a procedure by which a proposed insulating concrete be certified as meeting the claimed insulating properties as well as practicality for the intended use (such as load bearing requirements). Perhaps just as importantly, the present regulation does not speak clearly to a requirement for ensuring that the impoundments are kept dry (no water or snow accumulation), either of which have the potential for dramatic increase in the LNG evaporation rate following a spill – with increased vapor cloud dispersion exclusion distance.

Choice of Model

The differences shown here by the comparison of results obtained for the Freeport spill with DEGADIS and Flacs (the DEGADIS predicted exclusion distance less than the Flacs prediction by about a third) are important to consider:

- When a choice of alternate models is allowed, it must be anticipated that differences (in the model predictions) will result, raising questions:
  - Should a Standard such as 49 CFR 193 not specify a standard method for determining safety exclusion zones?
  - If not, how does one determine which one to use, or which one is right?[9]
  - Should the applicant be required to use the more conservative result?[10]
- Without full access to the model details (description and computer coding), it is extremely difficult, if not impossible, to determine the reasons for differences that will inevitably arise when the applicant has the choice of model – and therefore to judge equitably the suitability of the exclusion distance zones submitted by an applicant.
- The DEGADIS and FEM3A models, including source coding, are available to the public at no charge for purposes of evaluation. The DOT has not made any statements regarding

the availability to the public of such additional models as may approved. We believe that past experience suggests a need for models that are to be used for such purposes to be available to the public (non-proprietary) to ensure transparency.

Endnotes

---

[1] Havens, Jerry, "Consequence Analyses for Credible LNG Hazards", Second American Institute of Chemical Engineers/Canadian Society of Chemical Engineers LNG Technical Conference, 8th World Congress of Chemical Engineering, Montreal, Quebec, Canada, August 24-26, 2009.

[2] There apparently have been applications for peak-shaving and satellite facilities which have utilized such methods.

[3] Not to be confused with SOURCE5.

[4] To the author's knowledge, the area for heat transfer from the impoundment to the spilled LNG has heretofore been limited to the impoundment floor area; the area of the walls exposed to LNG were (are) not included. This assumption is not conservative, as inclusion of the wall area in the determination of vapor input to the dispersion model would increase the evaporation rate and, in turn, the vapor cloud exclusion zone distance.

[5] Materials Transportation Bureau, the responsible agency for LNG Import Terminal Permitting, prior to the promulgation of 49 CFR 193.

[6] Specified by Weavers Cove for the "Shoreside" facility in its application to FERC

[7] DEIS for Downeast LNG – Appendix T, Source Strength and Dispersion Calculations for the Marine Transfer Area Containment Basin.

[8] Specified by Freeport in its application to FERC

[9] This question warrants careful consideration. If the models being considered have already passed the "evaluation" tests in the model approval process, it is unlikely that differences in model input can be clearly determined to reflect anything except the degree of uncertainty remaining (in different model results). Even if different models could be shown to give identical results under all cases, which is not expected, it is important to recognize the fact that any of the atmospheric dispersion models under consideration necessarily predicts some kind of time average behavior of a turbulent atmospheric flow/dispersion process. It follows that repeats of the actual event that the model is attempting to predict are not expected to give the same results; instead there will be a distribution of results that reflects the natural variability in atmospheric (turbulent) flows.

[10] Given the inability to decide which model result is the more accurate, it is suggested that the more conservative result would be the more defensible. Consideration of past practice may be helpful. The purpose of the approval of FEM3A as an alternate model was to provide a model which would provide for more realistic consideration of site specific factors such as terrain features or obstacles to the flow. In the authors' opinion, consideration should be given to further model selection which involves a similar practice - of providing multiple models only as they are required to be used under circumstances which makes one the more appropriate. There seems to be little reason, for example, for FEM3A to be considered as an alternate to DEGADIS, unless circumstances dictate the use of a model which DEGADIS is not designed for.

Life time Prediction of 9% Ni Steel LNG Storage Tank Using Fatigue Analysis

Seung Rim Lee, Taek Seong Jin, Hyo Jung Bang
Institute of Gas Safety R&D, Korea Gas Safety Corporation, Korea

ABSTRACT

LNG storage tanks are the core facilities of which construction cost accounts for 45 to 65% of the total construction cost of an LNG receiving terminal. LNG started to be supplied in Korea in 1986 and the oldest storage tanks have been in operation for about 25 years and 270,000 ㎘ class tanks of which capacity is largest in the world are now being designed in step with consistent increase of LNG demand. Therefore, now is the time for us to enhance cost competiveness by establishing systematic life cycle management systems and reducing life cycle cost through the development of life time assessment technologies for the tanks approaching the end of their design life time (about 25 years) and for super tanks.

In this study, life time is estimated through structural analysis (ABAQUS ver. 6.10) and fatigue analysis (FEMFAT ver. 4.8b) for two elements i.e., tank shell and pump column with its supporting structure, in consideration of welds among the aging factors of the 9% Ni steel inner tank. The structural analysis results show that the maximum stress is 230.6 MPa for the inner shell when the maximum liquid level is 35.2 m and 124 MPa at the bottom of the piping for the pump column with the same maximum liquid level. In the case of fatigue analysis, the shell with welds shows a maximum damage of D=1.677e-014 at the bottom course weld when the tank is operated in a cycle of 4 time/month and the fatigue life time is 1.24e+16 years. In the case of the pump column, the max damage is 4.143e-016 at the pump bottom and the fatigue life time is 5.03e+17 years. In conclusion, it is demonstrated that the fatigue life time for the 9% Ni steel inner tank of an LNG storage tank is almost infinite and it is concluded that the fatigue life time of the concrete structure or roof structure which is more vulnerable should be assessed to estimate the overall life time of an LNG storage tank.

1. INTRODUCTION

In general, an LNG storage tank consists of an inner tank of a 9% Ni steel type or membrane type and a concrete outer tank consisting of a bottom, a wall and a roof.

All 9% Ni steel LNG storage tanks being operated in Korea can be classified as full containment LNG storage tanks (EN 1473). Figure1 shows the schematic drawing of a full containment type LNG storage tank. A full containment type LNG storage tank means a double structure tank of which inner tank and outer tank have the function to individually and independently contain LNG.

LNG storage tanks are the core facilities of which construction cost accounts for 45 to 65% of the total construction cost of an LNG receiving terminal. Korea started to supply LNG in 1986 and currently has a supply history of about 25 years. In the case of some countries with advanced LNG technology, it is reported that some LNG storage tanks constructed in 1960s when LNG started to be used are still in operation. This is possible because appropriate techniques for integrity assessment and residual life time assessment have been developed and applied, cost saving based on safety has been achieved in the matter of storage tank reconstruction and attempts have been being made to enhance the usefulness of national infrastructures.

Domestically, LNG storage tanks constructed in late 1980s are approaching their design guarantee life time (about 25 years) and now is the time for us to develop techniques for integrity assessment and residual life time assessment suitable for local environments and to enhance the cost competiveness of super LNG storage tanks which are still in design stage by reducing life cycle cost through systematic life cycle management.

Presently, life time assessment for LNG storage tanks is in the early stage of commercialization and there are some examples set by some companies by directly introducing life time assessment technique based on nuclear power plant cases. There are only 4 to 6 LNG storage tanks all over the world on which systematic life time assessment consulting has been performed. Therefore, on-site applicability is expected to be enhanced by developing systematic life time assessment and monitoring technologies and applying them to domestic LNG storage tanks through this task.

Even though LNG storage tanks are expensive core facilities and one 200,000 kℓ class storage tank costs about 102 billion won (about USD 90 million) for its construction, there is no established standard life cycle management system and this matter is recognized as a prerequisite to be solved for safe and economical operation and for making inroads into the world market.

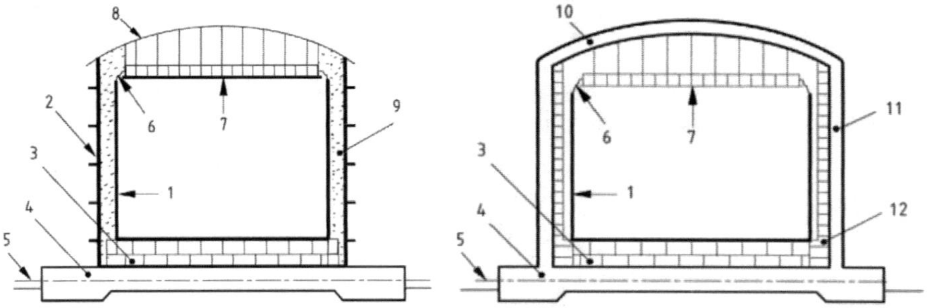

Key

1	primary container (steel)	7	suspended roof (insulated)
2	secondary container (steel)	8	roof (steel)
3	bottom insulation	9	loose fill insulation
4	foundation	10	concrete roof
5	foundation heating system	11	pre-stressed concrete outer tank (secondary container)
6	flexible insulating seal	12	insulation on inside of pre-stressed concrete outer tank

Fig 1. Schematic of full containment tank (EN 1473)

In this connection, the aging factors of LNG storage tanks have been derived through analysis of design data and overseas examples as shown in Table 1.

Table 1. Major Aging Factors for LNG Storage Tanks

Classification	Life Cycle Assessment Element	Description
Inner Steel Tank	Low-temperature thermal fatigue in inner tank	• 9% Ni steel / stainless steel membrane
	Fatigue in roof welds	• Dissimilar metal welds, roof fillet welds
	Fatigue in pump columns	• Pump column support structure / welds
Outer Concrete Tank	Loss of pre-stress	• Loss of pre-stress
	Durability of concrete	• Concrete strength • Covering depth • Chlorine ion concentration and salt damage diffusion factor (chloride content test, etc.) • Rebar corrosion • Creep and contraction of concrete • Carbonation
	Corrosion in piles	• Corrosion in piles
	Corrosion in tendons	• Corrosion in tendons
Insulation Materials	Deterioration of insulation performance and strength of bearing insulation materials	• PUF, cellular glass, etc.
	Deterioration of insulation performance of non- bearing insulation materials	• Perlite, glass wool, etc.

Therefore, in this study, fatigue life time is to be estimated using a commercial structural analysis program, ABAQUS and a fatigue life time estimation program, FEMFAT for 9% Ni steel inner tanks and pump column structures in accordance with the analysis procedure shown in Figure 2.

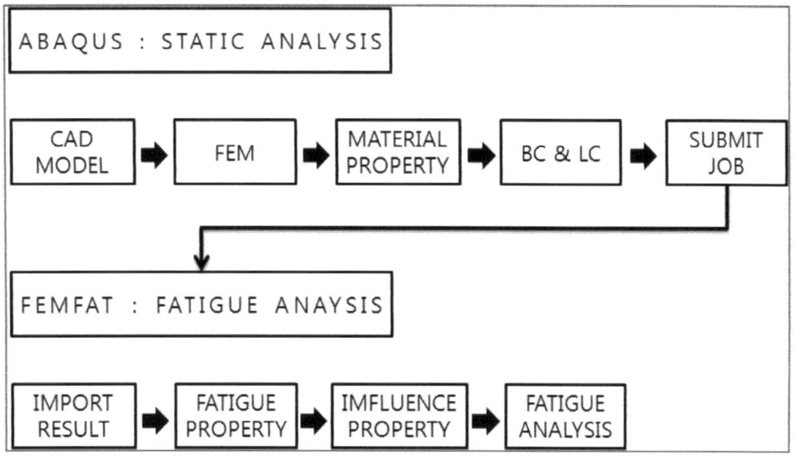

Fig. 2 Fatigue analysis process

2. ANALYSIS MODEL AND ANALYSIS CONDITIONS

2.1 Analysis Model

Traditionally, the inner tank of an LNG storage tank has been a self-supporting independent tank and is fabricated from 9% Ni steel plates. Since 9% Ni steel plates used in the inner tank structures have very small thermal expansion coefficients, they are designed to solve behavior problems due to thermal load variation and to play a structural role to support loads at the same time.

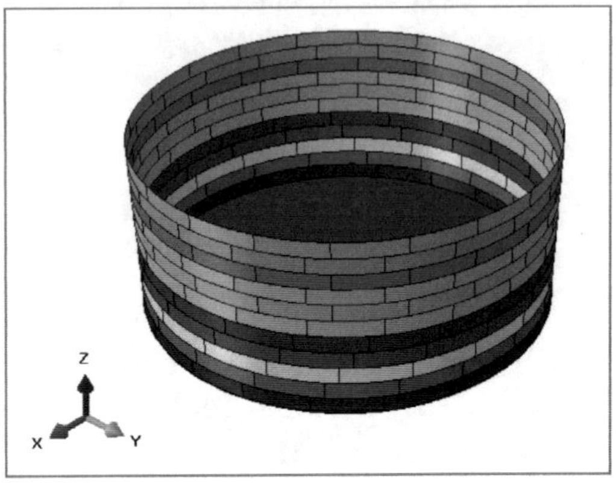

Fig. 3 Inner tank model

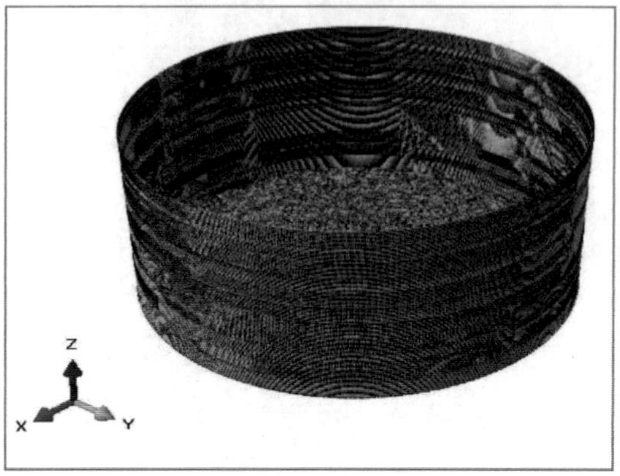

Fig. 4 Inner Tank FE model

An inner tank 3D CAD model and a FE model are shown in Figures 3 and 4 and physical property values of the material is indicated in Table 2.

Table 2. Inner Tank Material Property

Material Property	9% Ni Steel	Concrete
Elastic modulus, GPa	210	31.75
Poisson's ratio	0.3	0.3
Density, Kg/㎥	8000	2500

There are various pipes penetrating the roof of an LNG storage tank and LNG pipe and NG pipe ranging from 2 inches to 20 inches are included. Pump columns, a bottom fill column and LTD pipes are among those pipes 4 inches and over reaching the tank bottom and affecting the internal structure. Two or three pump columns are generally installed in an LNG storage tank and they have been chosen for analysis as the most important elements among the roof-penetrating pipes. A 3D CAD pump column model and a FE mode are shown in Figures 5 and 6, and the material properties of the pump column material are indicated in Table 3.

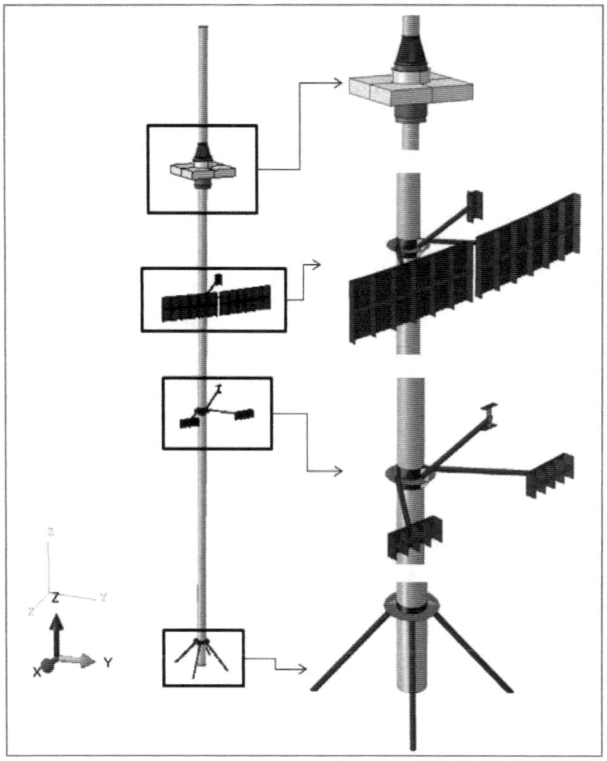

Fig. 5 Pump column model

Fig. 6 Pump Column FE model

Table 3 Material Properties for Pump Column

Material Property	Pump Column	Concrete
Elastic modulus, GPa	208	38.89
Poisson's ratio	0.3	0.2
Density, Kg/m^3	7850	2500

2.2 Loading Condition and Boundary Condition

The loads considered in the design of a 9% Ni steel inner tank include thermal stress due to cryogenic LNG temperature, static pressure applied on the inner tank wall depending on the stored LNG level, pressure of perlite transferred through the blanket and the self weight of the inner tank. In the case of thermal stress, the outside surface of the inner tank is wrapped with a blanket with spring action and this blanket restrains the expansion and contraction of the inner tank to prevent the generation of thermal stress due to cryogenic LNG temperature. Because of this reason, thermal stress due to temperature difference is not considered in the design of an inner tank, and only the static pressure due to LNG storage, self weight of the inner tank and pressure of perlite transferred through the blanket are considered. In addition, all 6 degrees of freedom at the bottom of the inner tank are restrained.

In the case of pump columns, if their sides are connected to the storage tank, they are subject to the stress caused by the thermal contraction of the tank interior due to temperature changes. The loads affecting the penetrating pipes include self weight, pressure and thermal load due to temperature difference. As the boundary condition of the pump column, all 6 degrees of freedom of the side supports and bottom supports are restrained.

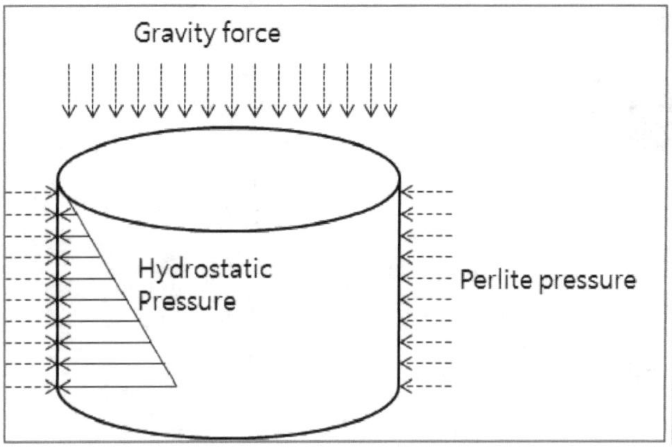

Fig. 7 Inner tank loading condition

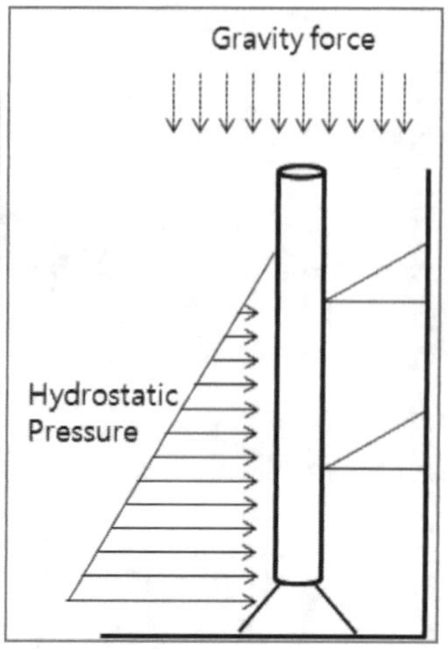

Fig. 8 Pump column loading condition

Figures 7 and 8 show the loads applied to the inner tank and the pump column. The uniform load around the tank shell in Figure 7 is the pressure on the inner tank from perlite, and the load applied to the pump column in Figure 8 is static pressure. The pump column is subject to its self weight and thermal stress as well as static pressure.

3. ANALYSIS RESULTS

3.1 Structural Analysis

The results of finite element analysis using ABAQUS for the inner tank indicate that a maximum stress of 230 MPa is applied to its bottom part as shown in Figure 9 when LNG is filled to a maximum liquid level of 35.2 m.

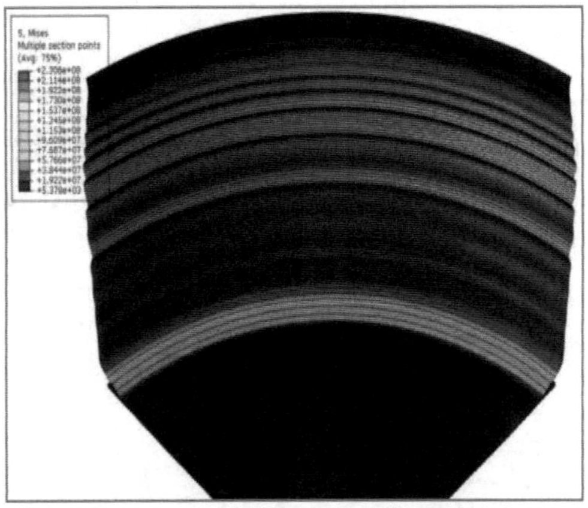

Fig. 9 Inner tank stress distribution

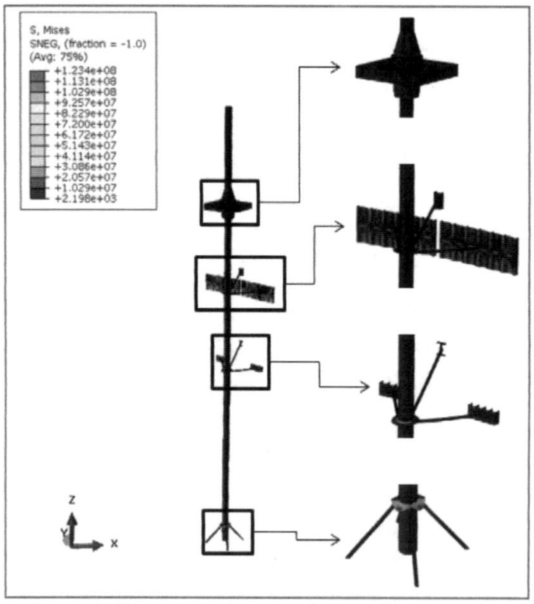

Fig. 10 Pump column stress distribution

The analysis results of the pump column show that a maxim stress of 123.4 MPa is applied to the bottom of the pump column as shown in Figure 10 when LNG is filled to a maximum liquid level of 35.2 m. Allowable stress values by structural members related to the LNG inner tank are indicated in Table 4.

Table 4. Maximum allowable stress

ASTM Specification		Stress (N/㎠)	
		Specified Minimum Values	
		Tensile strength	Yield strength
Plates and structural members	A353	68,947	51,711
	A553, Type 1	68,947	58,605
Pipes	A333, Grade 8	68,947	51,711
	A334, Grade 8	68,947	51,711
	B444(UNS-N06625), Gr.1	82,737	41,368
	B444(UNS-N06625), Gr.2	68,948	27,579
	A312, Grade 304	51,710	20,684
	A358, Grade 304	51,710	20,684

From Table 4, it is known that the inner tank and pump columns are safe.

3.2 Fatigue Analysis

In the case of LNG inner tanks, contraction and expansion due to static pressure and thermal load are repeated during their life time according as LNG is stored and discharged. Therefore, materials used in inner tanks shall have safe mechanical properties and their fatigue life time shall be guaranteed against cyclic loads.

1) Expected fatigue life time of inner tank

One of the loads being applied to inner tanks in LNG storage tanks is the thermal load which is applied according as the static pressure changes and the temperature changes from ambient temperature to -162℃ in conformity to the stored LNG quantity. The stress being applied like this is generated in the process of filling or discharging LNG, which is carried out once or three times a week. Fatigue test with low cycle like this has important significance for LNG storage tanks.

In this study, the fatigue life times of inner tanks and pump columns have been estimated using a common fatigue life time estimation program, FEMFAT.

The characteristics of the welds of inner tanks have been defined in FEMFAT as shown in Figure 11. The characteristics have been defined through welding factors of which physical meaning is an index indicating how precisely the grounding treatment of the weld has been carried out. The less this value is, the more perfect grounding treatment is. In general, the value of welding factor used for welds is taken as 0.89. In the analysis of inner tank welds, the maximum operable cycle for the inner tank was assumed to be 10,000 cycles and the weld factor was taken as 0.89. The analysis results indicated the max damage of the welds of the inner tank as D=1.677e-014 and the location of max damage was indicated as the bottom part of the inner tank as shown in Figure 12.

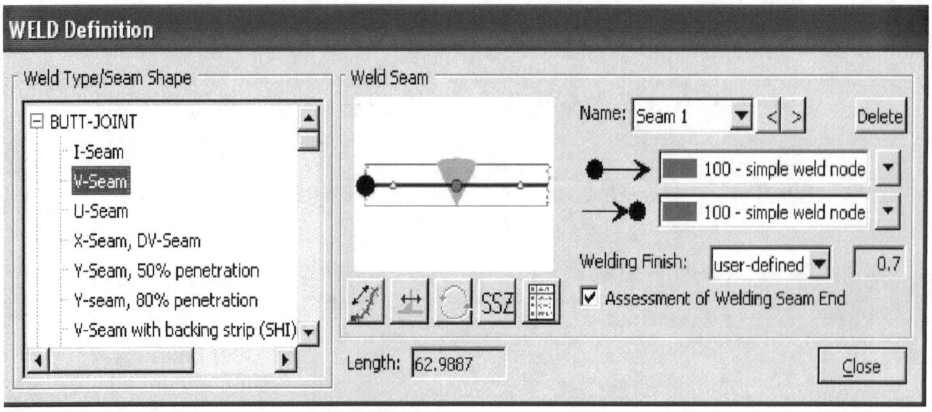

Fig. 11 Definition of welding finish

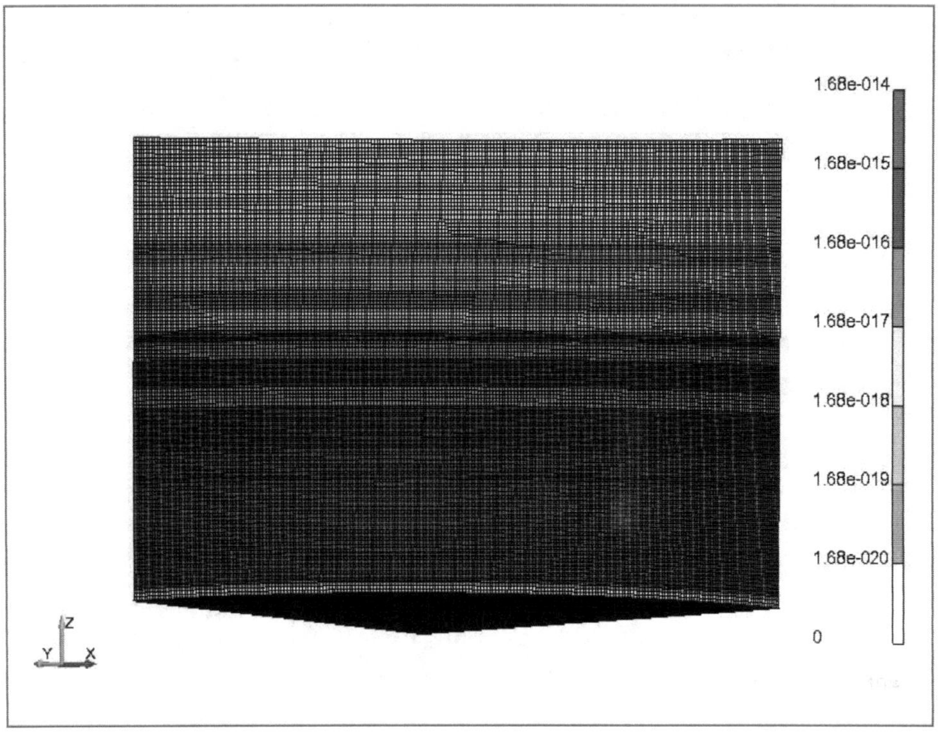

Fig. 12 Location of maximum damage

Accordingly, the number of actual cycles is the number of cycles divided by max damage i.e., 10,000cycle/1.677e-014 = 5.96e+17 cycles, and the life time 2.41e+17/48 (cycle/year) = 1.24e+16 years.

2) Expected fatigue life time of pump column

In order to calculate the fatigue life time of the pump column, the welds of the pump column was defined as shown in Figure 13, and the weld type was defined as T-joint type. The maximum operable cycle was assumed to be 10,000cycles. The analysis results indicated the max damage as 4.143e-016 and the location as the bottom guide of the pump column shown in Figure 14.

The number of actual operable cycles is calculated as 10,000 cycles/4.143e-016 = 2.41e+19 cycles, and the fatigue life time as 5.03e+17 years when the LNG storage tank is assumed to be operated 4 times a month.

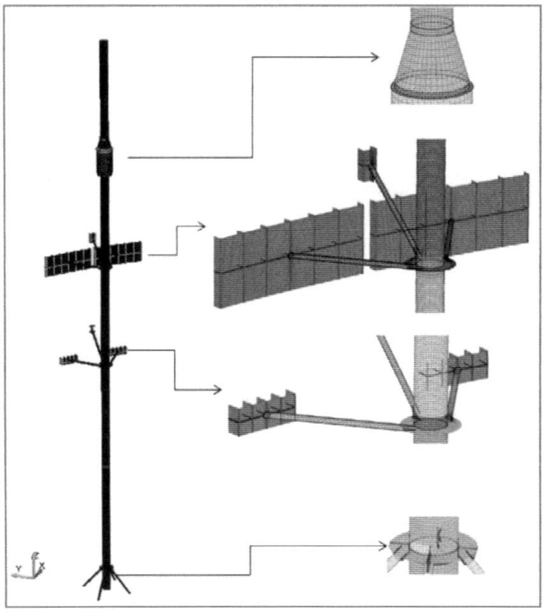

Fig. 13 Definition of weld joints

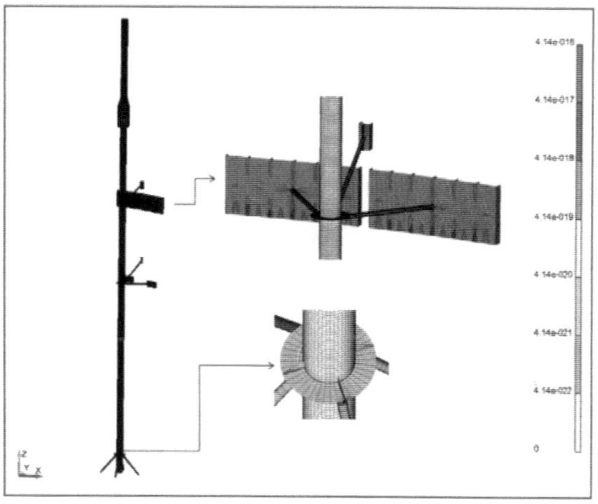

Fig. 14 Distribution of damage

4. CONCLUSION

In this study, fatigue analysis processes for the inner tank and pump column have been presented, and the fatigue life times of the inner tank and pump column have been estimated. The fatigue life time is calculated as 1.24e+16 years for the inner tank and 5.03e+17 years for the tank column.

In the subsequent study, the fatigue characteristics of the welds depending on their finishing treatment and the behavior characteristics of 9% Ni steel at cryogenic temperatures will be confirmed through experiments and the reliability of estimated fatigue life time will be secured through finite element analysis by obtaining the actual operation data base of LNG storage tanks.

ACKNOLWLEGEMENT

This research was supported by a grant from the LNG Plant R&D Center which belongs to the Plant Technology Advancement Program funded by the Ministry of Land, Transportation and Maritime Affairs (MLTM) of the Korean government.

REFERENCES

1. "Development of technologies and standards for the life time management of LNG storage tanks", Seung Rim Lee, 2010 AIChE Spring National Meeting(San Antonio), 2010
2. "EN 1473(Installation and equipment for liquefied natural gas-Design of onshore installations), CEN, 2007
3. "Cases of Diagnoses and Life Extension Measures of LNG Facilities at Tokyo Gas Negishi and Sodegaura Receiving Terminals", Tokyo Gas Co., Ltd, LNG 14, 2004
4. "Aging of installations at LNG Terminals", Takehiko Hasegawa, 22nd WGC, 2003
5. "API RP 579(Fitness for service)", API, 2000
6. "Korean Regulatory Approach to Aging Management of NPPs", KINS, 2003

AUTHOR BIOGRAPHY

Name: Seung Rim Lee
Position: Principal Researcher
Company: Institute of Gas Safety R&D, Korea Gas Safety Corporation
332-1, Deaya-dong, Siheung-si, Gyeonggi-do, 429-712, Korea
Phone: +82.31.310.1462, Fax: +82.31.315.4363
E-mail: srlee@kgs.or.kr
Cellphone : +82.10.5290.6668

Seung-Rim Lee is a Principal Researcher in the Institute of Gas Safety R&D at Korea Gas Safety Corporation. He is a registered professional engineer and has been in LNG industry for sixteen years. He has been involved in research and development projects related to domestic standardization and risk assessment in LNG industry. Especially, he has been participating in development of Korean Standards and international standards (ISO/TC67/WG10) related with LNG storage tanks and LNG terminals from 1999. He graduated from Seoul National University of Science & Technology with a Ph.D. degree in gas engineering in 2009.

Evaluation of Key Parameters for Designing Effective Water Curtain Systems for LNG Facilities using Computational Fluid Dynamics

Byung Kyu Kim, Dedy Ng, Ray A. Mentzer, M. Sam Mannan[*]

Mary Kay O'Connor Process Safety Center, Artie McFerrin Department of Chemical Engineering, Texas A&M University, College Station, TX, USA

**Corresponding author. Tel.: +1 979 862 3985. E-mail address: mannan@tamu.edu (M.S. Mannan)*

Abstract

Concerns over public safety and security of an accidental LNG spill have motivated the continued study of LNG mitigation measures. Forced dispersion of LNG vapors using a water curtain system has been proven to be effective in reducing the LNG vapor exclusion zone by enhancing the dispersion. Currently, no engineering criteria for designing effective water curtains are available mainly due to a lack of experimental data. Different experimental conditions and the discrepancy in defining the parameters make it difficult to understand the complex interaction of the water droplet-LNG vapor system.

This work applies computational fluid dynamics (CFD) modeling to evaluate different key parameters involved in dispersing LNG vapors using an upwards-conical water spray. A forced dispersion model based on an Eulerian-Lagrangian approach was used to model the water spray system. The Eulerian-Lagrangian approach calculates momentum and heat transfer by considering the various effects of the droplets (discrete phase) on the air-vapor mixture (continuous phase). The results were validated against the March 2009 Mary Kay O'Connor Process Safety Center (MKOPSC) outdoor LNG spill experiments conducted at the Brayton Fire Training Field. Consequence assessments on the influence of various properties of water curtain systems were conducted. On the basis of the findings, the effect of different droplet sizes, droplet temperatures, and installation configuration of water curtain systems are evaluated. Finally, the potential for applying the CFD application in providing guidance for setting up the design criteria for the water curtain systems as an integrated mitigation measure for LNG facilities are discussed.

1. Introduction

Growing concerns over carbon emissions and environmental impacts have led industries and governments to search for cleaner energy sources. Natural gas has been recognized as the cleanest fossil fuel and is posed to play a critical role in meeting the environmental criteria while providing a sustainable energy source [1]. The demand for natural gas is expected to increase its share an average rate of 1.7 % per year, catching up with coal consumption by 2035 [2]. The discrepancy between the gas supply and demand market has called for an alternative method of transporting natural gas, and Liquefied Natural Gas (LNG) has been a preferred alternative way of transportation for inter-regional gas trade. With the liquefaction capacity of natural gas expected to double by year 2035 [3], the safety and security concerns over the LNG facilities have motivated the need for continuous study on improving safety mitigation systems for the LNG industry.

The water curtain system has been recognized as one of the most effective and economic mitigation systems in the chemical and petrochemical industries [4]. Applying the water curtain directly to LNG spills has proven to enhance the dispersion and reduce the safety distance to a lower flammability limit zone [5]. However, there has been limited experimental work to draw conclusions and to establish engineering design criteria for water curtain applications in LNG facilities [6, 7]. This work applies computational fluid dynamics (CFD) to simulate the forced dispersion model on LNG vapors and evaluate various design parameters applicable for enhancing LNG vapor dispersion. The physical parameters for setting up the forced dispersion are calibrated using the Mary Kay O'Connor Process Safety Center (MKOPSC) outdoor LNG spill experimental data. In this work, the effects of different design parameters are evaluated using CFD forced dispersion modeling.

2. Forced Dispersion Modeling

Previous work by the authors discusses the physical parameters that are essential inputs in setting up a model for the forced dispersion of LNG vapor clouds using CFD [8]. A dispersion model of LNG vapors was initially set up and validated using the concentration and temperature data collected from March 2009 MKOPSC LNG spill experimental work [9]. The parameters for setting up the forced dispersion model were then calibrated using the dilution factors determined from the experimental data after application of the water curtain. After calibrating the essential parameters in simulating the forced dispersion of the LNG vapor cloud, the meteorological conditions from the Falcon-1 LNG spill test, which was conducted by Lawrence Livermore National Laboratories in 1987, was adopted to apply the worst-case scenario [10]. The atmospheric condition from the Falcon-1 experiment allows simulating an actual stable condition which allows the largest concentration to be observed at the furthest location from the LNG source.

The March 2009 MKOPSC outdoor LNG spill experiment used a total of 40 m^3 of LNG which was spilled on a small confined water pit. Point gas detectors and type-K thermocouples were used to collect experimental data [5]. Three different types of water nozzles, the conical upwards, downwards, and flat-fan upwards, were used to verify the effectiveness of each type of nozzle. The conical upwards nozzle, which had been verified as most effective in dispersing a LNG vapor cloud, has been modeled using the Eulerian-Lagrangian approach to simulate the complex interaction between the air-vapor mixture and the water droplets.

The Eulerian-Lagrangian approach has been recognized as a sound approach to provide flexible solutions in simulating a water curtain [11-13]. Detailed equations and application of the Eulerian-Lagrangian approach in simulating the water curtain application are discussed in previous work by the authors [8]. This work expands upon the results of earlier simulation work setup by the authors. The Discrete Phase Model (DPM) in ANSYS Fluent was applied to simulate the water droplets as a discrete particle travelling through the continuum. The DPM requires some important parameters, such as the water flow rate, initial droplet velocity, droplet sizes, size of nozzle, type of nozzle, droplet materials, and temperature, in simulating water sprays. The interaction between the droplets and the air-vapor mixture are simultaneously calculated based upon underlying principle of fundamental fluid dynamics of the multiphase flow [8].

The forced dispersion model was set up using Fluent 12.1 on ANSYS platform. The modeling work was carried out on Window Server 2008 on Chemical Engineering Department server at Texas A&M University, with a two-quad core Xeon processor (clock speed of 2.93GHz) and 18 GB RAM memory. The simulations were set to run for 200 seconds of natural dispersion and 200 seconds with the water curtain application. Previous results from the simulation showed that the dispersion scenarios reached steady state before 100 seconds in transient mode.

3. Designing An Effective Water Curtain System

Water curtains apply water droplets produced from the breakup of bulk liquid from the nozzles [14]. Some of important factors for designing effective water curtain systems include installation distance, nozzle configuration, flow rate, and droplet size.

The droplet size characterized using the Sauter Mean Diameter (SMD) can show performance of water curtain with the droplet's surface and volume [15]. The droplet size is an essential characteristic of water curtain application as smaller droplets entrain more air, but provide less momentum to the air-vapor mixture. The heat transfer effect, which is one of the dominant mechanisms in warming up the LNG vapor cloud, is also determined by the droplet size. The installation distance from the source is also important as dilution effects can dramatically increase as the mitigation measures are installed close to the source. Water curtains must be placed so that the droplets can effectively interact with the vapor cloud, while it must be placed far enough so that the water droplets do not enter the LNG spill. When water droplets contact LNG, a physical explosion known as a rapid phase transitions (RPT) may occur [16]. There is no current understanding of how effective the dilution becomes as the nozzles are moved closer to the LNG source. The configuration or the orientation of the water nozzles also plays a critical role in determining the effectiveness of water curtain application.

4. Results and Discussion

4.1 Droplet size

Water nozzles produce droplets due to the pressure difference across the liquid pressure at the nozzle inlet and the atmosphere. The droplet size is determined by the water pressure and the nozzle design, which is an intrinsic characteristic for a specific nozzle design. A total of eight conical nozzles (1" TF 48 NN BETE Fog Nozzle) were installed on a two separate 2" OD carbon steel pipeline. Four nozzles were installed on each pipeline in a V-shape, which produces a 60^0 full cone, spiral pattern. The size of water droplet produced from the nozzle is inversely proportional as the water pressure increases. The correlation between the water pressure and the droplet size for the 1" TF 48 NN BETE Fog Nozzle is provided by the vendor [17].

Table 1. Mass flow rate and droplet sizes

		S01	S02	S03	S04	S05	S06	S07	S08	S09
Mass flow rate	**Kg/s**	2.11	2.99	4.23	5.18	5.98	6.69	7.33	8.46	9.46
Droplet size	**mm**	1.43	1.16	0.94	0.83	0.76	0.71	0.68	0.62	0.58

A total of 9 different droplet sizes were simulated to evaluate the heat transfer rate from the droplet to the air-vapor mixture as shown in Table 1. As the water flow rate increases, the size of the droplet decreases as the pressure gradient across the nozzle increases.

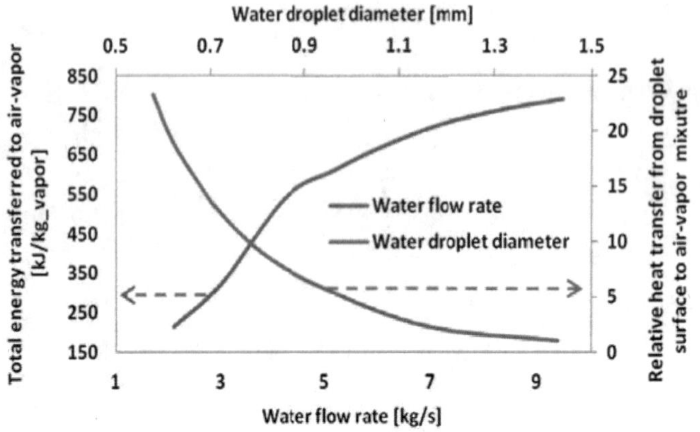

Fig. 1. Heat transfer rate from droplets and droplet surface to air-vapor mixture

Fig. 1 shows the total heat transfer rate from the droplets and droplet surface to air-vapor mixture. The results clearly indicate that the total amount of heat transferred from the water droplet to air-vapor mixture increased dramatically as the water flow rate increased. The total amount of heat transferred from the droplet surface to air-vapor mixture decreased as the droplet size increased. This is due to smaller droplets providing larger surface, where the heat transfer effect mainly takes place.

By normalizing the total amount of heat transfer from the water droplet to air-vapor mixture by the total amount of water used in the forced dispersion, the optimal droplet size which induces maximum heat transfer effect can be verified. Although this result is only applicable for the specific nozzle applied in this work, the approach taken in this analysis provides a way to evaluate the optimal droplet size which can effectively induce maximum heat transfer to warm up the LNG vapors.

4.2 Water Droplet Temperature

Three water droplet temperatures (293K, 303K, and 313K) were applied to compare the different heat transfer effects on behavior of the air-vapor mixture. The other variables, such as the water droplet size and flow rate, were set constant to observe only the thermal effects from various water droplet temperatures.

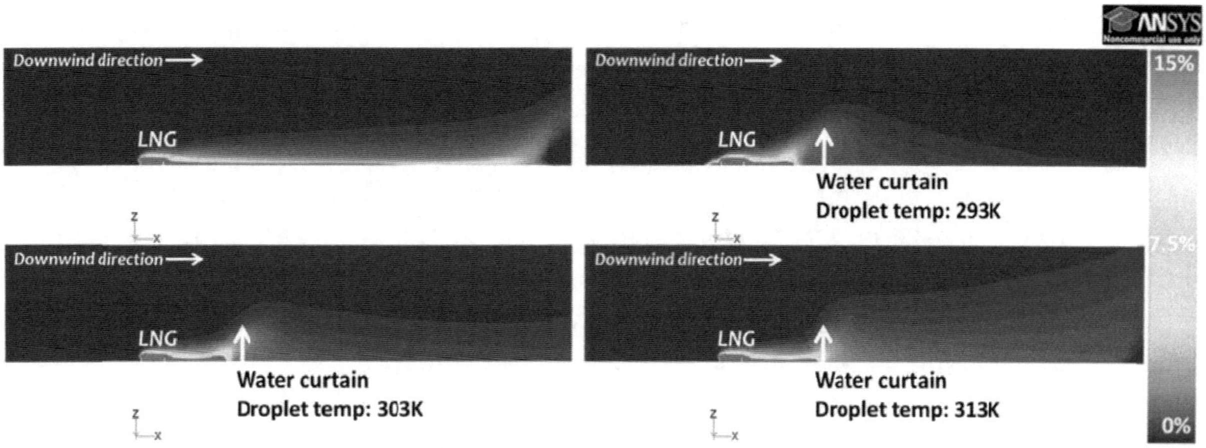

Fig. 2. Vapor concentration contour ([v/v] %) of LNG natural dispersion and forced dispersion with three different water droplet temperatures; 293K, 303K, and 313K

Fig. 2 shows the vapor concentration contour of LNG natural dispersion and with a water curtain applied. Three different droplet temperatures (293K, 303K, and 313K) were applied to assess its sensitivity to different vapor behavior behind the vapor-water curtain system. Dense gas behavior is observed for the natural dispersion of LNG vapors, dispersing in the downwind direction until sufficiently warmed to become positively buoyant. Water droplets at 293K show LNG vapor interaction around the water curtain region, however, the vapors are not sufficiently warmed to be dispersed into atmosphere. On the other hand, the droplets with 303K and 313K show vapor sufficiently warmed to become positively buoyant after the LNG vapors pass through the water curtain region. This indicates that water droplet must be above certain temperature to induce adequate thermal effects to make the LNG vapor positively buoyant, otherwise, the effectiveness of dilution is rather limited as observed from 293 K case. These results can be used to study the effect of various water droplet temperatures and determine the minimum temperature required to design an effective mitigation system for LNG facility.

4.3 Installment Configuration

The installation distance must be carefully determined as the effectiveness of the water curtain is strongly influenced by the location of the water curtain.

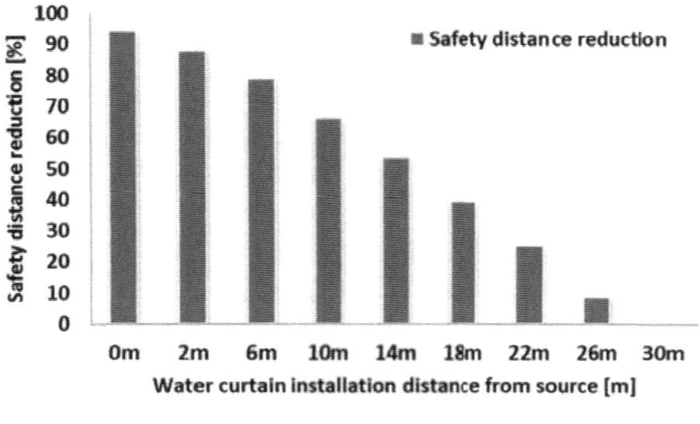

Fig. 3. Safety distance reduction depending on the water curtain installation distance from LNG source

Fig. 3 shows the safety distance reduction from placing the water curtain at different distances from the LNG source. The same water curtain settings were applied in order to compare the effectiveness of dilution for various locations. The safe distance, defined as where the LFL is reduced by 50% (2.5% v/v), reduced dramatically as the water curtain was installed closer to the LNG source. These results can be used to evaluate the impact of the distance factor in calculating the dilution effects.

A water curtain must be separated a certain distance from a potential spill to prevent the water droplets from entering the LNG pool. Most of experimental work in applying the water curtain in enhancing LNG dispersion mainly used water nozzles installed vertically upwards or downwards. Different configurations of the water nozzle were considered, which may allow water curtain systems to be installed closer to the LNG pit. Water curtain nozzles were installed 30°, 45°, and 60° tilted towards downwind direction (Fig. 4(a)). These configurations, in a real application will allow the water curtain system to be installed closer to the potential LNG spill without water droplets entering the LNG pit.

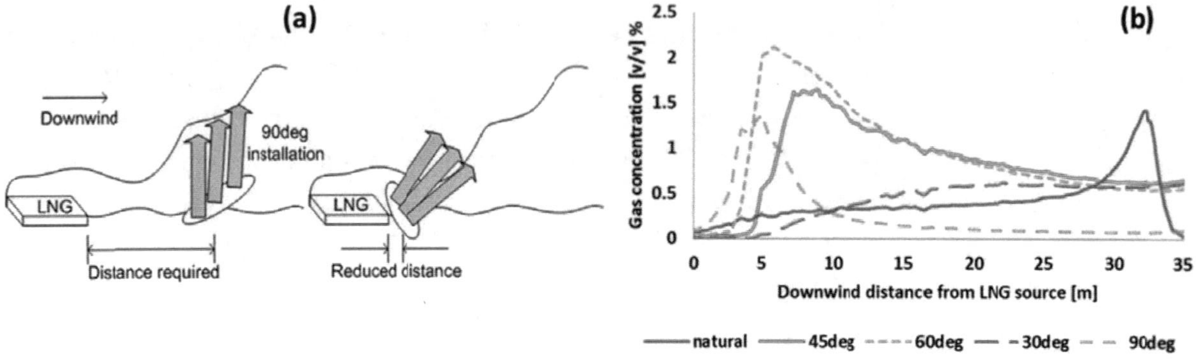

Fig. 4. Various configuration design (a) and vapor concentration data at z=2.1m elevation (b)

Fig. 4(b) shows the concentration data ([v/v] %) of the LNG vapors approximately 8 m away from the LNG source at 2.1 m high above the ground. Different vapor behavior behind the water curtain-vapor system for different nozzle configurations as observed from the vapor concentration. This may be due to differences in the momentum transferred from the water droplets to the air-vapor mixture. Also, the water curtains with different configurations can induce different turbulence behind the water curtain-vapor system, which also plays a critical role in enhancing the mixing effects. Further work is planned to verify different physical effects induced from various nozzle configurations.

5. Conclusion

The forced dispersion modeling using CFD has been applied to evaluate the key parameters required to design an effective water curtain system applicable to a LNG facility. The physical parameters for setting up the forced dispersion model were calibrated using March 2009 MKOPSC outdoor LNG spill experiment data. Sensitivity analyses were conducted to seek various parameters for designing effective mitigation systems.

The effects of various droplet sizes were verified by evaluating the amount of heat transferred from the water droplets to LNG vapors. An optimal droplet size can be verified by normalizing the amount of heat transfer by the amount of water applied. Also, the effects of various droplet temperatures on the vapor behavior were verified, which can provide guidance on determining the minimum water temperature required for effective dilution effects. The effects of different installation distances were verified by varying the points where the water droplets were introduced. The effectiveness increased as the installation distance decreased from the LNG source. These results can be used to find the correlation of direct impacts for various installation distances. Lastly, the water nozzles were tilted towards the downwind direction to evaluate the effects, and the concentration data showed different vapor behavior induced from some of the modified configurations. Further investigation is required to understand impacts of the modified configuration, such as the turbulence and mixing effects.

The investigation made in this work can only be applied for only a specific nozzle (1" TF 48 NN BETE Fog Nozzle) used in this study. However, this work provides a promising methodology which could be expanded to evaluate the key parameters for designing effective water curtain applications for different types of nozzles, or other flammable or toxic gas mitigation.

References

1. DOE, Balancing Natural Gas Policy: Fueling the Demands of a Growing Economy (Vol. I, Summary of Findings and Recommendations). 2003, National Petroleum Council: Washington, DC.
2. EIA, International Energy Outlook 2011. 2011, US Energy Information Administration: Washington, DC.
3. IEA, World Energy Outlook 2011: Are we entering a golden age of gas? 2011, International Energy Agency: Paris, France.
4. Uznanski, D.T., Buchlin, J. M., Mitigation of industrial hazards by water spray curtains, in proceeding of The International Emergency Management Conference 1998: Disaster and emergency management: International challenges for the next decade. 1998: Washington, DC. p. 153-168.
5. Rana, M.A., Mannan, M. S., Forced dispersion of LNG vapor with water curtain. Journal of Loss Prevention in the Process Industries, 2010. 23(6): p. 768-772.
6. Martinsen, W.E., Muhenkamp, S. P., Olson, J., Disperse LNG vapors with water. Hydrocarbon Processing, 1977: p. 260-267.
7. Heskestad, G., Meroney, R. N., Kothari, K. M., Neff, D. E. Effectiveness of water spray curtains in dispersing LNG vapor clouds. in proceeding of the American Gas Association Transmission Conference Paper. 1983. Seattle.
8. Kim, B.K., Ng, D., Mannan, M.S. Application of computational fluid dynamics modeling in forced dispersion of LNG vapor cloud. In proceeding of 14th Annual Symposium Mary Kay O'Connor Process Safety Center. 2011. College Station, TX.
9. Rana, M., Forced dispersion of liquefied natural gas vapor clouds with water spray curtain application, PhD dissertation, 2009, Texas A&M University. p. 232.
10. Brown, T.C., Cederwall, R.T., Chan, S.T., Ermak, D.L., Koopman, R.P., Lamson, K.C., McClure, J.W., Morris, L.K., Falcon Series Data Report 1987 LNG Vapor Barrier Verification Field Trials. 1990, Gas Research Institute: Livermore, CA.
11. Gant, S.E., CFD Modelling of Water Spray Barriers. 2006, Health & Safety Laboratory.

12. Alessandri, E., Buchlin, J.M., Cavallini, A., Patel, M.K., Galea, E., On the modelling of the thermal interactions between a spray curtain and an impinging cold gas cloud. 1996, Centre for Numerical Modelling and Process Analysis, University of Greenwich: London, UK: in Paper 96/IM/11, CMS Press.
13. Nijdam, J.J., Guo, B., Fletcher, D.F., Langrish, T.A.G., Challenges of Simulating Droplet Coalescence within a Spray. Drying Technology: An International Journal, 2004. 22(6): p. 1463 - 1488.
14. Jiang, X., Siamas, G.A., Jagus, K., Karayiannis, T.G., Physical modelling and advanced simulations of gas-liquid two-phase jet flows in atomization and sprays. Progress in Energy and Combustion Science, 2010. 36(2): p. 131-167.
15. CCPS, Guideline for postrelease mitigation technology in the chemical process industry. 1997, New York: Center for Chemical Process Safety of American Institute of Chemical Engineers.
16. Melhem, G.A., Saraf, S., Ozog, H., LNG Properties and Hazards: Understand LNG Rapid Phase Transition (RPT), ioMosaic Corporation Whitepaper. 2006, ioMosaic Corporation.
17. BETE, Nozzles for industry, pollution control and fire protection. 2007, BETE Fog Nozzle.

CFD simulation of the Falcon tests using the homogeneous equilibrium model in FLACS

Mathieu Ichard GexCon AS, Jayson Geiser GexCon US, and Filippo Gavelli GexCon US

GexCon

Abstract

The Falcon test series consisted of five large-scale LNG spill tests performed by the Lawrence Livermore National Laboratory in 1987 at Frenchman Flat, Nevada. The purpose of the tests was to evaluate the effectiveness of vapor fences as a mitigation technique for accidental LNG releases as well as to provide a dataset for model validation purposes. For this reason, LNG was spilled onto a specially designed water pond equipped with a circulating system to maximize evaporation of the LNG pool.

Previous attempts at simulating the Falcon test release scenarios utilizing computational fluid dynamics tools used either pure vapor releases or pool spreading models (assuming pure liquid spills) to introduce LNG vapors into the domain; in some of these simulations, additional contributions (e.g., turbulent kinetic energy) were introduced inside the fenced area in order to cause the vapor cloud to grow and overflow the impoundment in a manner approximately consistent with the experiments.

In this paper, the Falcon tests are simulated using the CFD software FLACS with a recently implemented two-phase modeling capability, which is based on the homogeneous equilibrium model (HEM). The HEM version of FLACS allows tracking of the vapor and aerosol streams from a flashing jet release, including heat and mass transfer, interaction with walls and obstructions, and droplet rainout. The paper compares the model predictions with the experimental data from the Falcon 1 trial and discusses the advantages of the two-phase model for safety studies in actual facilities.

Background

The Falcon test series consisted of five large-scale LNG spill tests performed by the Lawrence Livermore National Laboratory in 1987 at Frenchman Flat, Nevada (1). The purpose of the tests was to evaluate the effectiveness of vapor fences as a mitigation technique for accidental LNG releases as well as to provide a dataset for model validation purposes (see Figure 1). For this reason, LNG was spilled onto a specially designed water pond equipped with a circulating system to maximize evaporation of the LNG pool. LNG was delivered through a spill "spider" consisting of four pipes (each 11.6 m long) spaced at 90° intervals in order to distribute the LNG pool over the water pond. The ends of the spider arms pointed downward and were fitted with splash plates at the water surface to prevent the LNG from sinking into the water. The vapor fence, approximately 8.7 m high, extended upwind of the water pond, enclosing a total area 44 m wide × 88 m long. In addition to the fence, a 13.3 m tall billboard, 17.1 m wide, was located just upwind of the water pond. The purpose of the billboard was to simulate the effect of a storage tank or other obstruction on the dispersion of the vapor cloud. The terrain surrounding the fenced area was flat and the experiments were performed under stable or neutrally stable atmospheric stability conditions (Pasquill-Gifford classes D–F). The total volume of spilled LNG was such that the corresponding volume of saturated vapor was less than the volume of the impoundment. The Falcon tests were heavily instrumented to measure the following quantities: air temperature, pressure and humidity, wind speed and direction, turbulence intensity, heat flux from the ground, and gas concentration. The sensors were located both within the fenced area, to record the behavior of the constrained LNG cloud, and downwind of the fence, to observe the dispersion of the

LNG vapor that spilled over the fence. Sensors were located up to 250 m downwind of the trailing edge of the vapor fence. Several aspects of these tests are desirable for validating a CFD model for LNG spills:

- These tests used a realistically complex geometry that represents conditions likely to be present at or near an LNG release. The vapor fence simulates the effect of the impoundment walls, and the barrier was placed upwind of the spill to introduce additional turbulence, in a manner similar to the effect that containment tanks, buildings or even ships would have on an actual spill. These types of surface features and realistic flow complexity test the capability of the CFD model to accurately predict dispersion in the flow patterns formed in complex geometries.
- A predetermined amount of LNG was spilled into an impoundment at a prescribed rate, and the flammable vapor cloud was then tracked as it was carried downwind and dispersed in the environment. Therefore, the Falcon tests provide data on the entire LNG vapor cloud dispersion scenario.
- These tests included detailed and redundant instrumentation, and measured several quantities of interest that can be directly compared with the CFD model predictions. The array of measurement stations was distributed appropriately in the region adjacent to and downwind of the spill.

Figure 1. Photograph of the experimental setup for the Falcon test series (1)

The LNG Release Mechanism

A critical parameter in the successful simulation of an LNG vapor dispersion scenario is the vapor source term. When LNG (or any other liquefied gas) is released to the atmosphere, vapor can be formed by:

1. Flashing, when the fluid is at a temperature above the normal boiling point;
2. Droplet evaporation, when the fluid is atomized by flashing or by being ejected at high velocity through a breach or orifice;
3. Liquid pool evaporation, when the fluid spills onto a surface which is at a temperature above the fluid's normal boiling point.

The most frequent assumption when considering LNG releases is to neglect flashing and atomization (items 1 and 2 above), reducing the release to a pure liquid spill onto a non-porous surface (e.g., concrete, water or other substrate). Therefore, the only means for vapor generation is evaporation from a liquid pool (item 3 above). Following this assumption, several models have been used to simulate various LNG release tests, such as the Burro, Coyote and Maplin Sands trials (2) (3) (4) (5), generally with good results. When the same approach was applied to simulating the Falcon trials, however, the results were poorer: in the simulations, the LNG vapor cloud formed by pool evaporation tended to remain confined within the fenced area much longer than observed during the experiments, with consequent large underprediction of downwind concentrations. Figure 2 shows a snapshot of the flammable cloud isosurface after 60 s from the beginning of the LNG release, for a simulation performed by Hansen et al. (6) using only evaporation from the LNG pool for the vapor source term.

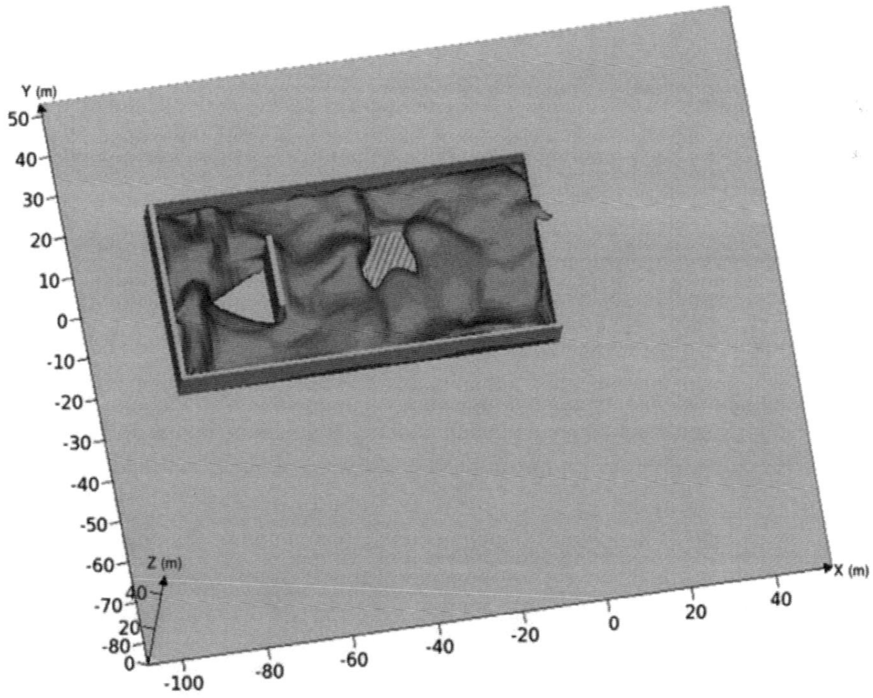

Figure 2. LFL isosurface for the Falcon 1 LNG vapor cloud forming by vaporization of the liquid pool. The snapshot is at 60 s after the start of the release. The simulations were performed with the FLACS pool model.

The main difference between the Falcon trials and the other LNG field test series is the tall vapor fence installed in the Falcon trials to shield the LNG pool from the ambient wind and turbulence. The characteristics of the LNG release, therefore, have a much stronger impact on the growth of the vapor cloud within the fenced area. Videos from the Falcon 5 test clearly show how the LNG release forms a high-momentum wall jet just above the water pond surface. Given the characteristics of the LNG delivery system (pressurized LNG flows through a long piping system, then passes through flow restricting orifices just before a 90-degree bend in the pipe, as shown in Figure 3) it is very likely that both flashing and aerosol formation occurred during the Falcon tests. Neglecting the effect of these phenomena on the growth of the LNG vapor cloud may lead to significant prediction errors.

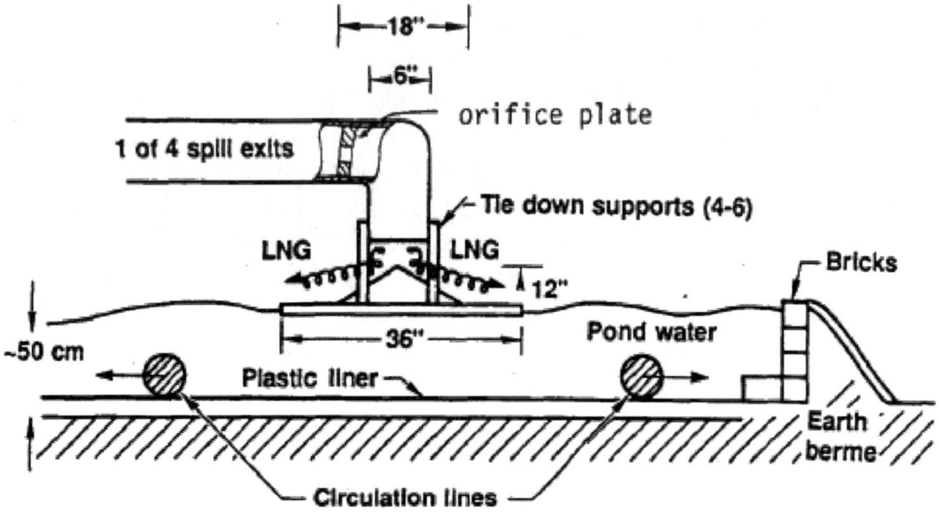

Figure 3. Falcon trial LNG delivery system (2)

Similar delivery systems were used in the other LNG trials; however, as mentioned by Gavelli et al (4), when an LNG pool is exposed to ambient wind, the ambient turbulence tends to dominate over the turbulence associated with the LNG release and the pool boiling. Therefore, for unobstructed releases, the effects of flashing and of the atomized jet are likely overwhelmed by natural turbulence. In the Falcon tests, however, the tall vapor fence shields the LNG pool from the ambient wind; in this case, the characteristics of the LNG release control the growth of the vapor cloud within the fenced area.

Previous Modeling Attempts

The importance of turbulence within the impoundment was recognized early on by Chan (7). In his CFD simulations of the Falcon tests, he imposed a uniform diffusivity both within and in proximity of the fenced region. Chan, however, did not discuss how he determined a suitable value of diffusivity (other than, presumably, by "tuning" it to best match the experimental results); therefore, this approach is not suitable for the application of CFD modeling to perform hazard analysess of LNG releases.

Gavelli et al. (4) focused on the analysis of Falcon trial videos to better describe the behavior of the LNG release and to extract release-related turbulence parameters that would improve a model's prediction of the cloud growth within the fenced region. While their analysis provides a scientific basis for the values of turbulence parameters used in the Falcon trial simulations, this approach is not applicable to scenarios (such as hazard analyses) where experimental data is not available from which such data may be extracted.

FLACS

FLACS, developed and maintained by GexCon AS in Norway, is a computational fluid dynamics (CFD) tool to model ventilation (i.e., natural or mechanical air flow), gas dispersion, gas/vapor cloud explosions and blast propagation in three-dimensional geometries, such as complex process areas. Because of its capabilities, FLACS is widely used for the quantification and management of explosion risks in the offshore petroleum industry and onshore chemical industries (8) (9) and in some cases is explicitly required by the regulations (10).

FLACS solves the compressible Reynolds-Averaged Navier–Stokes (RANS) equations on a 3D Cartesian grid using the finite volume method.[1] The conservation equations for mass, momentum, enthalpy and species closed by the ideal gas law are included. The RANS equations are closed using the k-ε turbulence model with the standard set of constants taken from Launder and Spalding (11).

Since even small objects can have a significant impact on gas dispersion and – even more so – on flame acceleration, the accurate representation of the 3D geometry has always been a focal point in the development of FLACS. A so-called "distributed porosity concept" (12) was developed to allow the efficient handling of complex geometries using a Cartesian grid: large objects and walls are represented on-grid, while smaller objects are represented as sub-grid elements. Sub-grid elements contribute to flow resistance, turbulence generation and flame folding in the simulation, which allows the effect of small obstacles to be accounted for accurately, while maintaining reasonable simulation times (13).

The Homogeneous Equilibrium Model

The Homogeneous Equilibrium Model (HEM) is a simplified approach for modeling two-phase flows which assumes that both phases are in local thermal and kinematic equilibrium. This implies that the heat and mass transport between the phases is infinitely fast. Therefore, this approach only distinguishes the two phases by their volume (or mass) fractions in a given mixed volume. The temperature, velocity, and other flow field parameters are the same for the two phases.

The HEM approach has two main advantages:

- It requires limited information about the aerosol characteristics;
- The conservation equations are similar to those for single phase flow.

Therefore, the HEM does not require information about the particle diameter and velocity at the source. This assumption can be advantageous because a lot of uncertainties remain in estimating the characteristics of droplets produced by the atomization of liquid sprays, despite recent efforts (14). The main disadvantage of the HEM approach is due to the equilibrium assumption, which tends to fail for jets characterized by large droplet diameters. Kukkonen (15) used the HEM approach to compute the dispersion of two-phase ammonia jets and found that the results were virtually identical (within 1%) to those from more sophisticated models when the ammonia droplets had a mean diameter smaller than 100 microns. Faeth (16) provided a thorough theoretical description of the HEM approach and reported comparisons between HEM predictions and experiments for evaporating and combusting sprays and concluded that the finite inter-phase transport rates account for much of the difference between the HEM approach and experimental observations.

The implementation of the HEM method within the FLACS CFD tool is relatively simple: in summary, it reduces to adding a species (the aerosol, or liquid fraction) and its conservation equations to the existing FLACS solver. In each grid cell, the mass fractions of air, gas phase, and liquid phase ("fuel") are tracked. Transport equations are solved for the two fuel mixture fractions (gas and liquid phases):

$$\frac{\partial \rho f_*}{\partial t}+\frac{\partial \rho u_j f_*}{\partial x_j}=\frac{\partial}{\partial x_j}\left(\frac{\mu_t}{\sigma_{f_*}}\frac{\partial f_*}{\partial x_j}\right)+S_{f_*} \qquad (1)$$

The source term Sf accounts for evaporation and eventual deposition of the liquid phase (i.e., rainout). The transport equation for the mixture enthalpy becomes:

$$\frac{\partial \rho h}{\partial t}+\frac{\partial \rho u_j h}{\partial x_j}=\frac{\partial}{\partial x_j}\left(\frac{\mu_t}{\sigma_h}\frac{\partial h}{\partial x_j}\right)+S_h \qquad (2)$$

where the source term Sh represents the loss of energy due to rainout. Evaporation of the liquid phase is taken into account by ensuring local thermodynamic equilibrium. Thermodynamic equilibrium implies that the partial pressure of the contaminant in the vapor phase is equal to the saturation pressure at the mixture temperature. The more challenging task is to couple the HEM with a rainout prediction model, since droplet deposition is strongly affected by the aerosol droplet size distribution, which is a parameter intentionally omitted in the HEM theory. Rain-out is due to jet impingement either directly on the ground or on obstacles. Consider a jet impinging on an obstacle located at a distance L in the positive x direction. In a grid cell (I,J,K) which is directly adjacent to the obstacle, the amount of liquid that rains-out at each time step is estimated via the following relation:

$$\begin{cases} m_{RO}(\vec{x})=\sum_{q,k} Y^l_{q,k} m_k \quad if \quad T(\vec{x})\le T_{NBP} \\ m_{RO}(\vec{x})=0 \quad if \quad T(\vec{x})>T_{NBP} \\ \vec{x}=\left[x(I),y(J),z(K)\right] \end{cases} \qquad (3)$$

Rain-out is allowed in FLACS only if the temperature of the mixture inside the grid cell (I,J,K) is less or equal than the normal boiling point of the mixture. The sum on the index k in Equation 7 is performed over all grid cells located at a distance less or equal than ℓ_{RO} from the wall. The characteristic length scale ℓ_{RO} for the rain-out process is defined as the product of the characteristic velocity (ϑ_{RO}) and characteristic time (τ_{RO}) scales for the rain-out process. The characteristic velocity for the rain-out process (ϑ_{RO}) is taken to be the local maximum velocity in regions of the flow where liquid is evaporating and thermodynamic equilibrium has been attained. This can be justified by the fact that the mass of liquid that deposits on the ground or on obstacles is expected to be contained inside the droplets having the largest diameters. These droplets have a large inertia, much larger than the inertia of fluid particles, and hence have an extended memory of their maximum velocities. Finally, the characteristic time τ_{RO} over which the rain-out process occurs is assumed to be the time needed by the flow to achieve a new thermodynamic equilibrium in the grid cell (I,J,K) adjacent to the wall. This time-scale is approximated by the time needed by the flow to bring new fluid particles to this grid cell.

Appropriate source terms are integrated inside the transport equations for the mixture fractions and mixture enthalpy. The amount of liquid that is removed from the flow is injected inside the pool model.

FLACS Model Validation

Model validation has been a critical component of FLACS development since its inception (17). As a result, a large database of FLACS validation examples currently exists which includes gas dispersion (18) (19) and vapor cloud explosion experiments (20), spanning from laboratory-scale to full-scale experiments performed by different groups. Several of the validation studies, particularly the most recent ones, consist of blind validation exercises (i.e., the simulations were performed prior to or without knowledge of the experimental results) and demonstrate the ability of FLACS software to accurately predict gas dispersion and explosion scenarios without "tweaking".

The atmospheric gas dispersion model in FLACS has been tested against a wide range of dispersion scenarios for different gases and release conditions, including releases of dense, passive and buoyant gases in open, obstructed, and enclosed spaces. Validation of the FLACS gas dispersion capabilities started in the 1990s with natural gas and propane releases and now includes hundreds of onshore and offshore experiments. Subsequently, FLACS was validated against several other field experiments involving both dense gas and passive gas releases, including CO2 releases in an array of flat "billboard"-shaped obstructions (the Kit Fox experiments), tracer gas releases in an array of rectangular storage containers in the Mock Urban Setting Test (MUST) experiments (3), and in intermodal comparisons, such as the simulation of tracer gas dispersion experiments in Manhattan (21). Recently, FLACS was evaluated according to the requirements in the Fire Protection Research Foundation's Model Evaluation Protocol (MEP) (22) whose purpose was to assess the suitability of dispersion models for predicting hazard ranges associated with large spills of LNG. The models need to be validated against key experimental data which are provided in a Model Validation Database, which consists of both large-scale field trials as well as wind tunnel tests. All 33 tests in the Model Validation Database were simulated with FLACS and the model predictions matched well with the experimental data, meeting the statistical performance requirements in the MEP (21). The MEP evaluation work was then submitted to the United States Department of Transportation to petition for approval of FLACS as an acceptable model for LNG vapor dispersion calculations under the Code of Federal Regulations; approval for such petition was granted on October 7, 2011 (23).

The homogeneous equilibrium model implemented in FLACS has been validated against experimental data sets involving different configurations and fluids, such as: the FLIE test series, which included propane and butane free and impinging jet experiments performed at INERIS (24); the Desert Tortoise large-scale ammonia releases (25); and liquid hydrogen jet releases performed by the UK Health and Safety Executive (26). Both the two-phase jet dispersion and the liquid rainout predictions using the HEM model were found to give accurate results when compared with the available experimental data

Falcon 1 Trial Simulation Results

The Falcon 1 trial is simulated with the HEM approach implemented in FLACS. The ambient conditions for the Falcon 1 trial are listed in Table 1.

Table 2: Meteorological data for test Falcon 1

Trial	U_{ref} (m/s)	Z_{ref} (m)	T_a (K)	Z_0 (m)	Stability class (-)	Monin-Obukhov length (m)
1	1.7	2	306	0.008	G	4.963

The release of LNG through the system illustrated on Figure 3 was simulated using four two-phase jets, located in the first grid cell above the water surface at the horizontal location corresponding to the ends of the four spill spider arms. The release configuration is shown in Figure 4.

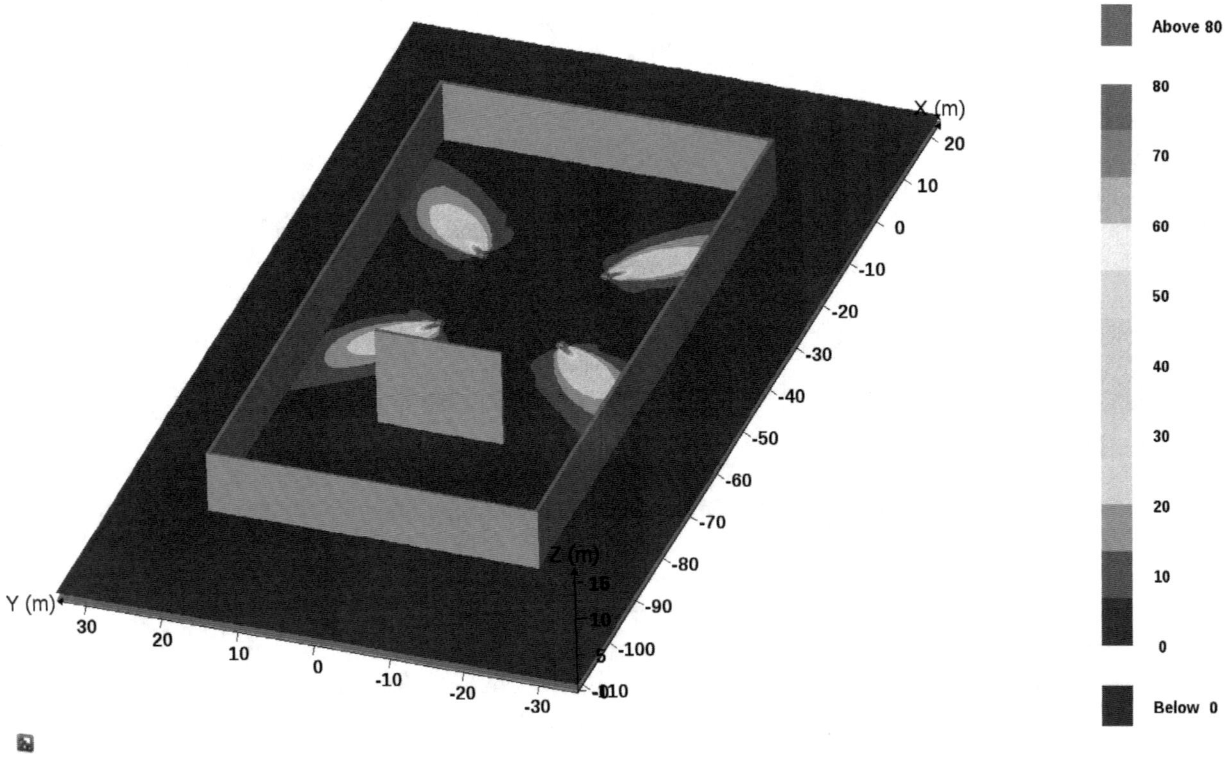

Figure 5: Cut-plane showing the mean velocity of the released LNG for the test Falcon 1

Since the Falcon trials did not record the information necessary to quantify the composition (vapor vs. aerosol) of the LNG release, a sensitivity study was performed on the mass fraction of vapor in the LNG release. Several values for the aerosol mass fraction were tested; for brevity, only three of the tested values are included in this paper. The LNG release flow parameters for the three scenarios here described are as follows:

a. Full flashing and/or vaporization prior to impingement on the water pond surface. This scenario did not result in pool formation;
b. Vapor mass fraction = 80.7%; therefore, only approximately 19.3% of the LNG released was assumed to remain in liquid form and contributed to the formation of a pool on the water pond;
c. Vapor mass fraction = 45.5%; therefore, approximately 54.5% of the LNG released was assumed to remain in liquid form and contributed to the formation of a pool on the water pond..

Table 3: The parameters of the LNG release flow for the Falcon 1 trial simulations

Scenario	x_g (-)
a	1.000
sb	0.807
sc	0.455

The size of the simulation domain goes from -150m to 400m in the x-direction, from -200m to 155m in the y-direction, and from 0m to 50m in the z-direction. The total number of grid cells is 630000.

Results

The predictions of the FLACS-HEM simulations were compared with experimental data from Brown et al. (1). Time series of gas concentration are shown in Figure 6 for two sensors 50 m downwind of the vapor fence and in Figure 7 for two sensors 150 m downwind of the vapor fence . The source term conditions that result in the closest match with experimental results are those of scenario "b" (approximately 80.7% vapor and 19.3% aerosol, by mass). For the scenario "b" simulations, the gas concentration time series at all four downwind sensors compare very well with the experimental results. The simulations performed with setups "a" and "c", on the contrary, largely underpredicted the experimental observations at all shown locations.

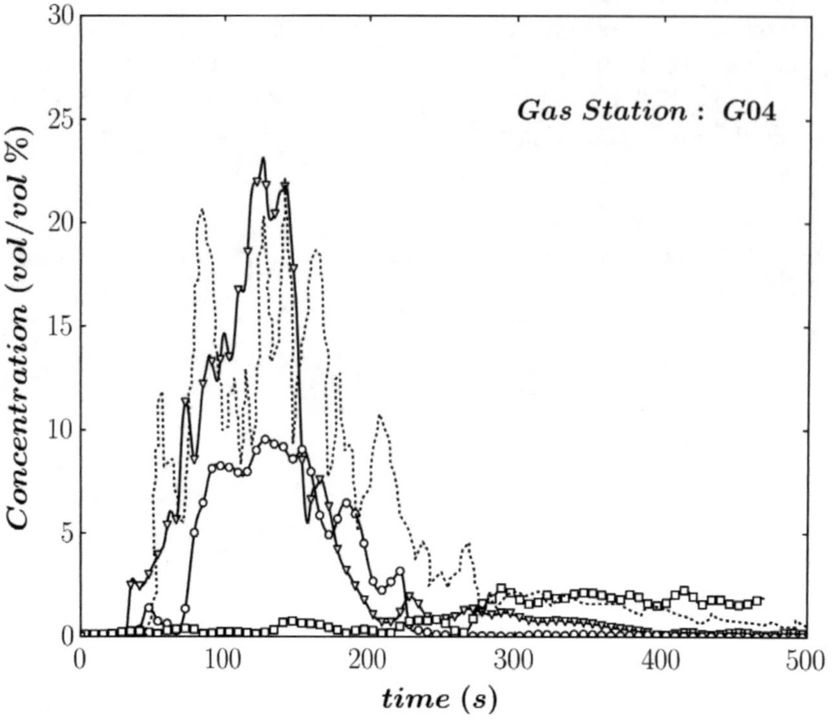

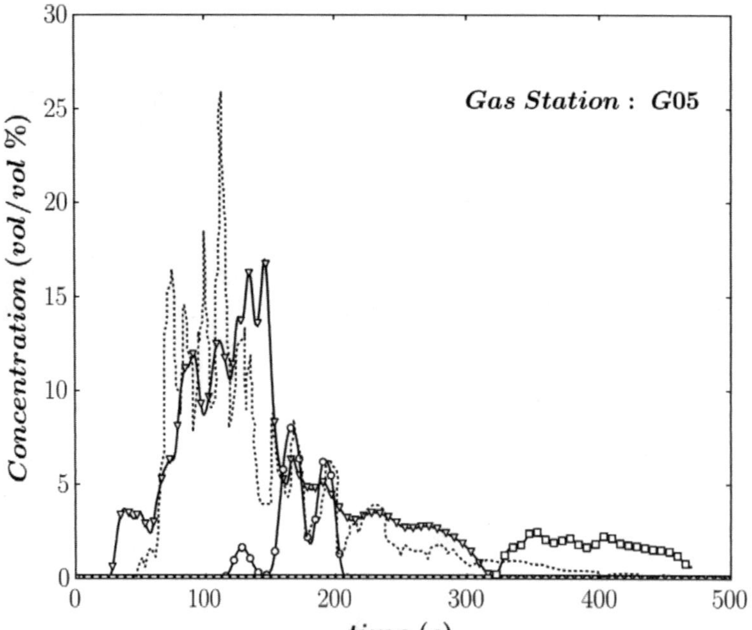

Figure 6: Time series of gas concentration for Falcon 1. Sensors G04 (top) and G05 (bottom) were located 50 m downwind of the vapor fence, 1 m above ground. The dashed lines represents the experimental data, (O) represents the FLACS-HEM simulation for scenario "a", (∇) represents the FLACS-HEM simulation for scenario "b" and (□) represents the FLACS-HEM simulation for scenario "c".

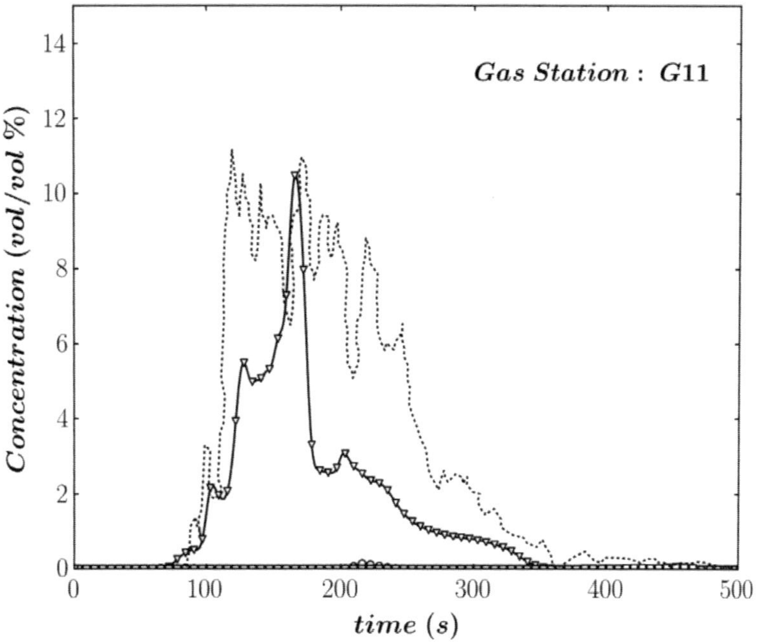

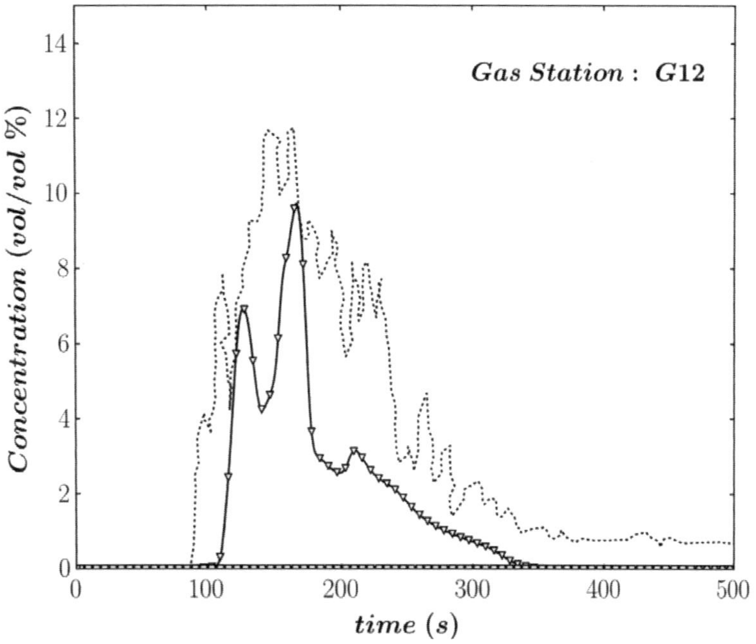

Figure 7: Time series of gas concentration for Falcon 1. Sensors G11 (top) and G12 (bottom) were located 150 m downwind of the vapor fence, 1 m above ground. The dashed lines represents the experimental data, (O) represents the FLACS-HEM simulation for scenario "a", (∇) represents the FLACS-HEM simulation for scenario "b" and (□) represents the FLACS-HEM simulation for scenario "c".

Figure 7 shows the time series of velocity recorded at sensor G04, which was located 1 m above the ground, 50 m from the fence. Both the experimental trace (left) and the FLACS simulation (right) show a marked increase of velocity as the vapor cloud reaches the sensor. It must be noted that wind speed remained relatively flat during the Falcon 1 trial, and that constant wind speed and direction were assumed for the FLACS simulations. Therefore, the approximately five-fold increase in velocity recorded at sensor G04 is a likely result of dense gas effects: the dense vapor cloud likely preserved some of the momentum from the high-velocity jet release, and also accelerates as it falls to the ground after "jumping" the vapor fence. The velocity readings at sensor G04 then return to the ambient wind speed value after the cloud has dissipated. Both effects are well captured in the FLACS simulations.

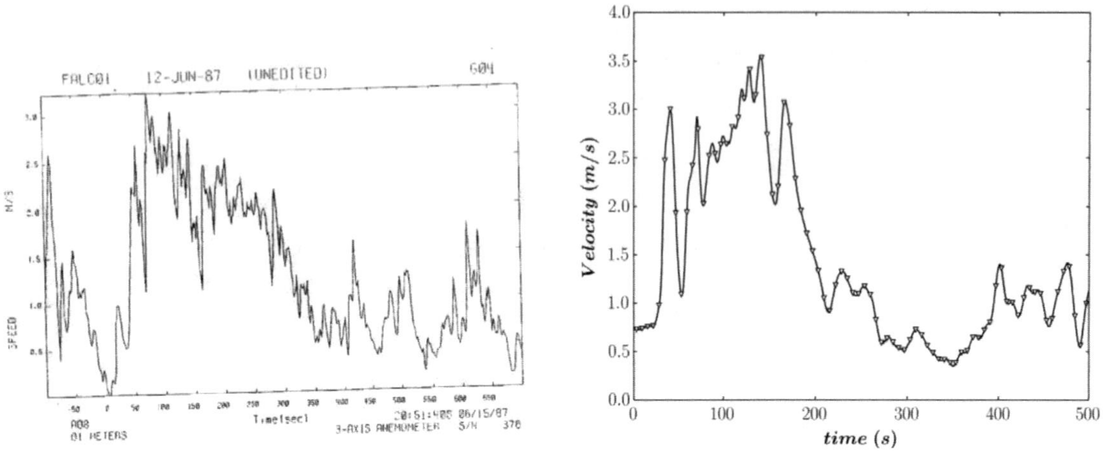

Figure 8. Time series of velocity for Falcon test 1: (left) experimental plot [Brown et al.]; (right) FLACS-HEM simulation using scenario "b". The location is that of sensor G04, which is located 50 m downwind of the fences at a height of 1m above the ground.

Discussion

The results presented in the previous section demonstrate the strong effect of both momentum and composition (vapor vs. aerosol) in the LNG source. The effect of source momentum is fairly intuitive: for a given cloud size and density, higher momentum (i.e., a stronger wall jet on the surface of the pond) will tend to result in earlier overflow of the vapor barrier. This effect is not presented in this paper but was demonstrated by Gavelli et al. (4). The effect of two-phase composition, however, is more complex and can be discussed by comparing the three scenarios examined in this paper.

In scenario "c", where almost half of the release (45.5% by mass) is assumed to flash and/or vaporize, the remaining aerosol contributes to reduce the source jet velocity, to increase the density of the cloud and to maintain the temperature of the mixture near the boiling point of methane (see Figure 9). As a result, a dense, cold vapor cloud grows to fill most of the fenced region and is late in overflowing the fence.In scenario "a", where the entire LNG release is assumed to flash and/or vaporize (i.e., no aerosol or pool formation occurs), the high jet momentum and turbulence lead to rapid mixing of the vapor cloud with air. Consequently, the vapor cloud rapidly begins to overflow the fence. However, as the vapors mix with ambient air, the cloud warms up and soon becomes buoyant, as shown in Figure 10.

In scenario "b", approximately 80.7% of the LNG release (by mass) is assumed to flash and/or vaporize. In this case, the presence of aerosol contributes to keeping the cloud colder than in scenario "a"; however, the smaller amount of aerosol results in a higher jet velocity and in a lighter vapor cloud. As a result, the momentum-driven cloud can overflow the fence within a few seconds, as well as maintain its dense gas behavior as it spreads downwind, as shown in Figure 11.

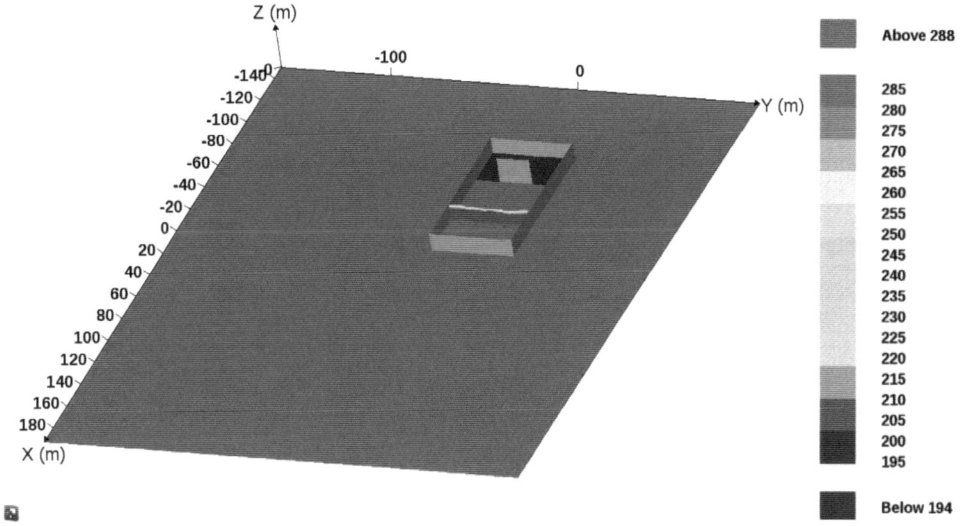

Figure 9: 3D cut planes of temperature for the FLACS-HEM simulation of Falcon 1 under scenario "c". Snapshot at time=140 s.

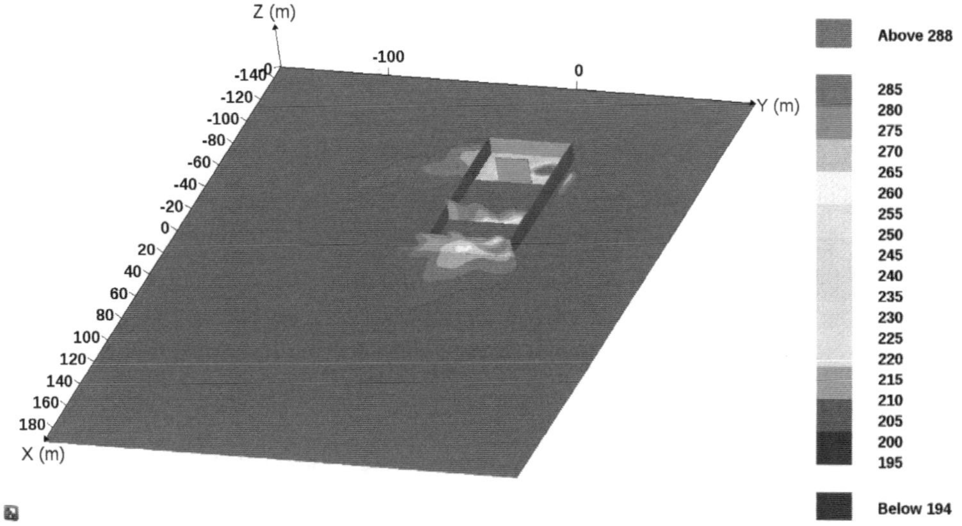

Figure 10: 3D cut planes of temperature for the FLACS-HEM simulation of Falcon 1 under scenario "a". Snapshot at time=140 s.

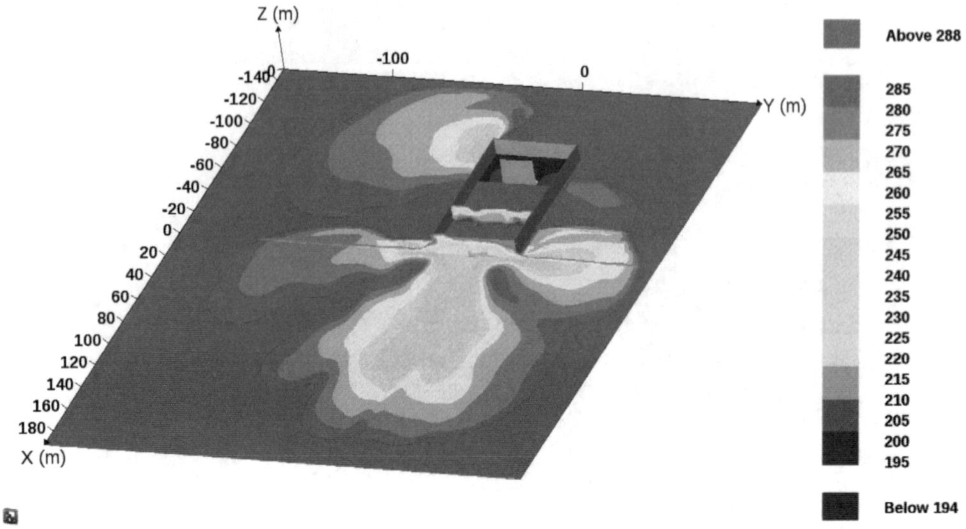

Figure 11: 3D cut planes of temperature for the FLACS-HEM simulation of Falcon 1 under scenario "b". Snapshot at time= 140s.

The sensitivity study performed on the vapor/aerosol mix at the source illustrates the critical importance of source characterization.

Conclusions

The paper discussed the basic principles of a recently developed, two-phase solver for the CFD model FLACS. The two-phase solver implements the Homogeneous Equilibrium Model to track the dispersion of both the vapor and aerosol phases frequently present during the release of a liquefied gas that is stored under pressure.

The paper also presented the application of the FLACS-HEM solver to the simulation of the Falcon 1 LNG trial. The purpose of the simulations was to demonstrate how a more accurate representation of the LNG source term could lead to a significant improvement in the model's prediction, without the need to impose artificial turbulence or momentum sources. The results of the simulations demonstrated a very close match between model predictions and experimental results. The simulations also identified a very strong effect of aerosol on the dense gas cloud behavior; the analysis of a sensitivity study performed on the vapor/aerosol composition of the LNG source suggested that this parameter could play a critical role on the consequences of a flashing jet release, particularly in partially confined and/or highly congested areas.

References

1. **Ermak, S. T. , Brown, T. C., Cederwall, R. T., Chan, R. T., Koopman, R. P., Lamson, K. C.** *Falcon series data report: 1987 {LNG} vapor barrier verification field.* s.l. : Gas Research Institute, 1990. GRI-89/0138.

2. **Brown, T. C., Blewitt, D. N., Yohn, J. F., Koopman, R. P.** *Conduct of anhydrous hydrofuoric acid spill experiments.* New York : Proc Int Conf on Vapour Cloud Modelling, 1987. 1-38.

3. **Hanna, S. R., Hansen, O. R., and Dharmavaram, S.** *FLACS CFD air quality model performance evaluation with Kit Fox, Prairie Grass, and EMU observations.* s.l. : Atmos. Environ., 2004, Vol. 38. 4675-4687.

4. **Gaveli, F., Bullister, E., Kytomaa, H.** *Application of {CFD} ({F}luent) to {LNG} spills into geometrically complex environments.* 1, s.l. : Journal of Hazardous Materials, 2008, Vol. 159. 158 - 168.

5. **Sklavounos, S., Rigas, F.** *Simulation of Coyote series trials—Part I:: CFD estimation of non-isothermal LNG releases and comparison with box-model predictions.* 5, s.l. : Chemical Engineering Science, 2006, Vol. 61. 1434-1443.

6. **Hansen, O.R., Gavelli, F., Ichard, M., Davis, S.G.** *Validation of {FLACS} against experimental data sets from the model evaluation database for {LNG} vapor dispersion.* 6, s.l. : Journal of Loss Prevention in the Process Industries, 2010, Vol. 23. 857 - 877.

7. **Chan, S. T.***Numerical simulations of LNG vapor dispersion from a fenced storage area.* 2, s.l. : Journal of Harzardous Materials, 1992, Vol. 30. 195-224.

8. **Hansen, O. R., Talberg, O., Bakke, J. R.** *Explosion risk assessment: how the results vary with the approach chosen.* s.l. : Proceedings of Mary Kay O' Connor Process Safety Symposium, 2001. 395–410.

9. **Hansen, O.R., Renoult, J., Bakke, J.R.** *Explosion risk assessment, how the results vary with the approach chosen.* Texas USA : Fall symposium Proceedings, Mary Kay O'Connor Process Safety Centre, 2001. 395-410.

10. **Z-013, NORSOK** *Risk and emergency preparedness analysis. Norsok standard* . Postboks 242, N-1326 Lysaker, Norway : Norsok standard. Available from Standard Norge, 2010.

11. **Launder, B. E. and Spalding, D. B.** *The numerical computation of turbulent flows.* 2, s.l. : Computer Methods in Applied Mechanics and Engineering, 1974, Vol. 3. 269-289 .

12. **Arntzen, B. J.** *Modeling of turbulence and combustion for simulation of gas explosions in complex geometries. PhD thesis.* Trondheim, Norway : NTNU, 1998. ISBN 82-471-0358-3..

13. **Bjerketvedt D, Bakke JR, van Wingerden K.** *Gas explosion handbook.* s.l. : J Hazard Mater, 1997, Vol. 52. 1–150.

14. **Witlox H.W.M, M. Harper, A. Oke, P. Bowen, P. Kay.** *Sub-cooled and flashing liquid jets and droplet dispersion I – Overview and model implementation/validation.* s.l. : J. Loss Prev. in the Process Industries, 2010, Vol. 23. 831-841.

15. **Kukkonen J, M. Kulmala, J. Nikmo, T. Vesala, D.M. Webber, T. Wren** *The homogeneous equilibrium approximation in models of aerosol cloud dispersion.* 17, s.l. : Atmospheric Environment, 1994, Vol. 28. 2763-2776.

16. **Faeth, G.M.** *Evaporation and combustion of sprays.* s.l. : Prog. Energy Comb. Sci., 1983, Vol. 9. 1-76 .

17. **Hjertager, B.H., Bjørkhaug, M., Fuhre, K.** *Gas explosion experiments in 1:33 scale and 1:5 scale; offshore separator and compressor modules using stoichiometric homogeneous fuel–air clouds.* s.l. : J. Loss. Prev. Process Ind., 1988, Vol. 1. 197-205.

18. **Venetsanos, A. G., Papanikolaou, E., Delichatsios, M., Garcia, J., Hansen, O. R., Heitsch, M., et al.** *An inter-comparison exercise on the capabilities of CFD models to predict the short and long term distribution and mixing of hydrogen in a garage.* 14, s.l. : International Journal of Hydrogen Energy, 2009, Vol. 34. 5912-5923.

19. **Middha, P., Ichard, M. & Arntzen, B.** *Validation of CFD modelling of LH2 spread and evaporation against large-scale spill experiments.* s.l. : International Journal of Hydrogen Energy, 2010, Vol. 36. 2620-2627.

20. **Hjertager, B.H., Bjørkhaug, M., Fuhre, K.** *Explosion propagation of nonhomogeneous methane-air clouds inside an obstructed 50 m3 vented vessel.* s.l. : J. Haz. Mater., 1988, Vol. 19. 139–153.

21. **Ivings, M. J., Jagger, S. F., Lea, C. J., & Webber, D. M.** *Consequence Analysis-Using a CFD Model for Industrial Sites.* **Dharmavaram, S., Hanna, S. R. and Hansen, O.R.** s.l. : Proc. Safety Prog., 2005, Vol. 24. 316-327.

22. *Evaluating vapor dispersion models for safety analysis of LNG facilities.* s.l. : The Fire Protection Research Foundation, 2007.

23. **PHMSA docket 2011-0101.** *Final decision, October 7, 2011.*

24. **Ichard, M., Hansen, O. R. and Melheim, J. A.** *Modelling of flashing releases around buildings.* Phoenix, AZ, USA : 90th AMS annual meeting, 2009.

25. **Ichard, M., Hansen, O.R., Melheim, J.A.** *Releases of pressurized liquefied gases: simulations of the Desert Tortoise test series with the CFD model FLACS.* Atlanta, GA, USA : 16th Conference on Air Pollution Meteorology, 2010.

26. **Ichard, M., Hansen, O.R., Middha, P. and Willoughby, D.** *CFD Computations of Liquid Hydrogen Releases.* San Francisco, California : Fourth International Conference on Hydrogen Safety, 2010.

VALIDATION OF PHAST DISPERSION MODEL AS REQUIRED FOR USA LNG SITING APPLICATIONS

Henk W.M. Witlox[*] and Mike Harper – DNV Software, London, UK
Robin Pitblado – DNV, Houston, Texas, USA

ABSTRACT

PHMSA in consultation with FERC has issued guidance relating to approval in the USA of atmospheric dispersion models for LNG siting applications. This guidance includes a Model Evaluation Protocol (MEP), and an associated experimental database against which the model needs to be validated. Final approval was obtained in October 2011 for the Phast dispersion model UDM, and this paper summarises the submission of this model according to the above PHMSA guidance.

The UDM model is the main model in the hazard assessment software package Phast. It is a Unified Dispersion Model (UDM) for two-phase jet, heavy and passive dispersion including droplet rainout and pool spreading/evaporation. The UDM model can deal with a wide range of scenarios including both pressurised jet releases and unpressurised releases.

The paper first summarises the overall verification and validation of the UDM (both for LNG and other chemicals), which includes validation against experiments included in the REDIPHEM database (as produced as part of the EU project SMEDIS) and the MDA database (as produced by Hanna et al.). Subsequently the paper outlines UDM validation against experiments as required by the PHMSA for the LNG model evaluation protocol (MEP). This includes large-scale LNG field experiments involving dispersion from a liquid pool (Maplin Sands, Burro, Coyote) and large-scale Freon/Nitrogen field experiments involving dispersion from ground-level vapour area sources (Thorney Island). Overall very good agreement was obtained for the field experiments.

The results of the validation were submitted to PHMSA including required statistics for model accuracy (mean ratio observed to predicted, variance, etc.). The PHMSA submission also included detailed technical documentation for the UDM dispersion model (theory, verification and validation), and a report relating to conformance of the UDM against the model evaluation protocol (MEP). The authors believe that this type of model evaluation exercise is essential to ensure sufficient model accuracy and to gain community support.

1. INTRODUCTION

The Pipeline and Hazardous Material Safety Administration (PHMSA) of the USA Department of Transportation (DOT) has issued standards (Regulation 49 CFR193) for safe design, siting, construction and operation of LNG facilities. These standards require that the operator or governmental authority control an 'exclusion zone' defined as the area that could be exposed to unsafe levels of thermal radiation or dispersion of flammable gas in case of a LNG release and ignition. In practice only one dispersion model was approved and available – DEGADIS, and this model has not been updated in a long time.

In conjunction with this standard, PHMSA in consultation with the Federal Energy Regulatory Commission (FERC) has issued guidance relating to approval of atmospheric dispersion models for LNG siting applications. This guidance is based on the Model Evaluation Protocol (MEP) developed by HSL (Coldrick et al., 2010), and an associated experimental database against which the model needs to be validated (Ivings et al., 2007). For further details see the FERC paper by Kohout (2012).

Final approval by the PHMSA was obtained in October 2011 for the dispersion model UDM contained in the hazard-assessment software package Phast developed by DNV Software. This paper summarises the submission of this model according to the above PHMSA guidance.

Section 2 provides an overview of UDM dispersion model. Section 3 summarises the overall verification and validation of the UDM (both for LNG and other chemicals). Section 4 subsequently outlines UDM validation against experiments as required by the PHMSA for the LNG model evaluation protocol (MEP). Section 5 summarises the overall submission of the Phast dispersion model UDM and its final approval by the PHMSA.

2. OVERVIEW OF PHAST DISPERSION MODEL UDM

The hazard-assessment package Phast (Witlox, 2010; Witlox and Oke, 2008) for consequence modelling of accidental releases of flammable or toxic chemicals to the atmosphere includes discharge, dispersion, toxic and flammable calculations. The flammable calculations include fireballs (instantaneous releases), jet fires (pressurised releases), pool fires (after rainout), and vapour cloud fires or explosion; see Figure 1 for the case of a continuous two-phase release of a flammable material with rainout. The UDM model is the main model in the hazard assessment software package Phast. It is a Unified Dispersion Model (UDM) for two-phase jet, heavy and passive dispersion including droplet rainout and pool spreading/evaporation.

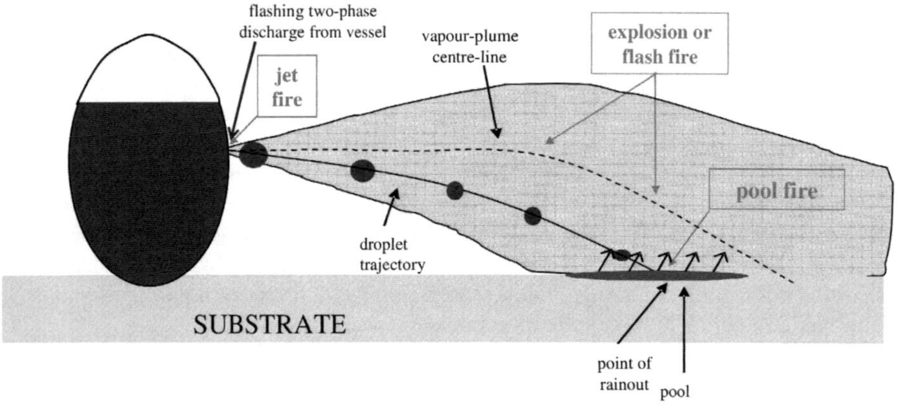

Figure 1. Continuous two-phase release of flammable material with rainout

The UDM can model a wide range of scenarios. Distinction can be made between momentum (un-pressurised or pressurised releases), time-dependency (steady-state, finite-duration, instantaneous or time-varying dispersion), buoyancy (buoyant rising cloud, passive dispersion or heavy-gas-dispersion), thermodynamic behaviour (isothermal or cold or hot plume, vapour or liquid or solid or multiple-phase, reactions or no reactions), ground effects (soil or water, flat terrain with uniform surface roughness), and ambient conditions (stable, neutral or unstable conditions).

The UDM models the dispersion following a ground-level or elevated two-phase pressurised release. It effectively consists of the following linked modules (see Figure 1):

- jet dispersion
- droplet evaporation and rainout, touchdown
- pool spread and vaporisation
- heavy gas dispersion
- passive dispersion

A single form of concentration profile is used to cover all stages of a release. This allows for anything from a sharp-edged profile in the initial stages of a jet release through to the diffuse Gaussian profile that would be expected in the final passive stage of spreading.

The UDM includes the effects of droplet vaporisation using a non-equilibrium model. Rainout produces a pool which spreads and vaporises. Vapour is added back into the plume and allowance is made for this additional vapour flow to vary with time.

The UDM allows for vertical variation in ambient speed, temperature and pressure. Another feature of the UDM is possible plume lift-off, where a grounded cloud becomes buoyant and rises into the air. Rising clouds may be constrained to the mixing layer if it is reached.

For further details the reader is referred to the UDM overview paper by Witlox and Holt (1999), the Phast 6.7 theory manual for the UDM (Witlox and Harper, 2011a), its thermodynamic sub-module THRM (Witlox and Harper, 2011b) and pool evaporation sub-module PVAP (Witlox, 2011).

3. OVERALL PHAST (UDM) MODEL VERFICATION AND VALIDATION

The overall UDM validation includes validation against experiments stored in the MDA database by Hanna et al. (1993). It also includes validation against experiments included in the REDIPHEM database, which was carried out partly as part of the EU project SMEDIS (Daish et al., 1999). The SMEDIS project did produce a protocol for evaluating heavy gas dispersion models, which was the basis of the LNG model evaluation protocol proposed by Ivings et al. (2007). The SMEDIS project also included an independent peer review of the UDM model by Britter (2002). He states in this model evaluation report (MER) that the UDM model is amongst the most extensively documented and validated models.

The verification and validation for the individual UDM submodels can be summarised as follows [see Witlox and Harper (2011a,b) and Witlox (2011) for further details and a detailed list of references]:

- Jet and near-field passive dispersion. For an elevated horizontal continuous jet (of air), the UDM numerical results are shown to be identical to the results obtained by an analytical solution. For vertical jets very good agreement has been obtained against both the "Pratte and Baines" and "Briggs" empirical plume rise correlations.

- Heavy-gas dispersion. The UDM numerical results are shown to be in identical agreement against an analytical solution for a 2D isothermal ground-level plume. The UDM has been validated against the set of three 2D wind-tunnel experiments (McQuaid, 1976) and the isothermal HTAG wind tunnel experiments (Petersen and Ratcliff, 1988). Furthermore the UDM model was verified against the HGSYSTEM heavy-gas dispersion model HEGADAS.

- Far-field passive dispersion. For purely (far-field) passive continuous dispersion, the UDM numerical results are shown to be in close agreement with the vertical and crosswind dispersion coefficients and concentrations obtained from the commonly adopted analytical Gaussian passive dispersion formula. The same agreement has been obtained for the case of purely (far-field) passive instantaneous dispersion, while assuming along-wind spreading equal to cross-wind spreading in the analytical profile.

- Finite-duration releases. The UDM "Finite-duration-correction" module has been verified against the HGSYSTEM/SLAB steady-state results, and shown to lead to finite-duration corrections virtually identical to the latter programs. Furthermore excellent agreement was obtained using this module for validation against the Kit Fox experiments (20-second releases of CO_2 during both neutral and stable conditions).

- Thermodynamics. The UDM dispersion model invokes the thermodynamics module while solving the dispersion equations in the downwind direction. This module describes the mixing of the released component with moist air, and may take into account water-vapour and heat transfer from the substrate to the cloud. The module calculates the phase distribution [component (vapour, liquid), water (vapour, liquid, ice)], vapour and liquid cloud temperature, and cloud density. Thus separate water (liquid or ice) and component (liquid) aerosols may form. The liquid component in the aerosol is considered to consist of spherical droplets and additional droplet equations may be solved to determine the droplet trajectories, droplet mass and droplet temperature. Rainout of the liquid component occurs if the droplet size is sufficiently large. The thermodynamics module also allows for more rigorous multi-component modelling (Witlox et al., 2006). The UDM homogeneous equilibrium model has been verified for both single-component and multi-component materials against the HEGADAS model. The UDM HF thermodynamics model (including effects of aqueous fog formation and polymerisation) was validated against the experiments by Schotte (1987).

- Release rate, initial droplet size and rainout validation. This relates to validation of both the Phast discharge model as well as the UDM dispersion model as carried out as part of a droplet modelling joint industry project (JIP). This includes validation of flow rate, initial droplet size and rainout against a wide range of experiments (Witlox and Harper, 2011c). This resulted in improvement of the droplet size correlation and rainout predictions in Phast.

- Pool spreading/evaporation. If the droplet reaches the ground, rainout occurs, i.e. removal of the liquid component from the cloud. This produces a liquid pool which spreads and vaporises (see Figure 1). Vapour is added back into the cloud and allowance is made for this additional vapour flow to vary with time. The UDM source term model PVAP calculates the spreading and vapour flow rate from the pool. Different models are adopted depending whether the spill is on land or water, and whether it is an instantaneous or a continuous release. The pool spreads until it reaches a bund or a minimum pool thickness. The pool may either boil or evaporate while simultaneously spreading. For spills on land, the model takes into account heat conduction from the ground, ambient convection from the air, radiation and vapour diffusion. These are usually the main mechanisms for boiling and evaporation. Solution and possible reaction of the liquid in water are also included for spills on water, these being important for some chemicals. These effects are modelled numerically, maintaining mass and heat balances for both boiling and evaporating pools. This allows the pool temperature to vary as heat is either absorbed by the liquid or lost during evaporation.

  PVAP was verified by Webber (2003) against the SRD/HSE model GASP for a range of scenarios with the aim of testing the various sub-modules, and overall good agreement was obtained. The PVAP spreading logic was first validated against experimental data for spreading of non-volatile materials. Subsequently the PVAP evaporation logic was validated against experimental data in confined areas where spreading does not take place. Finally comparisons were made for simultaneously spreading and vaporising pools. The above validation was carried out for both spills on water and land, and a wide range of materials was included [LNG, propane, butane, pentane, hexane, cyclo-hexane, toluene, ammonia, nitrogen, water, Freon-11].

The validation of the overall UDM model was carried out against large-scale field experiments selected from the MDA and REDIPHEM databases, including the following:

- Prairie Grass (continuous passive dispersion of sulphur dioxide).
- Desert Tortoise and FLADIS (continuous elevated two-phase ammonia jet)
- EEC (continuous elevated two-phase propane jet)

- Goldfish (continuous elevated two-phase HF jet)
- Maplin Sands, Burro and Coyote (continuous evaporation of LNG from pool)
- Thorney Island (instantaneous un-pressurised ground-level release of Freon-12)
- Kit Fox (continuous and finite-duration heavy-gas dispersion of CO_2 from area source)
- BP Spadeadam (pressurised CO_2 release: cold steady-state releases and supercritical hot vapour releases)

Each of the above experimental sets was statistically evaluated to determine the accuracy and precision of the UDM predictions with the observed data. Formulas adopted by Hanna et al. (1993) were used to calculate the geometric mean bias MG (under or over-prediction of mean) and mean variance VG (scatter from observed data) for each validation run. A perfect result would have both MG and VG = 1. This was carried out for centre-line concentrations, cloud widths, and (for the SMEDIS experiments) also off centre-line concentrations. The overall performance of the UDM in predicting both peak centreline concentration and cloud widths was found to be good for the above experiments. A graphical presentation of the overall MG,VG validation results for arc-wise concentration (excluding CO_2 experiments) is shown in Figure 2. The figure also shows the factor of 2 band ($0.5 < MG < 2$) expected from good quality similarity models. See the UDM validation manual for further details.

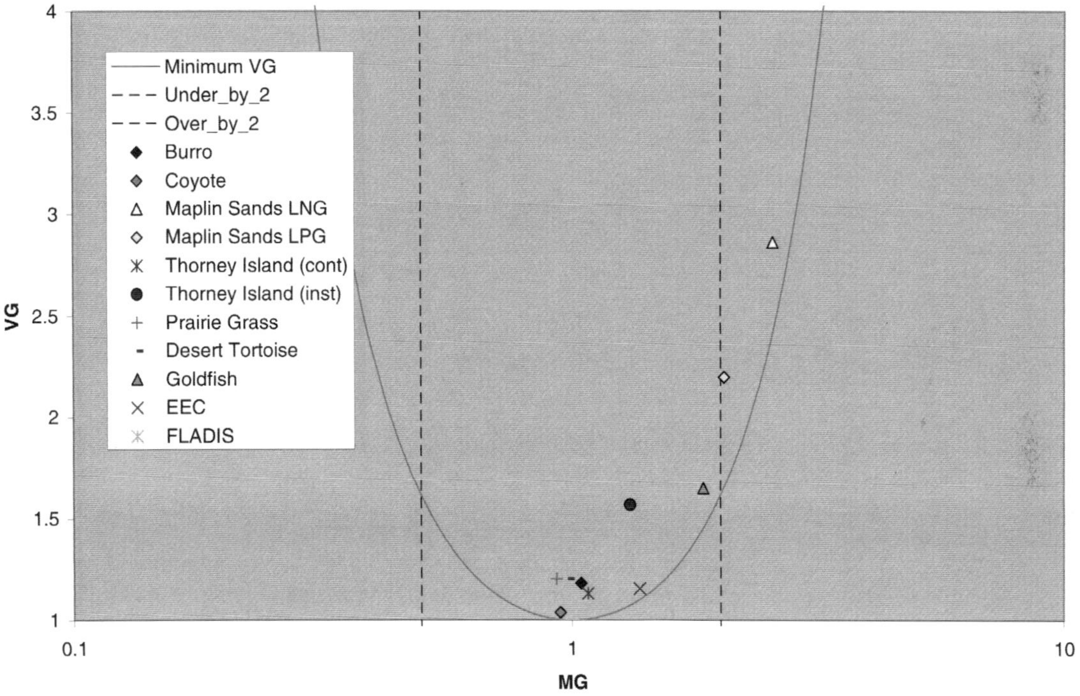

Figure 2. **Phast UDM Version 6.7 validation statistics (MG, VG) for maximum arc-wise concentration**

The overall UDM model was also verified by means of comparison against other models (HGSYSTEM, SLAB, TRACE, ALOHA, SCIPUFF) for three US chlorine accidents involving elevated two-phase chlorine jet releases, and the Phast predictions were found typically in the medium range of the predictions; see Hanna et al. (2007) for full details.

4. PHAST (UDM) VALIDATION AGAINST PHMSA EXPERIMENTS

This section outlines UDM validation against experiments as required by the PHMSA for the LNG model evaluation protocol (MEP). Full details are provided in the UDM validation document by Witlox and Harper (2011d) submitted to the PHMSA.

4.1 Selection of experiments

Table 1 lists the experiments against which the UDM model has been validated and also lists how each experiment has been modelled by the UDM:

- The large-scale LNG field experiments involve dispersion from a liquid pool (Maplin Sands, Burro, Coyote). These experiments have been modelled as low-momentum elevated horizontal releases (with immediate virtually 100% rainout).
- The large-scale Freon/Nitrogen field experiments involve dispersion from a ground-level vapour area sources (Thorney Island), and have been modelled as a low-momentum ground-level horizontal release.
- The CHRC, BA-Hamburg and BA-TNO scaled wind-tunnel experiments were modelled at full scale as a ground-level vapour pool source

Experiment	trial nr.	Type	Material	Modelled by UDM as
Maplin Sands	27,34,35	Field	LNG	Low momentum elevated horizontal release
Burro	3,7,8,9	Field	LNG	Low momentum elevated horizontal release
Coyote	3,5,6	Field	LNG	Low momentum elevated horizontal release
Thorney Island	45,47	Field	Freon&N_2	Low momentum ground-level horizontal release
CHRC	A	Wind tunnel	CO2	Ground-level vapour pool source
BA-Hamburg	DA0120,DAT223	Wind tunnel	SF_6	Ground-level vapour pool source
BA-TNO	TUV01,FLS	Wind tunnel	SF_6	Ground-level vapour pool source

Table 1. List of experiments for UDM validation

4.2 UDM input and results

After rainout, the UDM model invokes the PVAP model for pool calculations and divides the time-varying pool evaporation rate into a number of segments (with constant evaporation rate during segment duration). The PHMSA includes both experimental maximum concentrations (one-second averaged), and (for Burro and Coyote) longer averaging-time measurements. For the short-averaging times, the pool segment is applied which produces the highest concentration. For the long averaging times, the pool segment most likely to be active in the given time-averaging window has been selected.

In line with the model evaluation protocol, the following UDM output data were produced:

- o arcwise maximum concentration at measurement elevation and downwind distance
- o distance to measured arcwise maximum concentration at measurement elevation

- arcwise cloud width at downwind distance where concentrations were measured
- point-wise concentrations at measurement location

The following UDM validation statistics were derived from the above results:

- MG (mean) and VG (variance) for above data [ratio observed to predicted; for each experiment and each group of experiments]
- MRB (mean relative bias) and MRSE (mean relative square error) [relative difference; for each group of experiments]
- FAC2 [fraction within factor of 2; for each groups of experiments]
- CSF (Concentration safety factor) [ratio predicted to observed; for each groups of experiments]
- LFL safety factors for LNG experiments (arcwise data only; LFL = 4.4%):
  - Concentration safety factor to LFL, CSF_{LFL} [ratio of predicted concentration (at observed distance to LFL) to LFL]
  - Distance safety factor to LFL, DSF_{LFL} [ratio of predicted to observed distance to LFL]

Table 2 lists the UDM input data for the example case of the Burro experiments (trials 3, 7, 8, 9). In this table the 'BU03'column, lists all the input data for BU03 experiment, while the subsequent columns indicate the input data of the trials BU07, BU08 and BU09 as far as they differ from BU03. Table 3 lists the observed and predicted results for these experiments. This includes UDM validation statistics (MG, VG for concentration and width; CSF).

Table 4 includes a list of MG, VG, CSF_{LFL}, DSF_{LFL} values for the individual experiments in the LNG Model Validation Database. The same data are plotted for the field experiments in Figure 3. Figure 4 includes the combined MG and VG values for each group of experiments. The following is concluded from these tables and figures:

- Field experiments – short averaging times
  - Excellent results are obtained for the Burro and Coyote experiments
  - Maplin Sand under-predicts the concentrations.

- Field – long averaging times
  - Thorney Island gives excellent results
  - Burro give good results for both concentrations and cloud widths, though with slightly higher variance than for short averaging times
  - Concentrations are over-predicted for the Coyote experiments

- Wind-tunnel experiments
  - Concentrations are consistently under-predicted, while the cloud widths are slightly over-predicted. To maintain conservation of mass this appears to imply that either the cloud depth is over-predicted (too much heavy-gas entrainment at top of cloud) and/or the cloud speed is over-predicted
  - The above may be partly caused by inaccurate scaling. To further evaluate the cause an in-depth study of the un-scaled experiments would be desired first. It is recommended to do this as part of further work.

As previously indicated modelling Maplin Sands releases tends to produce large duration pool segments which will underestimate the actual peak evaporation rate. This will in turn lead to concentrations that are too low. The combination of significant time-varying effects and long averaging times is difficult to model with the Phast 'segment' approach, as it is difficult to choose a segment with an evaporation rate representative of the time-averaging window.

According to verbal communication with PHMSA/FERC, the above UDM under-prediction for the Maplin Sands experiments and the wind-tunnel experiments appears to be in line with other model predictions, and as such this may be caused by the quality of experimental data (Maplin Sand experiments) or inaccuracy of scaling (wind-tunnel experiments).

Description	Units	BU03	BU07	BU08	BU09	Notes
RELEASE DATA						
Duration		167	174	107	79	
Material		Methane				Assume LNG = pure methane
Release rate	kg/s	87.98	99.46	116.93	135.98	
Initial state [-1: saturated liquid at boiling point]		-1				Saturated liquid at its boiling point
Droplet size	m	0.01				Assume maximum permitted in Phast
Release height	m	1.5				
Release angle [0 = horizontal]	radians	0				
Release velocity (continuous only)	m/s	0.1				Assume minimum release velocity
AMBIENT DATA						
Pasquill stability class (5-C,7-D,8-E)		5	7	8	7	
Wind speed at reference height	m/s	5.58	8.75	1.94	5.94	
Reference height for wind speed	m	3				
Temperature at reference height	K	307.75	306.96	306.02	308.52	
Pressure at reference height	N/m2	94840.2	94029.6	94130.925	94029.6	
Reference height for temperature and pressure	m	1				
Atmospheric humidity (fraction)	-	0.052	0.074	0.045	0.144	
SUBSTRATE DATA						
Surface roughness length	m	0.0002				
Dispersing surface type (1-land,2-water)		1				
POOL DATA						
Surface [5:deep water (no ice), 8:shallow water]		8				
Temperature of pool surface	K	307.75	306.96	306.02	308.52	
Bund diameter (= 0: no bund)	m	0				
AVERAGING TIME						
Averaging time	s	100	140	80	50	

Table 2. Burro experiments - UDM input data (long averaging time)

	Experiment ID	Downwind distance m	Height of interest m	Concentration observed mol%	Concentration predicted mol%	Width observed m	Width predicted m	Concentration Mean MG	Variance VG	Width Mean MG	Variance VG	CSF (cp/co)
Burro (long)	BU07	57	1	14.19	17.01	-1.00	14.00	0.81	1.35	1.14	1.02	1.20
		140	1	4.40	10.30	20.50	18.03					2.34
		400	1	2.29	1.56	-1.00	26.17					0.68
	BU08	57	1	30.67	16.03	28.80	85.08	2.40	2.23	0.56	1.80	0.52
		140	1	16.36	5.52	-1.00	87.81					0.34
		400	1	3.50	1.71	87.04	93.37					0.49
		800	1	2.08	0.73	-1.00	101.02					0.35
	BU09	140	1	6.52	12.47	30.90	24.21	1.10	1.36	1.41	1.13	1.91
		400	1	2.79	2.07	49.20	32.46					0.74
		800	1	1.16	0.61	61.60	42.20					0.53
	BU03	57	1	7.89	15.56	20.86	18.76	0.55	1.45	1.11	1.01	1.97
		140	1	6.11	10.35	-1.00	24.34					1.69
Burro (short)	BU07	57	1	17.90	20.10	-1.00	26.04	0.97	1.24	-1.00	-1.00	1.12
		140	1	7.13	12.41	-1.00	33.23					1.74
		400	1	3.86	2.17	-1.00	43.91					0.56
	BU08	57	1	55.87	22.62	-1.00	178.27	1.91	1.56	-1.00	-1.00	0.40
		140	1	16.49	10.07	-1.00	194.10					0.61
		400	1	4.27	2.51	-1.00	200.50					0.59
		800	1	2.12	1.09	-1.00	210.01					0.52
	BU09	140	1	10.56	17.17	-1.00	55.76	0.93	1.11	-1.00	-1.00	1.63
		400	1	3.96	3.83	-1.00	72.20					0.97
		800	1	1.40	1.09	-1.00	85.56					0.78
	BU03	57	1	22.40	18.25	-1.00	35.28	0.95	1.07	-1.00	-1.00	0.81
		140	1	8.98	12.24	-1.00	45.25					1.36

Table 3. Burro experiments – UDM validation against arcwise concentration & width

Type experiment	Experiment	Trial nr.	Arcwise Concentration		Width		Pointwise Concentration		CSF_LFL	DSF_LFL
			MG	VG	MG	VG	MG	VG		
Field - Short Averaging time	Maplin Sands	27	3.89	7.15	-	-	-	-	0.23	0.36
		34	2.20	1.88	-	-	-	-	0.47	0.58
		35	3.10	3.83	-	-	-	-	0.41	0.55
	Burro	3	0.95	1.07	-	-	1.09	1.08	0.79	0.91
		7	0.97	1.24	-	-	0.82	4.01	0.78	0.88
		8	1.92	1.57	-	-	0.95	1.35	0.6	0.62
		9	0.93	1.11	-	-	1.02	1.17	1.1	1.04
	Coyote	3	0.79	1.08	-	-	1.36	1.37	1.4	1.15
		5	1.05	1.02	-	-	1.47	2.05	1.13	1.03
		6	0.98	1.03	-	-	0.62	1.75	1.02	1.01
Field - Long Averaging time	Burro	3	0.55	1.45	1.11	1.01	0.31	6.23	0.22	0.51
		7	0.81	1.35	1.14	1.02	0.47	9.82	2.34	1.69
		8	2.40	2.23	0.56	1.80	1.06	1.31	0.49	0.51
		9	1.10	1.36	1.41	1.13	1.14	1.81	1.54	1.23
	Coyote	3	0.46	1.87	1.46	1.15	0.64	1.63	2.05	1.47
		5	0.33	3.52	-	-	0.40	3.79	3.88	3.44
		6	0.77	1.11	1.07	1.14	0.38	6.49	1.43	1.19
	Thorney Island	45	1.15	1.12	-	-	-	-		
		47	0.97	1.15	-	-	-	-		
Windtunnel - Pool Source	CHRC	A	2.83	3.16	0.60	1.33	1.94	2.69		
	Hamburg	DA0120	3.89	6.78	-	-	-	-		
		DAT223	1.51	1.48	-	-	1.92	1.79		
	TNO	TUV01	-	-	-	-	2.74	3.72		
		FLS	3.49	5.23	0.84	1.07	3.34	6.34		

Table 4. List of MG, VG, CSF_{LFL} and DSF_{LFL} values for experiments

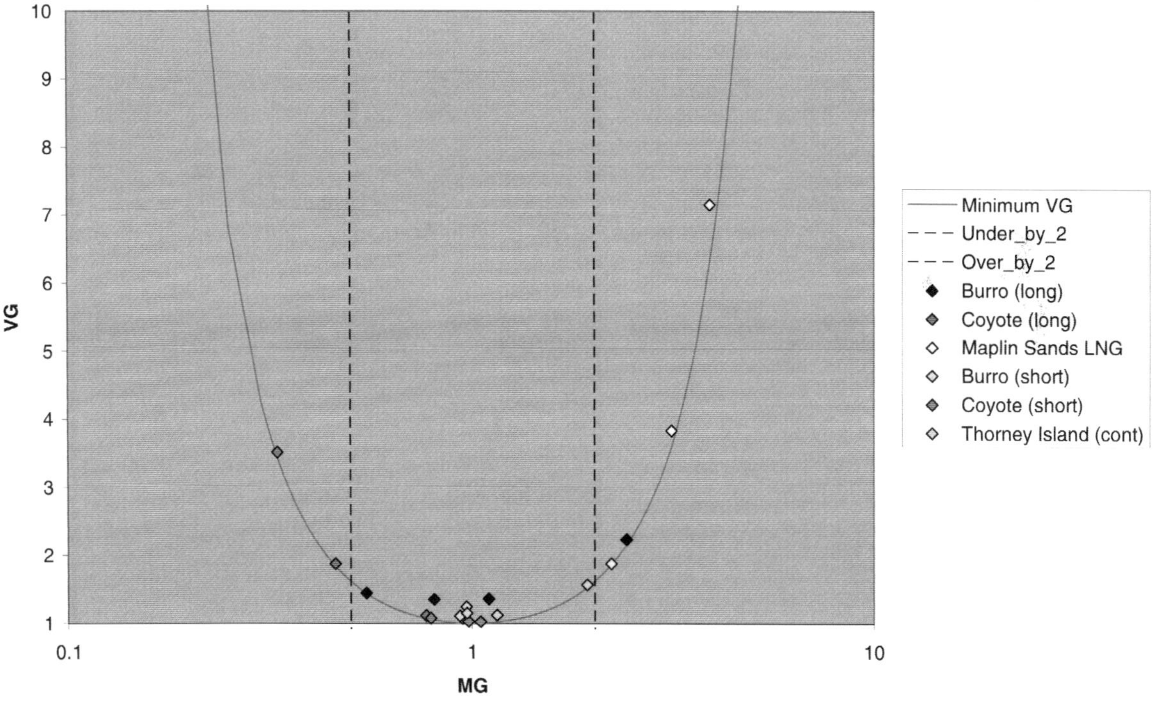

Figure 3. Plot of MG and VG values (arcwise concentrations) for individual field experiments

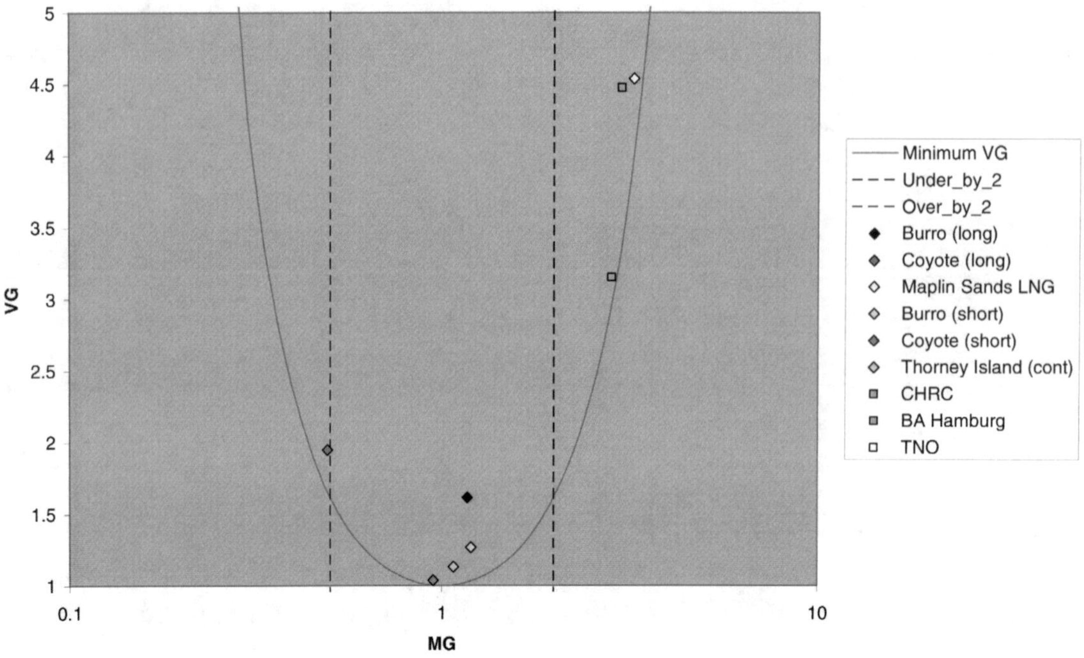

Figure 4. Plot of MG and VG values (arcwise concentrations) for each group of experiments

4.3 Sensitivity analysis

This submission also included a sensitivity analysis. This includes sensitivity to the experimental uncertainty of the input parameters (wind speed, stability class, surface roughness, ambient pressure, humidity, LNG mixture composition) and a sensitivity analysis to deviations to the measured maximum arc-wise concentrations.

This sensitivity analysis is illustrated in Figure 5 for the case of Burro 7 (short averaging time). Here the observed value and absolute deviations from the mean (downwards and upwards) were provided by FERC/PHMSA (2011). Figure 5 illustrates the inaccuracy in measuring the concentrations (blue curves/markers), versus the spread in the model predictions (pink curves/markers) resulting from the range of input variations as given by FERC/PHMSA. For this specific case, the maximum prediction concentration corresponded to the base-case, while the minimum prediction concentrations corresponded to a higher value of surface roughness (10mm versus base-case value of 0.2mm) and mixture composition (exact composition versus base-case assumption of pure methane).

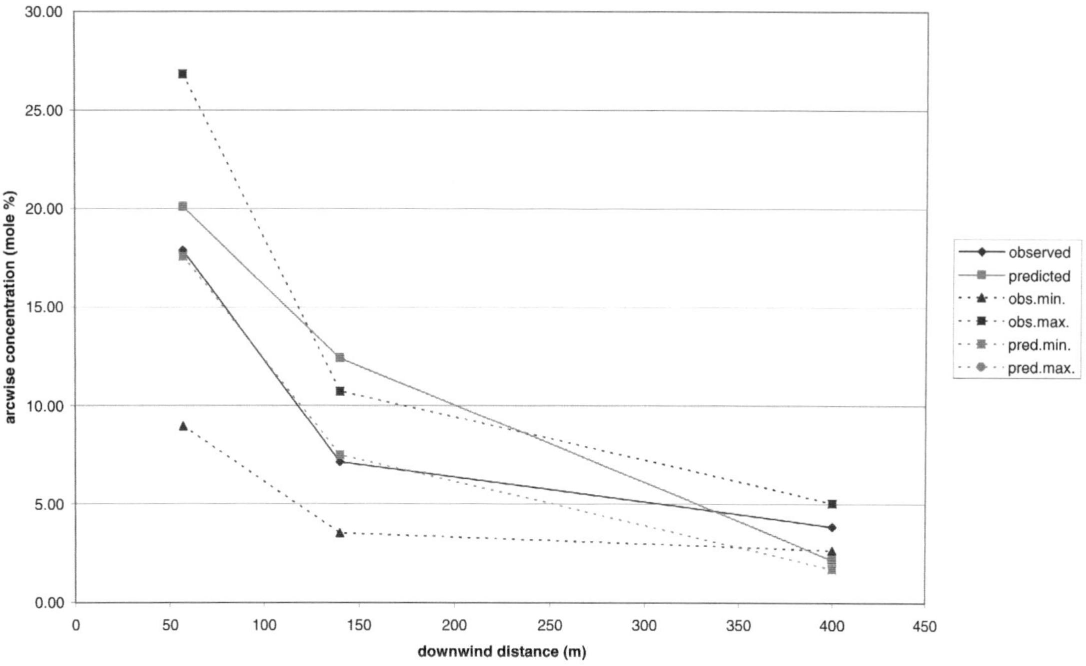

Figure 5. **Burro 7 (short) - Measured versus observed concentrations**

5. PHAST (UDM) SUBMISSION AND PHMSA APPROVAL

The results of the validation presented in the previous section were submitted to PHMSA including all required UDM validation statistics for model accuracy. The PHMSA submission also included detailed technical documentation for the UDM dispersion model (theory, verification and validation), and a report relating to conformance of the UDM against the model evaluation protocol (MEP).

Final approval was obtained in October 2011 for the Phast dispersion model UDM by the PHMSA. The approval was obtained for both versions 6.6 and 6.7 of the Phast software. Both versions produce virtually identical results for dispersion from ground-level LNG pools (using new UDM 'Version 2' solver), but the new version 6.7 includes more advanced rainout modelling for elevated two-phase releases (Witlox and Harper, 2012).

The approval was obtained for scenarios involving dispersion from circular shaped LNG pools, dispersion from LNG pools in impoundments with low aspect ratios, and dispersion from releases in any direction (including releases from flashing, venting and pressure relief). Although the Phast dispersion model UDM has been validated against line sources (as describe in Section 3; validation against McQuaid experiments), this feature has currently not yet been made available in Phast. Furthermore Phast currently presumes dispersion over terrain with a uniform surface roughness. Thus the PHMSA decision acknowledged that the current Phast may not be appropriate for dispersion from high aspect-ratio pools (e.g. trenches), across highly varying terrain, or around large obstacles. Furthermore PHMSA recommended that the UDM is used with a safety factor of 2 (i.e use 0.5 LFL) to account for turbulent fluctuations and model uncertainties. This is in line with the Phast default settings for flammable materials.

While DNV is pleased to note approval of Phast, we have commented to PHMSA that real releases involve complex source term aspects (flashing, drop size distributions, rain-out, pool formation and re-evaporation) not included in the PHMSA list of validation experiments. Thus other CFD models, which may appear superficially to be superior to Phast (UDM) in the dispersion over varying terrain aspect, may in fact be deficient in their less sophisticated source-term modelling and overall predicted results may not be superior. Issues related to a CFD model may also include numerical accuracy/convergence, appropriate selection of grid, turbulence model (to account for e.g. heavy-gas-dispersion effects) and boundary conditions (e.g. wall laws etc.). Thus many different CFD results could be obtained depending on the chosen input. Furthermore the extensive CPU time does often not allow repetitive runs in order to obtain more confidence in the CFD results.

ACKNOWLEDGEMENTS

Steven Zhang assisted in carrying out the Phast calculations. Also the input is acknowledged from PHMSA (Charles Helm, Keith Coyle) and FERC (Andrew Kohout, Terry Turpin) while submitting the UDM for PHMSA approval.

REFERENCES

Britter, R., 2002, Model Evaluation Report on UDM Version 6.0, Ref. No. SMEDIS/00/9/E, Version 1.0, 21 January 2002, Prepared by Cambridge Environmental Research Consultants Ltd, Cambridge, UK

Coldrick, S., Lea, C.J. and Ivings, M.J., 2010, "Validation database for evaluating vapour dispersion models for Safety Analysis of LNG facilities", Guide to the LNG model validation database, Health and Safety Laboratory, Revision May 2010

Daish, N.C, Britter, R.E., Linden, P.F., Jagger, S.F. and Carissimo, B., 1999, "SMEDIS: Scientific Model Evaluation Techniques Applied to Dense Gas Dispersion models in complex situations"., *Int. Conf. and workshop on modelling the consequences of accidental releases of hazardous materials*, San Francisco, California, CCPS, New York, 345-372

FERC/PHMSA, "Sensitivity bounds and experimental uncertainty", Memo provided by FERC/PHMSA in email from Charles Helm dated 3 March 2011

Hanna, S.R., Chang, J.C. and Strimaitis, D.G., 1993, "Hazardous gas model evaluation with field observations", *Atm. Env.*, 27a: 2265-2285

Hanna, S., Dharmavaram, S., Zhang, J., Sykes, I., Witlox, H. W. M., Khajehnajafi, S. and Koslan, K., 2007, "Comparison of six widely-used dense gas dispersion models for three actual chlorine railcar accidents", *Proceedings of 29th NATO/SPS International Technical Meeting on Air Pollution Modelling and its Application*, 24 - 28 September 2007, Aveiro, Portugal

Ivings, M.J., Jagger, S.F., Lea, C.J. and Webber, D.M., 2007, "Evaluating vapor dispersion models for safety analysis of LNG facilities", Contract by HSL for Fire Protection Research Foundation, Quincy, Massachusetts

Kohout, A., 2012, "Evaluation of dispersion models for LNG siting applications", Session LNG Plant Safety and Protection, *12th Topical Conference on Gas Utilization*, AICHE Spring Meeting, April 2012

McQuaid, J., 1976, "Some experiments on the structure of stably stratified shear flows", Technical Paper P21, Safety in Mines Research Establishment, Sheffield, UK

Nedelka, D., Moorhouse, J., and Tucker, R. F., 1990, "The Montoir 35m diameter LNG pool fire experiments", *Proc. 9th Intl. Cong and Exposition on LNG*, LNG9, Nice, 17-20 October 1989, Published by Institute of Gas technology, Chicago, 2-III-3: 1-23

Petersen, R.L. and Ratcliff, M.A., 1988, "Effect of homogeneous and heterogeneous surface roughness on HTAG dispersion", CPP Incorporated, Colorado. Contract for API, Draft Report CPP-87-0417

Schotte, W., 1987, "Fog formation of hydrogen fluoride in air", *Ind. Eng. Chem. Res.*, 26: 300-306; see also Schotte, W., "Thermodynamic model for HF formation", 31 August 1988, Letter from Schotte to Soczek, E.I. Du Pont de Nemours & Company, Du Pont Experimental Station, Engineering Department, Wilmington, Delawere 19898

Webber, D., 2003, "PVAP Theory Document", distributed as part of Phast 6.7 software, DNV Software, London

Witlox, H.W.M. 2010, "Overview of consequence modelling in the hazard assessment package Phast", Paper 7.2, *6th AMS conference on applications of air pollution meteorology*, Atlanta, USA, 17-21 January 2010

Witlox, H.W.M., 2011, "PVAP Theory Document", UDM Version 6.7, distributed as part of Phast 6.7 software, DNV Software, London

Witlox, H.W.M. and Harper,M. 2011a, "UDM Theory Document,", UDM Version 6.7, distributed as part of Phast 6.7 software, DNV Software, London

Witlox, H.W.M. and Harper, M. 2011b, "UDM Thermodynamics Theory Document,", UDM Version 6.7, distributed as part of Phast 6.7 software, DNV Software, London

Witlox, H.W.M. and Harper, M., 2011c, "Two-phase jet releases, droplet dispersion and rainout, I. Overview and model validation", *14th Annual Symposium, Mary Kay O'Connor Process Safety Centre*, Texas A&M University, College Station, Texas, 25-27 October 2011, pp. 642-655 *also accepted for publication in Journal Loss Prevention*

Witlox, H.W.M. and Harper, M. 2011d, "Validation of Phast dispersion model UDM against experiments in LNG Model Validation Database" (submission to PHMSA/FERC), DNV Software, London, April 2011

Witlox, H.W.M., Harper, M., Topalis, P. and Wilkinson, S., 2006, "Modelling the consequence of hazardous multi-component two-phase releases to the atmosphere", *Hazards XIX Conference*, Manchester, 250-265

Witlox, H.W.M. and Holt, A., 1999, "A unified model for jet, heavy and passive dispersion including droplet rainout and re-evaporation", *International Conference and Workshop on Modelling the Consequences of Accidental Releases of Hazardous Materials*, CCPS, San Francisco, California, September 28 – October 1, 315-344

Witlox, H.W.M. and Oke, A., 2008, "Verification and validation of consequence models for accidental releases of hazardous chemicals to the atmosphere", *Hazards XX Conference*, Manchester, 14-17 April 2008

Modeling of LNG Spills into Trenches and Troughs

Nicolas F. Ponchaut, Ph.D., P.E., CFEI
Exponent, Inc.
9 Strathmore Road
Natick, MA 01760
nponchaut@exponent.com

Alfonso Ibarreta, Ph.D., P.E., CFEI
Exponent, Inc.
9 Strathmore Road
Natick, MA 01760
aibarreta@exponent.com

Harri Kytömaa, Ph.D., P.E., CFEI
Exponent, Inc.
9 Strathmore Road
Natick, MA 01760
hkytomaa@exponent.com

Abstract

Trenches and troughs are commonly used in LNG storage and transfer facilities to contain potential LNG spills. The trenches are concrete-lined rectangular channels dug into the ground. They are typically slightly inclined from the horizontal to direct the flow of LNG towards the sump and they are often configured with sharp turns and dead ends.

In sizing the trench, it is important to ensure that it can contain the intended design spill. The trench design also influences the rate of wetting of the trench floor and walls during a design spill, which in turn controls to the rate of vapor generation. The latter dictates the dispersion of vapors and associated exclusion zones. Despite the regulatory importance associated with trenches, there is little formal guidance for proper trench design in LNG facilities, or guidance for the tools that are needed for modeling the liquid propagation.

This study examines the use of Computational Fluid Dynamics (CFD) to model the flow of LNG within trenches and troughs during a design spill. Detailed calculations of liquid flow are obtained using Star-CCM+, a state of the art, commercially available, broadly used, generally accepted and validated CFD software package. Calculations performed with this software reveal the dynamics of design spills and show how sharp turns and dead ends in trench systems can result in overspilling of LNG. These software tools are useful in sizing trenches and troughs to contain worst-case scenario design spills.

In this work, transient 3-dimensional calculations are compared to a transient 1-dimensional model that uses the shallow-water equations to predict liquid propagation in trenches and troughs. The study also shows how these liquid propagation models can be used to determine the LNG evaporation rates during the propagation of the LNG in the trench.

1. Background

Trenches and troughs are used in LNG receiving or liquefaction terminals to direct the flow of LNG into containment sumps in the event of a line rupture or leak. The terminals are designed to contain any potential LNG spills in a confined sump, limiting the extent of the resulting methane cloud.

The trenches are concrete-lined rectangular channels dug into the ground. They are placed next to, or under, areas containing LNG processing equipment and run parallel to LNG transfer lines. The role of the containment trench is to minimize the amount of LNG evaporated during the conveying of the liquid spilled onto the ground. They are, in many instances, lined with insulated or aerated concrete in order to minimize the heat transfer to the liquid from the substrate.

Figure 1 Aerial view of sample LNG trench and containment sump at an LNG receiving terminal.

LNG conveying trenches are typically slightly inclined from the horizontal (0.2 to 1% typical slopes) to direct the flow of LNG towards the sumps. The trenches tend to follow the placement of the LNG transfer lines, normally running parallel to the pipes. They are generally designed with long straight runs, and many times configured with sharp turns, sudden changes in elevation and dead ends.

The LNG trenches and sump are normally one of the last elements included in the design of the facility. Their placement and geometry tends to depend on the location of LNG piping and available space in process areas. Trenches vary widely in width, from about a foot to over 10 feet in width. The trench depth also tends to vary from just a few inches to over 10 feet. There seems to be little guidance in the industry as to the optimum design of the trench geometry. The trench design, however, has a big impact on the resulting LNG spill propagation speed, concrete surface wetting rate, and maximum conveying capacity of the trenches. Wide trenches, for example, normally result is slower moving LNG spills, capable of conveying large amounts of liquid, but tend to cause higher surface wetting rates and therefore higher evaporation rates. Narrow trenches reduce the evaporation rate, but are more susceptible to liquid over-spilling at high leak flowrates [1].

DOT and PHMSA currently require that the vapor dispersion due to evaporation of LNG travelling through trenches be modeled at LNG terminals. The required vapor dispersion analysis is typically performed in two phases. First, a model evaluates how the LNG will travel along the trench or trough, and how fast it vaporizes, commonly referred to as a vapor source term model. A second model is then used to estimate how the evaporated LNG disperses in the atmosphere, commonly referred to as a dispersion model. The emphasis in 49 CFR 193 [2] and NFPA 59A [3] has been on the tools that model the behavior of vapors, and providing little attention to the source term model. The FLACS modeling software from Gexcon allows simultaneous modeling of the liquid spill and the resulting vapor dispersion, but it requires the use of a Cartesian grid and does not allow the modeling of inclined trenches._ The proper modeling of liquid flow through realistic trenches is essential for several reasons:

1) The evaporation rate of LNG is directly dependent on the trench wetting rate, and therefore the liquid spread velocity. Any source term model, therefore, must properly model the conveying of LNG through trenches.

2) It is necessary to ensure that the trenches are properly designed to contain the maximum LNG flowrate expected during a worst-case spill scenario, without over spilling or flooding surrounding areas.

The spread of LNG spills on sloped ground have been modeled in the past using steady-state assumptions [4] or integral models [5]. These models, however, are not able to properly model the unsteady propagation of LNG through an inclined trench as a function of time. It is also important to take into account hydraulic jumps in the liquid as it encounters a sharp turn, elevation change, or a dead end within the trench. Hydraulic jumps are sudden changes in liquid height during free surface flow, caused by localized changes in liquid velocity. These frequently occur just upstream of sudden turns or dead ends, resulting in a sudden increase in liquid level height, potentially leading to over spilling.

A previous Exponent study showed the effect of trench geometry and substrate properties on the propagation speed and resulting evaporation rate in a simple straight run of trench [1]. That work provided several rules of thumb to reduce the rate of evaporation during transfer of LNG in trenches. The current work analyzes in greater detail the liquid spreading behavior in trenches with turns and dead ends. This paper compares three separate liquid propagation models of increasing complexity:

1) Chézy equation (steady-state free surface flow)
2) Exponent's Shallow Water Equations (SWEs) solver
3) Star-CCM+ CFD software (from CD-Adapco)

These three methods are used to calculate the LNG propagation rate and maximum predicted liquid height of LNG traveling down a trench encountering a 90-degree turn or a dead end.

2. LNG Spill Spread Modeling in Trenches

The following describes the three models compared in this work.

2.1 Chézy Equation (Steady-State Free Surface Flow)

The velocity of liquid flowing down a uniform channel of small slope can be computed for steady flow using generally accepted empirical formulas. One of these is the Chézy equation [6], which expresses the velocity of the liquid (*U*) as follows:

$$U = \sqrt{\left(\frac{2g}{f}\right) \cdot D \cdot S} \qquad \textbf{Equation 1}$$

Where *g* is the acceleration of gravity; *f* is the surface friction factor; *D* is the hydraulic diameter; and *S* is the slope of the trench. The friction factor, *f*, can be approximated in the fully turbulent regime as a function of the hydraulic diameter [6]:

$$f = \frac{0.25}{\left[0.57 - \log_{10}\left(\varepsilon/D\right)\right]^2} \qquad \textbf{Equation 2}$$

In this equation, ε represents the surface roughness (set to 1 mm for a concrete surface).

Equation 1 is only valid for steady-state channel flow, where the flow velocity is uniform along the channel and the liquid surface is at a constant height from the channel floor. In the early stages of an LNG spill, the liquid depth changes as a function of time and location. The steady-state equations cannot capture this early behavior nor can they model the propagation of liquid in a trench with little or no slope. The steady-state Chézy equation predicts a zero flow velocity for a trench that is not inclined; this means that the steady state Chézy equation predicts an infinite liquid depth for such a trench.

The hydraulic jump at a dead end in the trench can be approximated using the free-surface flow equation by assuming the flow suddenly stops at a location downstream. This results in a hydraulic wave that travels from the location of the stoppage towards the source. Across the hydraulic wave, mass and momentum are conserved, allowing for the calculation of the wave height and traveling velocity. Using the Chézy equation for the flow upstream of the hydraulic jump, the maximum liquid height achieved for a spill in a sloped trench reaching a dead end or sudden turn can then be estimated using the equations of mass and momentum conservation.

2.2 Exponent's Shallow Water Equations (SWEs) Numerical Model

A model was developed by Exponent to accurately describe the unsteady LNG liquid propagation through trenches and troughs [1]. This model solves the full quasi one-dimensional Shallow Water Equations (SWEs), thereby capturing the details of the pool at every location. This allows for a scientifically accurate representation of the LNG spill source and features, such as hydraulic jumps, to be captured by the model. The quasi one-dimensional spill evolution can be described by two one-dimensional equations that correspond to the mass and momentum conservation equations:

$$\frac{\partial}{\partial t}(h)+\frac{\partial}{\partial x}(hU)=S_1 \qquad \textbf{Equation 3}$$

$$\frac{\partial}{\partial t}(hU)+\frac{\partial}{\partial x}\left[hU^2+g\frac{h^2}{2}\right]=-F\frac{\tau}{\rho}+S_2 \qquad \textbf{Equation 4}$$

where h and U are the local liquid depth and liquid velocity, respectively, and where F is a shape factor that depends on the trench width and on the liquid depth. F is introduced to take both the floor and the wall friction into account. The source terms S_1 and S_2 include the effects of the spill source at the origin, of the pool evaporation, the effect of an inclined floor, and of the friction between the LNG and the concrete surface.

A standard friction model for turbulent flow was used in our model:

$$\tau=\frac{f}{4}\frac{1}{2}\rho U^2 \qquad \textbf{Equation 5}$$

where τ is the shear stress, ρ is the LNG liquid density and U is the local liquid velocity. In Exponent's model, the friction factor, f, varies according to Equation 2. The Shallow Water Equations (SWEs) were solved using a second order finite volume discretization scheme.

2.3 Star-CCM+ Computational Fluid Dynamics (CFD) Model

The Star-CCM+ CFD software is commercially-available and can model two phase flow and dispersion scenarios. The CFD model provides a time-dependent solution to the Navier-Stokes equations. In the model, conservation equations for mass of LNG, mass of air, momentum and energy are solved on an unstructured grid using a finite volume method. This allows the user to refine the mesh only where fine spatial resolution is needed without any constraints imposed by the coordinate system used by the code. The equations are closed using the $\kappa-\varepsilon$ equations for turbulence.

The free surface two-phase flow is solved using the Volume- Of- Fluid (VOF) method. This numerical technique is used to track and locate the liquid free surface (or gas-fluid interface).

3. LNG Trench Geometry

Three trench geometries, with varying slopes, were modeled in this work. The basic trench geometry and spill scenario have the following set of parameters:[1]

- Length of main trench = 254 ft
- Width of trench = 4 ft
- Flowrate of liquid spill = 10,000 m^3/hr
- Liquid density = 424.1 kg/m^3
- Slope of trench
  - No slope
  - 0.5 % slope
  - 1.0 % slope

In the steady-state free surface flow and shallow water equation calculations, the 254-ft trench terminates in a dead end. In the Star-CCM+ calculations, the trench has a 90-degree left-hand turn at the end of the 254-ft trench.

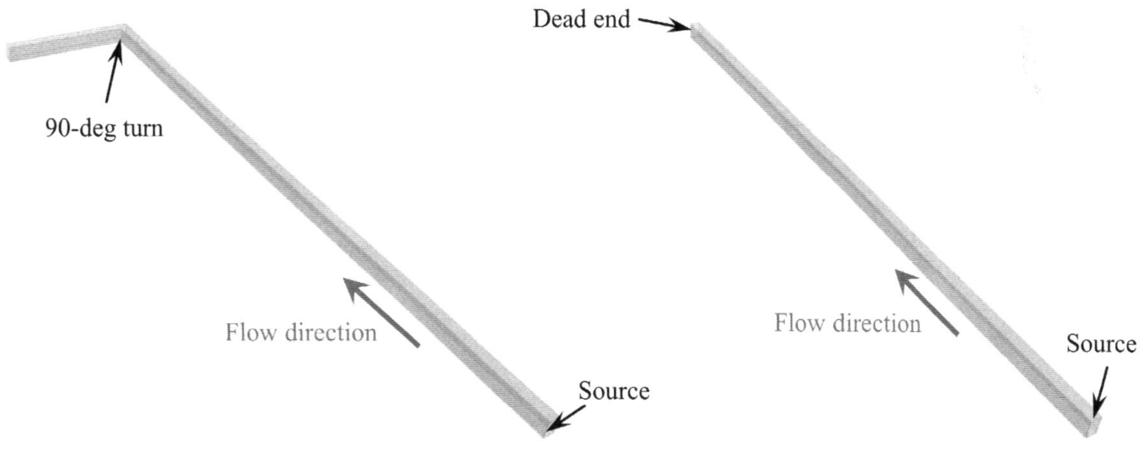

Figure 2 Image of trench geometry. A 90 degree turn was modeled in Star-CCM+ (left). A straight trench with a deadend was modeled with the SWEs, the Chézy equation and the hydraulic jump analysis (right)

[1] Values for base case were chosen based on trench characteristics at existing LNG facilities

The evaporation of LNG is not included in our calculations. This simplifying assumption allows the decoupling of the liquid flow solution from the evaporation rate calculations. In most LNG spill scenarios, the amount of LNG evaporated during its conveying to a sump accounts for only a small percentage of the spill flowrate, making this a reasonable, practical and a conservative assumption, as it results in a more rapid wetting rate of the concrete surfaces.

Figure 3 top shows a 3-D view of the Star-CCM+ solution when LNG has just passed the 90-degree corner, while the Figure 3 bottom shows the same location one minute after the start of the spill. The high 3-dimensionality and local complexity of the flow structure is clearly observed in the image.

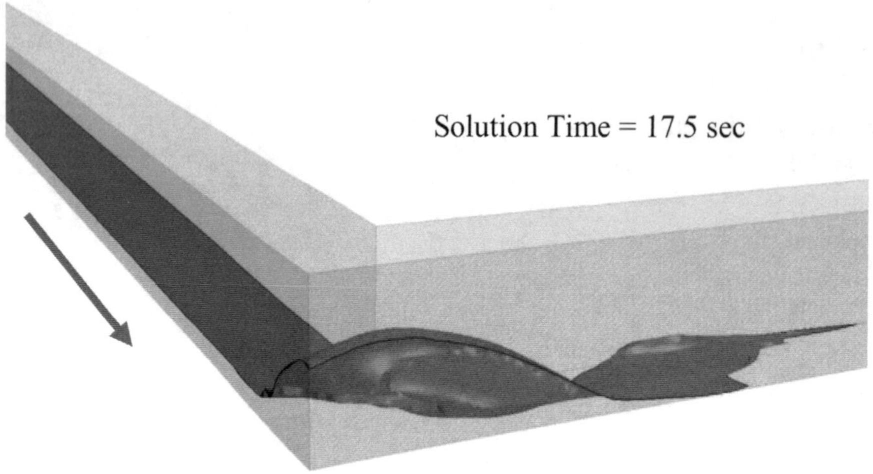

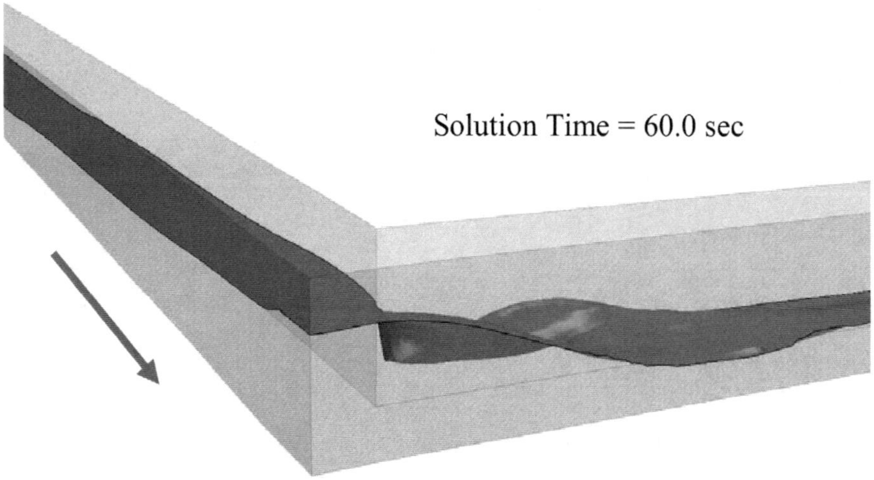

Figure 3 3D Image of liquid surface just after it reaches 90-degree turn (top), and 60 seconds into the spill (bottom). The results were obtained using a flow rate of 10,000 m^3/hr and a 1% grade trench in Star-CCM+. The arrows show the flow direction.

4. Liquid Flow Profiles

The following sections compare the results of the three models for the three trench spill scenarios with varying slopes.

4.1 Flat Trench Scenario Propagation

A sequence of 4 snapshots of the LNG spill liquid surface for the 0% slope scenario is shown in Figure 4. In the figure, the time dependent results from Exponent's SWEs model is compared to the 3-D CFD solution from Star-CCM+. Although the liquid height in the Star-CCM+ is 3-dimensional, only the maximum liquid height at any trench cross-section is shown. Although the Star-CCM+ geometry contains a turn, the liquid stagnates at the corner away from the turn, and the maximum liquid height reached at this point is almost as high as that reached in a dead end geometry.

There is no steady-state solution for the 0% slope scenario. In Figure 4, the liquid source is at the left end of the trench domain and the trench terminates or turns at a distance of 254 ft, at the right end of the plot. The LNG liquid is introduced with an initial velocity of 4.1 ft/s, which corresponds to a drop of about 2 ft from the breached pipe to the floor of the trench.

The images show the highly unsteady character of the event. Near the source, the liquid quickly speeds up, rapidly reducing in height away from the source. As the liquid propagates, its surface levels off over time. As the liquid reaches the corner, it hits the trench back wall, suddenly stopping the forward momentum, and forming a hydraulic jump wave that reflects back towards the source.

Figure 4 shows good correlation between the 1-D SWEs model and the 3-D Star-CCM+ calculation, except near the source, where the inlet conditions are slightly different for the two models. The slope of the liquid free surface away from the source is almost identical. The Star-CCM+ calculation shows a leading edge that extends very slightly farther downstream.

The maximum height of liquid against the back wall is initially identical (20 seconds). Once the hydraulic wave reflects, however, the SWEs dead-end geometry continues to fill, increasing the maximum liquid level height. The Star-CCM+ geometry models the flow with a 90-degree turns at the end of the trench, so the liquid does not accumulate as much in the trench.

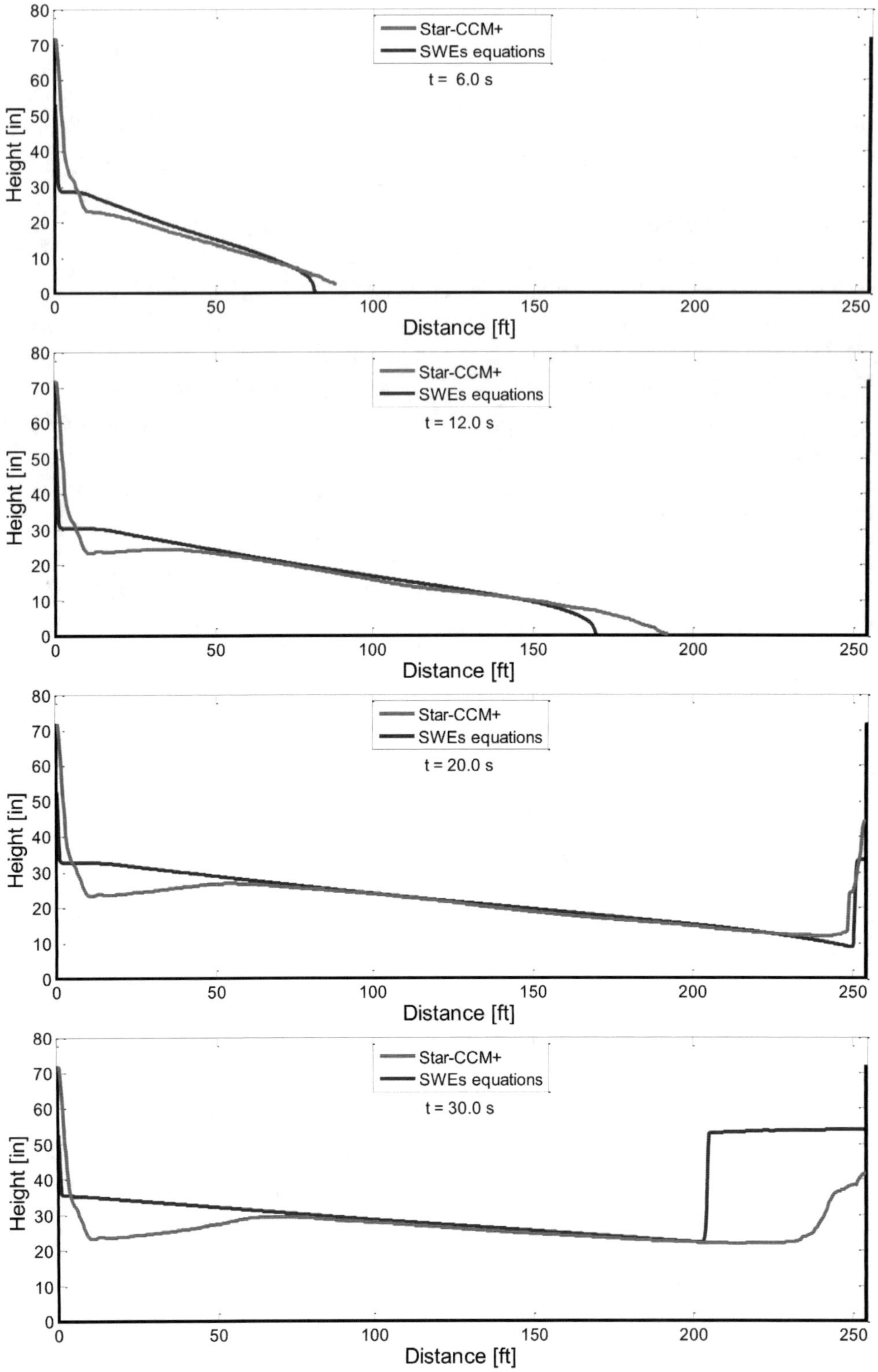

Figure 4 LNG liquid profiles at 4 different times for the 0% slope scenario. Note that the SWE solution shows a flow into a dead-end while the Star-CCM+ solution describes the flow with a right angle turn at the end of the trench. The computations were performed with the conditions in Section 3.

4.2 0.5% Slope Scenario

A sequence of 4 snapshots of the LNG spill liquid surface for the 0.5% slope scenario is shown in Figure 5. In the figure, the time dependent results from Exponent's SWEs model for a flow into a dead end is compared to the 3-D CFD solution from Star-CCM+ for a flow with a 90° left angle turn at the end of the trench. The liquid source is again at the left end of the trench and the trench terminates or turns at a distance of 254 ft to the right.

The steady-state approximation is shown as the dashed green line. After the initial flow transient, the liquid depth for the unsteady-calculations becomes nearly constant. The depth of the liquid surface also becomes nearly constant prior to the formation of the the hydraulic jump and is slightly below but close to the value predicted by the steady-state Chézy equation. After the liquid has hit the back wall (20 seconds and 30 seconds), a second dotted line shows the approximated hydraulic jump based on the Chézy equation and conservation of mass and momentum.

Again, the 1-D SWEs model and the 3-D Star-CCM+ calculations compare well away from the source. The slope of the liquid free surface away from the source is almost identical. The maximum height of liquid against the back wall is initially identical (20 seconds). The wave that forms at the end of the trench, however, behaves differently for the two models at later times due to the different geometries (dead-end for SWE and right angle turn for Star-CCM+). The hydraulic jumps calculated using the 1-D SWEs and 3-D Star-CCM+ models remain below the level approximated by the steady-state assumptions at all times.

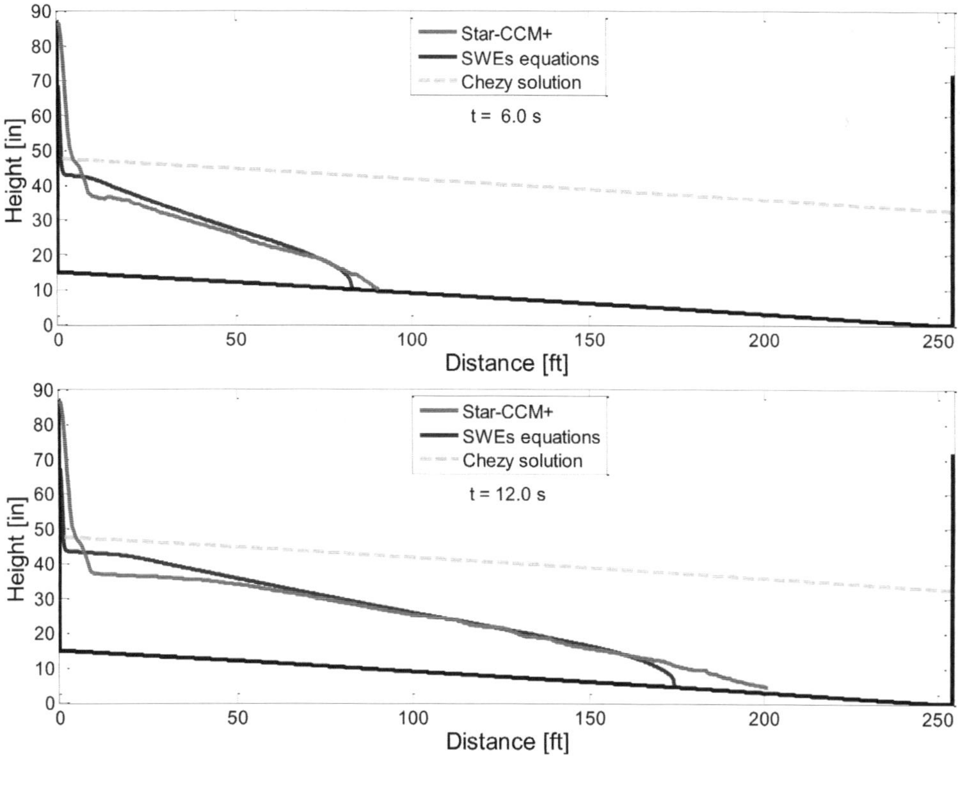

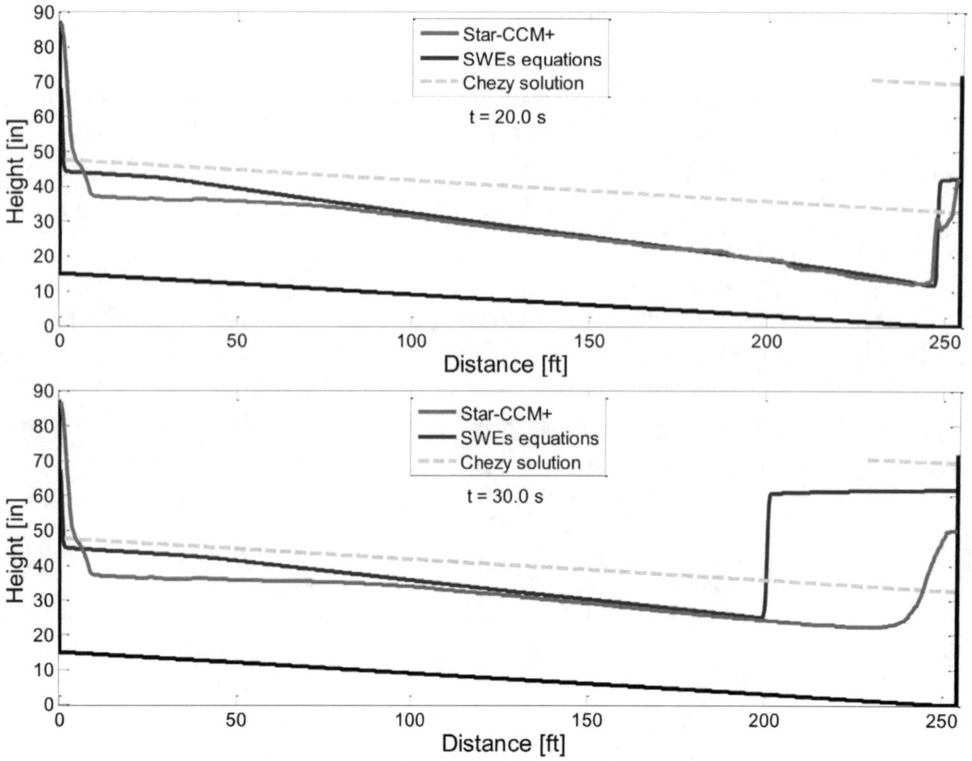

Figure 5 LNG liquid profiles at 4 different times for the 0.5% slope scenario. Note that the SWE solution shows a flow into a dead-end while the Star-CCM+ solution describes the flow with a right angle turn at the end of the trench. The computations were performed with the conditions in Section 3.

4.3 1% Slope Scenario

A sequence of 4 snapshots of the LNG spill liquid surface for the 1% slope scenario is shown in Figure 6. In the figure, the time dependent results from Exponent's SWEs model are compared to the 3-D CFD solution from Star-CCM+. The steady-state approximation is again shown as the dashed green line, with a second dashed line showing the predicted hydraulic jump at the end of the trench.

Again, the 1-D SWEs model and the 3-D Star-CCM+ calculations compare well away from the source. The slope of the liquid free surface away from the source is almost identical. After the initial flow transient, the liquid depth for the unsteady-calculations becomes nearly constant at later times (see 30 second solution) upstream of the hydraulic jump. At this time, the LNG liquid height prior to the hydraulic jump is equal to the value predicted using the steady-state Chézy equation for a free surface flow in an inclined channel.

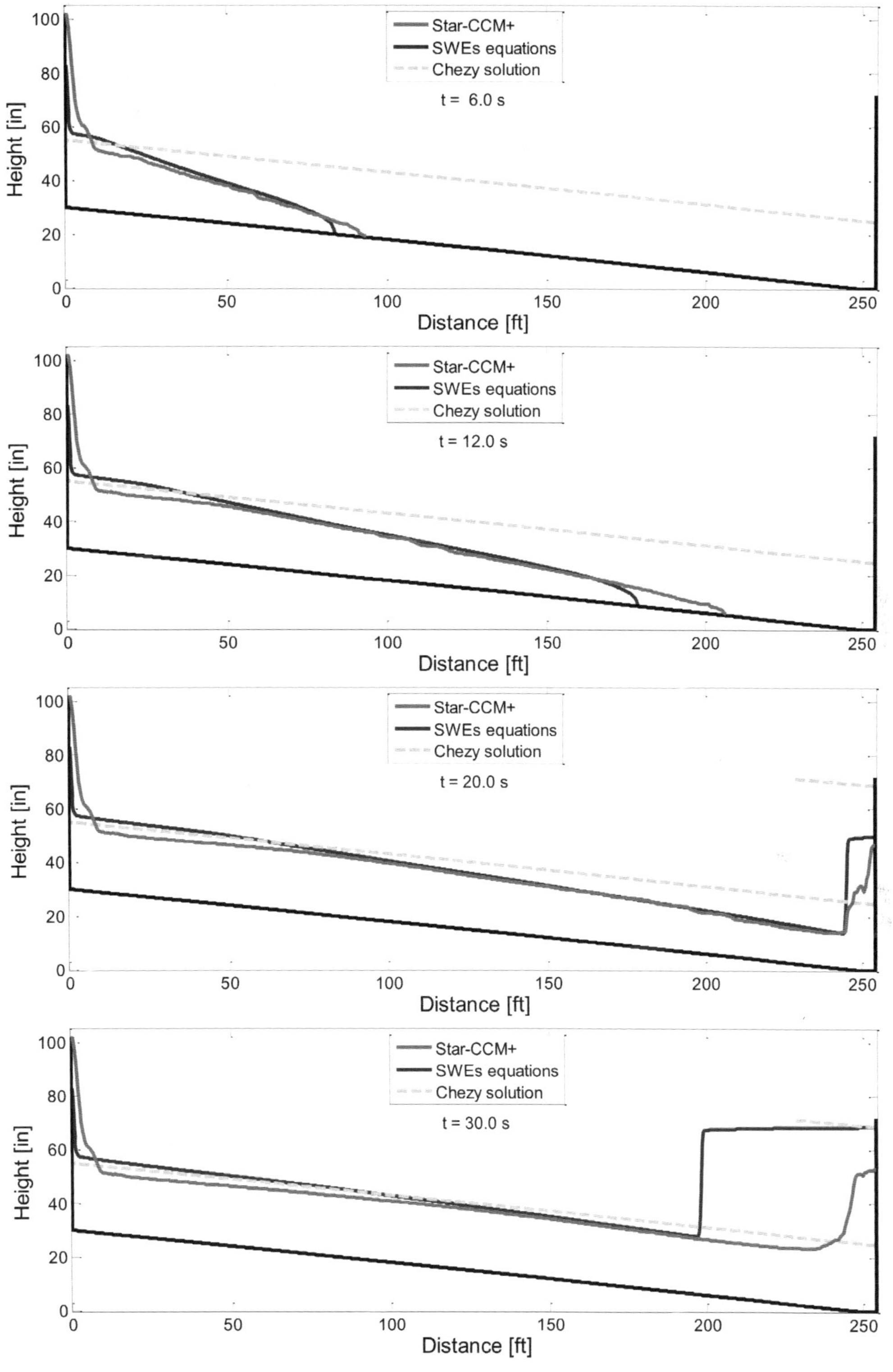

Figure 6 LNG liquid profiles at 4 different times for the 1% slope scenario. Note that the SWE solution shows a flow into a dead-end while the Star-CCM+ solution describes the flow with a right angle turn at the end of the trench. The computations were performed with the conditions in Section 3.

The maximum height of liquid against the back wall is initially identical for the 1-D and 3-D models (20 seconds). The wave that forms at the end of the trench, however, behaves differently for the two models at later times due to their different geometries. The hydraulic jumps calculated using the 1-D SWEs and 3-D Star-CCM+ models remain below the level approximated by the steady-state assumptions at all times shown. It is worth noting that the positions of the leading edge at the 12 second mark are very close to one another for all three cases.

5. Liquid Propagation Rate and Maximum Liquid Height

The choice of the slope influences the depth of LNG in the trench as well as the speed at which it spreads, both of which also influence the rate of vapor generation and trench flow capacity. The following section summarizes the key results obtained for all three scenarios.

5.1 LNG Propagation Velocity

Figure 7 shows the propagation velocity of the LNG leading edgedcalculated using the 1-D SWEs model and the 3-D Star-CCM+ model as a function of time for the three trench slopes explored in this work. The steady-state Chézy equation solution for the 0.5% and 1% cases are shown by the two arrows.

The figure shows how immediately after the beginning of the spill the flow rapidly accelerates to a maximum. The peak liquid velocity values are relatively high, up to 20 ft/s, when compared to the expected steady state solutions. After the maximum velocity is achieved, the LNG propagation slows down but does not reach a steady-state condition before reaching the end of the trench at (between 14 and 19 seconds after the start of the spill). The 1-D SWEs model shows a clear trend in the LNG propagation rate with increasing trench slope, although the values are very similar for all 3 cases. The 3-D Star-CCM+ model does not exhibit a clear trend of LNG front velocity versus trench slope, due in part by the relative short length of the trench not allowing the flow to reach steady-state.

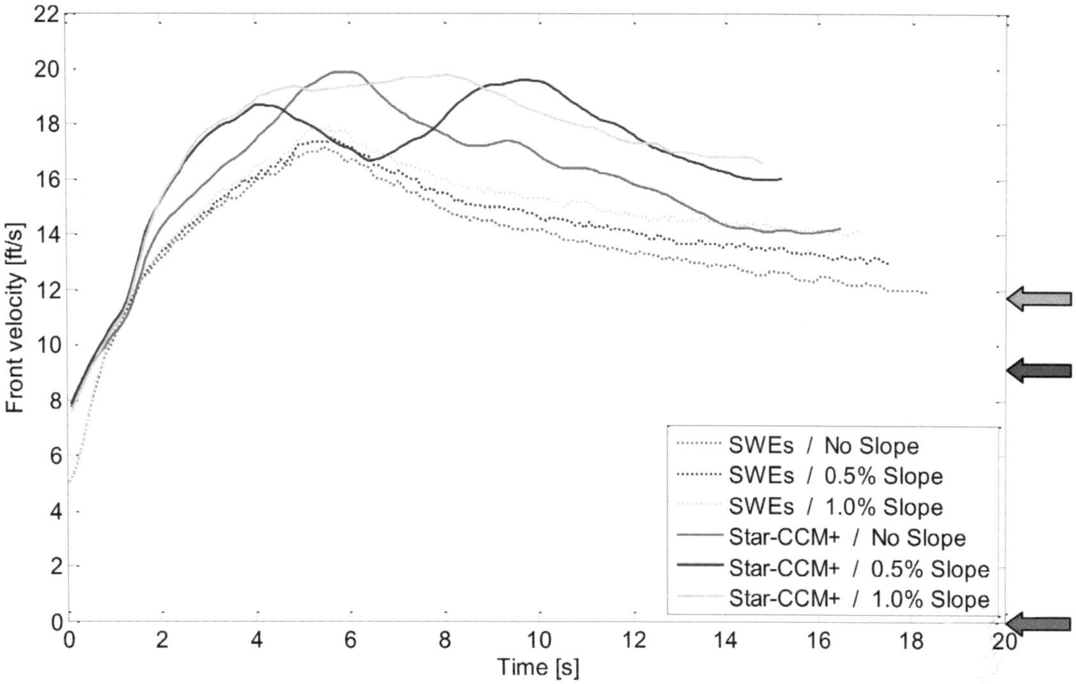

Figure 7 LNG spill propagation velocity as a function of time for 3 different sloped trenches. The Chézy steady-state solutions are represented with the arrows on the right.

5.1 LNG Liquid Height

Figure 7 shows the maximum liquid height of the LNG front (not including the area near the source) calculated using the 1-D SWEs model and the 3-D Star-CCM+ model as a function of time for the three trench slopes explored in this work. The steady-state Chézy equation solution for the 0.5% and 1% cases are shown by the two arrows.

The figure shows how the maximum liquid level height is relatively constant after the initial transient near the source and prior to the liquid reaching the far wall. Once the liquid reaches the dead end or sharp turn, the maximum liquid height suddenly jumps due to a stagnation point forming at the end wall. In the 3-D Star-CCM+ calculations with the 90-degree turn, the maximum liquid level height stabilizes shortly after the formation of the hydraulic wave. The 3-D Star-CCM+ solutions show very similar jumps in liquid height, irrespective of the trench slope. In the 1-D SWEs model solutions, the maximum liquid level height continues to increase after the hydraulic wave formation because the dead-end geometry continues to fill the trench with liquid.

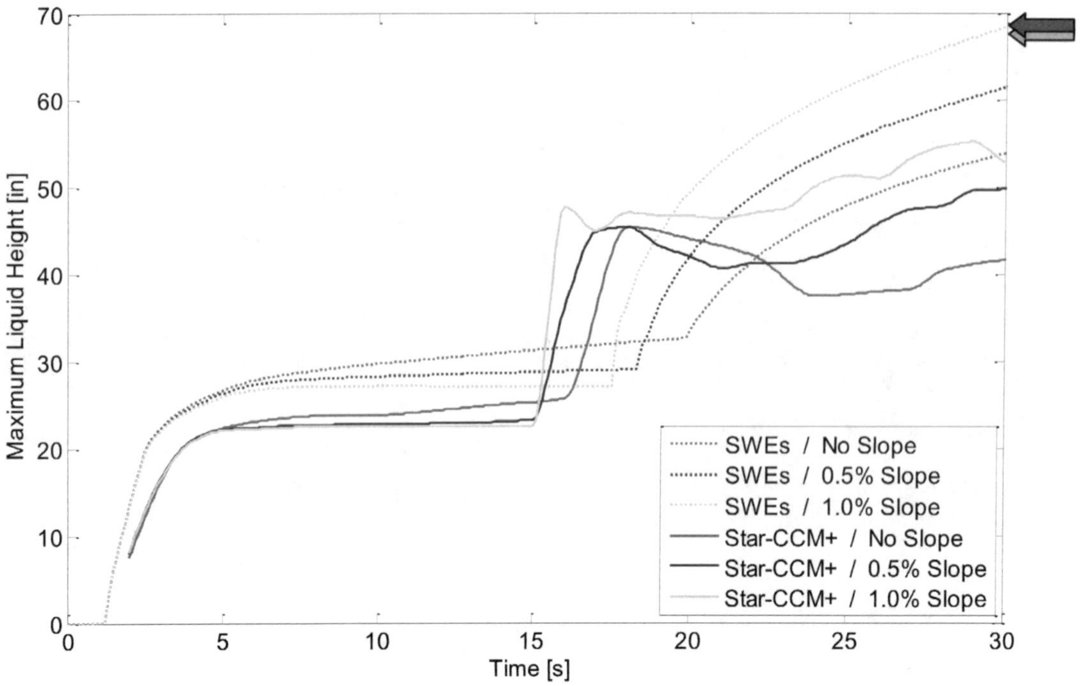

Figure 8 LNG maximum liquid height above the trench floor as a function of time. Arrows indicate the Chézy steady-state solutions for the maximum height of the hydraulic jump.[2]

The data in Figure 8 shows that the LNG maximum height is greatly increased by the presence of a dead end or turn in the trench. This is an important factor that needs to be considered in the design of an LNG trench system. The trench walls must be high enough to contain the LNG, even in the event of a hydraulic jump that could be several times the height of the liquid level flowing down the trench (steady-state scenario). The Chézy steady-state solutions for the maximum height of the hydraulic jump seems to provide a conservative value for the maximum liquid level height after a 90-degree turn (at least for the 0.5% and 1% sloped scenarios considered).

6. Summary and Conclusions

Three different models to predict liquid level propagation and heights during an LNG spill into a trench were compared for varying trench slopes. The solutions provided the following set of conclusions:

1) The propagation of LNG in inclined trenches is a highly transient event that needs to be properly modeled to predict the LNG propagation speed, wetting rate and liquid level height. These types of calculations are critical to properly model the substrate wetting

[2] To avoid the flow features of the source, the first 10 ft downstream of the source are neglected in the maximum height calculation.

rate and resulting LNG evaporation rates.

2) Exponent's Shallow Water Equations (SWEs) Solver provides similar results to the Star-CCM+ CFD software during LNG propagation down a flat or inclined trench.

3) The trench slope had a very small influence on the LNG propagation velocity and maximum liquid height when encountering a sharp 90-degree turn.

4) The liquid response to a sharp 90-degree turn is similar to flow encountering a dead end in the trench, forming a sudden rise in liquid level and a wave that propagates back towards the source.

5) Sharp turns in the LNG trench system can result in an increase of the maximum liquid height by as much as 2.5 times (for the cases considered), when compared to the liquid level height achieved prior to the turn.

6) The 3-D Star-CCM+ solutions indicate that the trench slope (for the cases examined) has little effect on the height of the hydraulic jump, for the limited trench lengths considered.

7) The propagation velocity of the spill leading edge is almost always larger than the steady-state liquid propagation velocity predicted using the Chézy steady-state solution, especially for trenches with little or no slope, and near the source. The liquid level prior to the hydraulic jump is also smaller than the Chézy steady-state solution for sloped trenches.

5. References

[1] Ponchaut, NF, Kytömaa, HK, and Ibarreta, AF, *Modeling the Vapor Source Associated with Spills into Troughs and Trenches*. American Institute of Chemical Engineers (AIChE) Spring National Meeting (2011).

[2] DOT Code of Federal Regulations - 49 CFR 193 *Pipeline and Hazardous Materials Safety Administration, DOT*

[3] *NFPA 59A Standard for the Production, Storage, and Handling of Liquefied Natural Gas (LNG)* (2009)

[4] Gavelli, F, Chernovsky, MK, and Kytömaa, HK. *Modeling of LNG spills into trenches.* Journal of Hazardous Materials 2010; 180(1–3):332–339.

[5] Raj, PK. *Evaporating Liquid Flow in a Channel (an Integral Model)* Mary Kay O'Connor Process Safety Center International Symposium, College Station, TX, October 27 (2010)

[6] Blevins, RD. *Applied Fluid Dynamics Handbook.* Krieger Publishing Co., Malabar, FL (2003) Pg. 164 – 170.

[7] Ponchaut, NF, Kytömaa, HK, Morrison, DR, and Chernovsky, MK, *Modeling the vapor source associated with the spill of LNG into a sump or an impoundment area.* Mary Kay O'Connor Process Safety Center International Symposium (October 2010).

AUTHOR INDEX

AUTHOR INDEX